Ecological Studies, Vol. 182

Analysis and Synthesis

Edited by

M.M. Caldwell, Logan, USA
G. Heldmaier, Marburg, Germany
R.B. Jackson, Durham, USA
O.L. Lange, Würzburg, Germany
H.A. Mooney, Stanford, USA
E.-D. Schulze, Jena, Germany
U. Sommer, Kiel, Germany

Ecological Studies

Volumes published since 2001 are listed at the end of this book.

H. Schutkowski

Human Ecology

Biocultural Adaptations
in Human Communities

With 36 Figures and 7 Tables

 Springer

Dr. Holger Schutkowski
Biological Anthropology Research Centre
Department of Archaeological Sciences
University of Bradford
Bradford, BD7 1DP
West Yorkshire, UK

Cover illustration: Picture of month December, by Stephan Kessler and fellows (Brixen, 1675/80). Benediktbeuern monastery, Germany

ISSN 0070-8356
ISBN-10   3-540-26085-4 Springer Berlin Heidelberg New York
ISBN-13   978-3-540-26085-1 Springer Berlin Heidelberg New York

Springer is a part of Springer Science+Business Media
springeronline.com

© Springer-Verlag Berlin Heidelberg 2006
Printed in Germany

The use of general descriptive names, registered names, trademarks, etc. in this publication does not imply, even in the absence of a specific statement, that such names are exempt from the relevant protective laws and regulations and therefore free for general use.

Editor: Dr. Dieter Czeschlik, Heidelberg, Germany
Desk editor: Dr. Andrea Schlitzberger, Heidelberg, Germany
Cover design: *design & production* GmbH, Heidelberg, Germany
Typesetting and production: Friedmut Kröner, Heidelberg, Germany

31/3152 YK – 5 4 3 2 1 0 – Printed on acid free paper

To Helen

Preface

Human Ecology comes in various guises and is addressed at very different levels; from the narrow focus of human household economics to the global issue of humans as the agencies of unconcerned and irresponsible consumption of landscapes and resources. Yet, Human Ecology comprises what it says: the study of human populations and their interrelationships with the characteristics and properties of their environment. But this is where the analogy with general ecology stops. Human interaction with their habitats is almost entirely shaped by their cultural characteristics. Culture, this complex entity of technological inventiveness, social institutions, belief systems and idiosyncratic identities, is an integral part of the human condition and forms a unique amalgamation with our evolutionary biological and behavioural heritage. Human Ecology, therefore, has to integrate both aspects into a study of biocultural adaptations of human communities. This is an almost all-encompassing, gargantuan brief that needs to be broken down into comprehensible units.

The attempt presented here takes its starting point from a very basic area of life-support strategies: the procurement of food and other resources. The topics explored in this book are therefore both immediately and indirectly connected to resources: subsistence strategies, subsistence changes, resource allocations, population dynamics and reproduction. These are areas which either directly refer to possibilities and patterns of tackling the challenge of securing survival, to satisfy the energetic or caloric requirements from existing options within the habitat, or indirectly refer to those options which are affected by these possibilities and patterns and which reflect the results of such efforts, both at the individual and population level. The starting point for such a resource-focused approach follows from the central ecological principles of material and energy flows in a given habitat, which humans are a necessary part of. Resources and their use are, as it were, the interface where humans integrate in the flows of material and energy and where their variable exploitation by means of cultural solutions become the hallmark of human niches.

After addressing theoretical and historical aspects of human ecological enquiry, the book will therefore examine modes of production and food acquisition and how they are influenced by the characteristics of different habitats and their limiting factors. This is followed by a consideration of subsistence change and explores its conditions and contexts against a range of likely causes. Resource use and social organisation explore the question of unequally distributed possibilities of access to resources within mainly historic societies. Finally, it will be attempted to synthesise all three topics into an appreciation of population development and regulation and to analyse the consequences of different forms of resource use on the size and composition of a population, i.e. in which form subsistence affects the reproductive options within a population.

The organising pattern of this book follows function or functional areas of life. This structure deliberately deviates from 'classic' approaches that take the division of the earth into biomes as a starting point and explore which kinds of adjustment enabled humans to survive under specific and typical habitat conditions. In contrast, an arrangement focussed on function has the advantage that, where thematically necessary and useful, a habitat-oriented view can be taken, for example with regard to strategies of food acquisition or modes of production. At the same time, however, such a system keeps the option of being able to show not only the differences but also the commonalities of strategies for coping with quite different habitats. The variability of biocultural responses is characterised by exactly this duality of the, as it were, universal functional areas and their local shapes and forms. Foragers, pastoralists or farmers do or did roam nearly all continents; and the respective habitat factors led to quite different adaptations. It is nevertheless the local and regional responses to the unchanging fundamental challenges of the use, availability and control of resources which can be examined for their conditions and effects and whose vital connection with structures in the social organisation of communities can be recognised. In this regard, this book pays tribute to the tradition of Steward's Cultural Ecology. At the same time, however, it draws from the systemic approach of Ecological Anthropology, as cultural achievements will be regarded as those components or system properties which act as determinants for coping with habitat constraints and the concrete problems of daily life resulting from such constraints. This approach will be complemented by reverting to concepts of Behavioural Ecology, thereby allowing individual decisions to be considered in the context of sociocultural institutions, which are the basis of observable characteristics on the population level.

The central areas of life and function are characterised by their mutual dependence on the biological and cultural characteristics of the populations which were chosen as examples. Humans do not exist in a pristine state of natural being. No human community is able to survive without cultural developments and inventions, without technology, or without traditions and institutions. Their long-term survival is secured because they live through human thoughts and actions and require to be translated by humans. Biological processes and cultural structures are tied together in human life and human evolution. Therefore humans in their ecosystems are the hinges between socio-cultural, biotic and abiotic components. It is this role that makes them unique.

Whilst the comprehensive demand behind this statement can only be competently redeemed in a cross-disciplinary discourse, the material used for this study must be necessarily limited. Certain areas usually connected with the topic of biocultural adaptations are therefore missing, e.g. the entire complex of material culture and its meaning. Devices, tools or handicraft, in other words the technological skills and achievements of a society, are undoubtedly important as part of life-support systems and are one key factor in analysing survival strategies. Major lines of argument, for example the connection of population growth and the intensification of food production have in the past referred to technological innovations as the driving force behind cultural evolution. To describe and analyse these is one thing. To understand the meaning of material culture in the context of its function in a certain culture is a different matter entirely. Both sides however are inseparable, if material culture is not to be reduced to its bare functional content.

For example, the houses built by the Toraja of Sulawesi (Indonesia) with their characteristic roofs, saddleback-shaped and the roof ends upswept, can most likely be understood by appreciating the laws and principles of physics and then be esteemed for their possible advantages and benefits in ambient climatic conditions. At the last, however, when one attempts to reconcile these aspects with their obvious meanings as part of cosmological concepts, with the integration of the material environment deeply rooted in the belief systems of the community, the compelling necessity of discourse and the unification of aspects of different subject matters becomes evident. However, the disciplinary leadership would then no longer be with scientific anthropology, but with subjects of the humanities that are close to anthropology. In this regard, the restriction on topics revolving around resource-related areas of life in human communities takes into account disciplinary competence.

The formal structure of the book sees general theoretical aspects of the topic at the beginning of the respective chapters, which are prepared by brief examples. In the more specific parts, different aspects of the topic are developed by case studies. The chapters close with a summarising view and a comparison of the differences and commonalities of biocultural adaptations derived from the case studies, as appropriate. The case studies may appear to be portrayed in some detail from time to time, yet their purpose is only to serve as illustrations of the underlying principle, notably the development of biocultural strategies of adaptation to changing and specific ecological conditions. Such examples are neither limited to certain geographical regions nor to certain time-horizons and will, as far as sub-recent examples are concerned, employ the ethnographic present tense. Thus, on the one hand this opens the possibility of being able to represent the versatility of biocultural adaptations in different natural conditions and different subsistence modes and, on the other hand this demonstrates that ecological questions and their systemic treatment are not subject to historical temporal delimitation or limited to synchronic analyses, but rather attempt to integrate reference to the course of historic events and developments.

This book owes its development and completion to the encouragement, stimulation, advice and support of colleagues and institutions, to whom I feel privileged to extend my gratitude.

Thanks are due to Bernd Herrmann, my long-time and influential academic teacher, also to Michael Casimir, Silvana Condemi, Randolph Donahue, Alexander Fabig, Gisela Grupe, Clark Larsen, Gabriele Macho, Donald Ortner, Britta Padberg, Mike Richards, Martin Rössler, Henry Schwarcz, Eva Stauch, Joachim Wahl and Philip Walker. Last, but not least, I should like to thank the many students for their inquisitiveness and interesting questions.

The completion of this book has greatly benefited from a teaching buy-out granted by the Department of Archaeological Sciences in the autumn term of 2004.

Springer-Verlag has admirably put up with the duration of this project and I am greatly indebted to Drs Andrea Schlitzberger and Dieter Czeschlik for their continuous support. Thanks also go to Alan Mill for congenial copy editing.

Bradford, July 2005   Holger Schutkowski

Contents

"...the use of human populations in ecological analyses preserves a view of humanity as a part of nature at the same time that it recognizes the uniqueness conferred upon the species by culture. As such it preserves the terms defining the condition of a creature that can live only in terms of meanings, largely culturally constituted, in a world to which law is intrinsic but meaning is not."

(p. 57 in Rappaport 1990)

1 History, Concepts, and Prospects

Humans are the ecologically dominant component in their respective habitats, certainly ever since they accomplished the transition from an extracting to a producing mode of food procurement. Humans attempt to appropriate nature, the living and inanimate environment, for their purposes by intervening in natural processes. They accomplish this within the framework of their personal interests and collective goals; and in order to reconcile and pursue these interests and goals, they develop strategies to facilitate the use their environment and to secure their survival. But at the same time humans are subject to change in the context of evolutionary processes, both biological and cultural. Human adaptation thus always embraces the twin aspects of cultural strategies and biological conditions and outcomes. The diversity and correspondence of solutions humans develop to co-ordinate these two sides of life support systems, to maintain and change them, is at the heart of human ecology.

Humans share their habitats with a multitude of other life forms. Like these, humans have to adapt to the given or, in the course of time, changing basic environmental conditions to achieve long-lasting use of their habitat. These basic conditions are largely provided by default settings in terms of certain natural factors, e.g. climate, geomorphology, soil properties or species diversity. However, humans themselves change these basic conditions by applying survival strategies, e.g. techniques, agreements, rules and modes of organisation, which they develop to facilitate survival in their habitat; and they manipulate and shape their environment as part of their adjustment strategies.

Humans first of all – and this again they share with other organisms – have to be suitably equipped with a set of biological traits to be adequately adapted to the respective environmental conditions. Such adaptive responses can be genetically coded, e.g. certain physiological reactions to cold or heat stress, and they provide a reaction norm of genetic plasticity from which different phenotypes emerge, are selected as successful responses to environmental variability and eventually lead to differences in allele frequencies between human populations (Cavalli-Sforza et al. 1994). But, more importantly, adaptive solutions for coping with ecological constraints are also found as culturally coded survival strategies. They take the form of acquired information that is constantly modified, increased and passed on across generations in a nongenetic, as it were socio-genetic, way. Although incipient cultural traditions can be observed in non-human primates (e.g. Tomasello 1999), humans are, above all, distinct in their ability to adapt culturally to ecological basic conditions. It is this flexible and lasting cultural adjustment, based on skills, knowledge and experience in achieving success or dealing with failure of environmental utilisation, which explains why humans have been so successful in spreading across the planet.

Their cultural, and to a certain extent also their phenotypic, variability allowed anatomically modern humans to seize a durable foothold, even far away from the habitat in which their natural history began. From an evolutionary point of view, such adaptations serve for nothing but survival in the first place and it is hard to tell whether this is about overcoming seasonal shortages or safeguarding long-term survival in a locality once it is inhabited. The two are inseparable. Human adaptability encompasses biological adaptations and, particularly, cultural responses and the concomitant behavioural reactions.

The variety of these cultural responses to different abiotic and biotic environmental conditions will be explored in the following. But it has to be kept in mind that environmental conditions perceived as natural today are in fact quasi-natural only, because they are uniquely shaped by humans, from subtle modification to radical change and interference, and may require new adjustments if necessary. The natural foundations of human existence are culturally mediated and the outcomes impact upon biological operational sequences: humans are genuinely biocultural by definition. These dynamic interrelations between humans and their environment are interconnected by functions and processes that are mutually dependent and affect each other. Therefore, the true meaning of biocultural refers exactly to a situation where biological conditions need to be explained as a result of the establishment and perpetuation of cultural strategies. But how and by which means do human populations succeed in long-term survival in a certain habitat? Can general patterns be identified or is the success of cultural variability due to characteristics in local strategies? In other words: How are biology and culture intertwined?

A common link for these questions comes naturally with the topic of food procurement, as the basic prerequisite for humans to engage with their environment and the resulting biocultural adjustments. Food acquisition is the trivial condition of physical survival. It is at the same time a basic way in which organisms participate in material and energy flows in their habitat. This immediate connection between subsistence and human culture has been an emphasis of research into human/environment relations and ecological studies in anthropology from the outset (see Sect. 1.1) and has *mutatis mutandis* not lost anything of its relevance. Therefore, this intertwining of biological and cultural characteristics of human communities will be dealt with (Chaps. 2–5) in topics either directly related to subsistence, i.e. the conditions of food acquisition, or those closely connected with the determinants or consequences of resource use.

Using ecological principles and ecosystem theory in anthropological research on human local populations has substantially contributed to a better understanding of the role of humans in nature, their influence on the environment and their being shaped by the environment. Therefore, in the course of this chapter, fundamentals of ecology are briefly introduced. Subsequently, and building on it, the suitability of the ecosystem concept for an application to human populations is discussed and subtopics relevant for further discussion are addressed with regard to possibilities of their application and modifications necessary to suit the interpretation of the complex non-linear or multiple stable-state conditions in human ecosystems. This will be preceded by a concise overview of theoretical positions and methods underlying the observation and analysis of interrelations between humans and their environment. This overview is really only meant to provide a very brief account of major relevant ideas and concepts, flashlights as it were, and the reader is referred to the much broader coverage of anthropological theory given, for example by Layton (1997), Barnard (2000) or Harris (2001). Primarily, only those trends will be presented here from which theoretical conditions for a cross-cultural and ecosystemic approach can be derived. The time frame, therefore, emphasises concepts developed in the second half of the twentieth century. The scientific starting points are numerous and their development, temporal succession and mutual pervasion demonstrate both the change in evaluating the role of nature and environment in their effects on humans and likewise the change in evaluating the effects humans and human culture have on the environment.

1.1 Human Ecological Concepts: A Brief Overview

1.1.1 Environmental Determinism

Towards the end of the nineteenth and the beginning of the twentieth century, in the wake of colonial expansion, a constantly growing amount of information about foreign worlds had been accumulating. In order to classify exotic artefacts and ethnographic knowledge gathered through expeditions and voyages of discovery an attempt was made to systematically structure the evidence according to provenance. The observation that similar cultural characteristics were connected with certain geographical locations led to the assumption that the material culture and technology of a society was caused by the specific makeup of the environment – the habitat in ecological terms (p. 3 in Hardesty 1977). The environment, in one way or another, was considered to have a determining effect on the possibilities of human cultural development. The view that similar environmental conditions would lead to similar forms in the political organisation of a society became popular in the emerging field of human geography, notably with the prominent proponent Friedrich Ratzel (1844–1904), but continued to be popular into the mid-twentieth century (e.g. Huntington 1945). Natural conditions such as climate or landscape were thus ascribed a strong formative power on human populations and their institutions, resulting in the notion of culture areas whose environmental make-up would define socio-economic expression of the human societies. Nature with its areas of different layout and composition provided, as it were, the default settings for paths of least resistance, by which humans reacted to the characteristics of their environment (p. 22ff in Moran 2000). It was this tradition of environmental determinism that called into play explanations of, for example, why dry areas were used to breed cattle rather than lay out irrigation fields. While the simplicity of this approach may have been attractive, its major assumptions have been proven wrong. The environment is not fixed and unchanging, nor are cultural responses to certain environmental conditions static. Rather, human culture kits are the result of flexibility, resilience and the ability to come up with alternative solutions even under the same or similar environmental conditions.

1.1.2 Possibilism

As a reaction to this strictly deterministic concept, a tradition of thought developed whose most eminent representative was Franz Boas (1858–1942). It was termed possibilism or historic particularism. According to possibilist thought, nature did not directly influence humans, but provided a framework and thus facilitated different possibilities of human development. Nature, as it were, offered the raw material from which traditions, belief systems or theories could develop. The role of nature was thought to be passive and any decision on the actual expression of culture traits, i.e. a realisation of the respective options under given environmental conditions, was due to the historic and cultural particularities and the selectivity by which societies made their choices. Human culture was not shaped by nature, but cultural decisions were thought to be subject to their own dynamics, so that cultural differences between populations would also be found in their respective particular cultural history. In the context of possibilism, it was not important to explain the origin of culture traits. Characteristics of the environment were not required in order to explain the presence of culture traits, but rather served as an explanation for their absence, i.e. the reason why they did not evolve. The absence of stone houses, for example, would be explained as a consequence of a lack of appropriate raw materials in the habitat. Thus certain characteristics would not emerge, simply because they or the means to produce them were not available (p. 4 in Hardesty 1977).

This culture-centred view of humans within nature left little space for a dynamic role of the environment, but rather reduced it to a generally limiting element of human cultural development. At the same time, the emphasis on historical specificity precluded that similar environmental conditions could also lead to similar selectivity (p. 33ff in Moran 2000), i.e. the possibility of a cross-cultural comparative view was handicapped from the start by the primarily case-by-case nature of the possibilist assumption.

1.1.3 Cultural Ecology

Following the comparatively extreme theoretical positions of environmental determinism and historical possibilism, with their respective exclusive emphasis on either nature or culture, a quite different concept was developed during the 1950s. It broke with both traditions and instead postulated interrelations between humans and their environment, i.e. it proposed a dynamic view. In a seminal study, Julian Steward (1902–1972) developed the idea that causal connections would exist between natural environmental conditions, subsistence and the social structures of a population or society (Steward 1955). It was further postulated that those social and political structures which developed in societies under comparable environmental conditions and comparable subsistence patterns ought to show similar causal connections among themselves. This notion of a Cultural Ecology thus searched for regularities and common grounds in human behaviour, social structure and belief systems which would develop as responses to certain environmental situations. Steward's method was culture-comparative in time and space and designed with the aim to search for generalisations in the function and emergence of human behaviour. Conditions and modes of food acquisition constituted the most immediate link between environment and behaviour. The underlying mechanisms leading to the development of such behaviour were believed to represent a human universal, whose impetus would arise from the necessity to use the naturally available resources, such as food.

According to the concept of Cultural Ecology, social institutions possess an internal functional connection, e.g. as certain modes of production occur in combination with certain modes of social and political organisation or the division of labour in a society. On this condition, the effect of one variable on a limited number of further variables can be examined within the system, rather than having to examine the much more complex system of social organisation in its entirety. By emphasising diachronic comparison Cultural Ecology differs from classic functionalism (e.g. Malinowski 1960) in that it puts an emphasis on the investigation of change and its causes and less so on the question of mechanisms by which equilibrium states can be maintained or basic and derived needs be met.

Central aspects of the culture-ecological approach refer to the question whether specific behavioural responses are necessary for the adaptation of human populations to their environmental conditions, or rather whether a broad behavioural repertoire would suffice, i.e. whether adaptation occurred through specialisation or generalisation of abilities. In this context, adaptation would be understood as the ability to find ever better solutions for the possibilities of habitat use. The method chosen to examine these questions first of all aimed at investigating the relationship between the environment and the subsistence system practised. Subsequently, those behavioural patterns were analysed which were connected with a certain subsistence technology, in order to eventually be able to study the effects of the respective behaviour patterns on other aspects of culture in the population. Thus Cultural Ecology attempted to support the basic assumption that there is a causal relationship between natural resources, subsistence technology and those behaviours in a population that facilitate the use of resources at a given level of technological development (p. 48 in Moran 2000).

Yet the environment would only affect certain elements of culture, the so-called culture core, while other elements would develop in the course of an autonomous culture history (Orlove 1980). Steward's (1955) term of the culture core comprises all those social, political and religious behaviour patterns which can be empirically determined and associated with subsistence activities and economic operational sequences. The aim of Cultural Ecology was and still is the investigation of adaptive processes which, under comparable environmental conditions, bring about cross-cultural regularities. Hence, principles of cause and effect between culture, technology and environment were implied which consequently led to the formulation of a deterministic model of intrinsic cultural development (Steward 1949). It was based on the idea that certain regular transitional culture stages would appear in the process of the evolution of human social structures, the sequence of which could be regarded as a universal model.

Studies in which the principles of Stewardian Cultural Ecology were applied and which dealt with the functional connections of subsistence and characteristics of social organisation were, indeed, able to confirm evidence of such intercultural commonalities. Among the well known examples is Lee's work with !Kung, which allowed general principles to be derived with regard to, for example, mobility patterns, food-sharing behaviour, co-operation or rules over access to resources (see Sect. 2.3.1), all of which were later found, too, in studies of other forager societies across the world.

It is true: most culture-ecological investigations concentrate on behaviour connected with human subsistence. Besides this, however, other aspects such as the effect of resource change, population dynamics or the influences of other cultures on the cultural evolution of a certain population were also examined (p. 441 in Baker 1988). Cultural ecology claimed to have developed an approach that would enable an explanation of origins of cultural characteristics. What the method of Cultural Ecology is actually capable of showing, however, is first of all only that the occurrence of certain types of resources, typical subsistence modes and behaviour traits are covariant, i.e. functionally interconnected, and furthermore that the same correlations occur in different regions and historical times. Yet this does not necessarily mean that characteristic culture traits are in fact caused by certain environmental conditions (p. 9 in Hardesty 1977).

Leslie White, another proponent of Cultural Ecology, shared with Steward both his emphasis on culture as the actually interesting and significant unit of investigation and his interest in the evolution of culture. Similar, too, was the way he organised culture into technological, social and ideological components. In contrast to Steward, however, White put the emphasis on the use of energy as the determinant of cultural evolution (White 1943). In this sense, he aimed at a linear and monocausal explanation for cultural evolution, in which both the efficiency of energy utilisation and the application of certain technological achievements was seen as the prime driver of cultural evolution. This is in contrast to the classic Stewardian view, which would permit different causes to bring about different lines of cultural development (Orlove 1980).

1.1.4 Ecological Anthropology

White can indeed be seen as a forerunner of an ecological anthropology that favoured the use and inclusion of the principles of general ecology. As humans are regarded as parts of the ecosystem, human adaptability could now, in contrast to the previous emphasis on culture-typical behaviour, be examined in terms of physiological, biological and behavioural reactions arising from interrelations with environmental conditions.

Accordingly, the term Ecological Anthropology (Vayda and Rappaport 1976) was coined, in order to show that such a discipline would aim at the analysis of culture characteristics by applying the principles of biological ecology. This was also expressed by the fact that biological terms such as population or ecosystem were increasingly being used now, thus allowing better comparability with other ecological studies (see p. 58ff in Moran 2000). With methods empirically more precise than those of Cultural Ecology, the role of other organisms could now also be factored into a description of shaping environmental characteristics and, in particular, the interaction with other human groups. Accordingly, Ecological Anthropology is defined as "the study of the relations among population dynamics, social organization, and culture of human populations and the environments in which they live" (p. 235 in Orlove 1980).

A systems approach offers a very attractive concept to anthropologists in that it possesses great similarity with the holistic notion of anthropology: a system is the great whole, where the individual parts are connected with one another in such a way that they cannot be understood in isolation and without knowledge of the others. Interpreting and understanding the variability of human reactions to given environmental conditions would thus have to consider the interrelation between the realms of cultural, biotic and abiotic factors within the system or habitat.

Subscribing to the principles of general ecology also meant adopting the concept of homoeostasis, according to which ecosystems would in principle have the property of self-regulation. The notion fostered one of the most influential investigations that promoted homoeostatic regulation of a population to its local resource situation. Taking the Tsembaga of highland New Guinea as an example, Rappaport (1968) believed it was possible to prove that, within the society through the cyclical nature of certain feasts, self-adjusting mechanisms were established as a culture trait. On the occasion of these celebrations, large numbers of pigs kept and accumulated over many years would be slaughtered as part of the ritual and thus removed from the system as competing with humans for food, since both relied on horticultural crops as their main staple. The discrepancies, which in hindsight arose from the idealistic interpretation of systemic correlations, consequently led to a redefinition of equilibrium states and homoeostasis was now seen as referring to the maintenance of general system properties, such as resilience.

Accordingly, human ecosystems are considered open and characterised by positive feedback, non-linear oscillating processes and intentional intervention. Homoeostasis and dynamic equilibrium are not equivalent to the absence or the impossibility of change. On the contrary: they require the constant adjustment of parts of the system or complete structures. Systems therefore possess low-level mechanisms which aim at the maintenance of stability and other mechanisms which have an effect at a higher general level and which maintain the system as a whole. In order to be able to follow and analyse such processes, Ecological Anthropology is thus explicitly oriented towards a diachronic method with an emphasis on the study of change, in order to identify adaptation as a process (Orlove 1980).

Not surprisingly, due to the inherent potentially high complexity, the application of an ecosystem perspective to human populations seems to work most satisfactorily at the micro level, in the comprehensible and moderately complex situations of local populations (e.g. Kemp 1971; Thomas 1976; Little et al. 1990). It remains contentious, however, whether and to what extent the results of such studies can be used in the overall context of human behaviour and human adaptability. However, only seemingly is there a contradiction in the incompatibility of interpretational levels. Increasingly, individual decisions and strategies are being included in ecological studies as factors whose effects on the population and the system can be analysed (see below, behavioural ecology; p. 62 in Moran 2000). Thus, there is a possibility of linking both levels and interpreting observable structures at the population level as the sum of individual behaviours.

1.1.5 Human Adaptability

Parallel to Stewardian Cultural Ecology and Ecological Anthropology, an approach was developed during the 1960s that was concerned with the influence of the natural and cultural environments on the biological characteristics of human populations. From the start, this concept of Human Adaptability was supported by several, predominantly scientific disciplines whose fundamental agreement was based on the realisation that humans are a product of natural evolution and in their genetic make-up would reflect the outcome of adaptations to their respective environments. Prominent representatives, particularly in England, are e.g. G.A. Harrison, Weiner or Tanner. The approach is largely based on a notion of adaptability, which regards individuals as being equipped with a set of possibilities encompassing physiological reactions and a behavioural repertoire that provides means of survival within certain limits. Well studied examples are, for example, thermoregulation or balanced polymorphisms, as in sickle cell anaemia. Adaptation is thus understood as any kind of biological reaction which reduces environmental stress and/or increases tolerance against the stressor (Baker 1988) and it may take the form of population-specific genetic characteristics, physiological acclimatisation and learned behaviours. In the context of integrative concepts such as Ecological Anthropology (see above) these kinds of biological/physiological adaptations blend naturally into the holistic framework. With a slightly different emphasis, fields like prehistoric anthropology have picked up on the concept of adaptability by extending it beyond the strictly Darwinian sense and including any kind of successful adjustment to ecological stress. The notion of environment is not limited to natural conditions alone, but embraces the cultural, social, political and economic reality of humans. Recently renewed interest in such biocultural connections are particularly fostered by anthropologists concerned with the effects of environmental factors on health and well being (e.g. Larsen 1990, 1997; Powell et al. 1991; Steckel and Rose 2002).

1.1.6 Cultural Materialism

A second major line of research, which developed from the ideas of Cultural Ecology, is called Neo-Functionalism, perhaps better known as Cultural Materialism, named after Marvin Harris' (1927–2001) influential great narrative (Harris 2001). It refers to Steward and White by adopting their concepts of functional connections between subsistence and culture. In contrast,

named much earlier

however, the social organisation and culture of a certain society are seen as functional adaptations themselves, which enable the successful exploitation of the environment, implicitly without overexploiting the habitat. Not the social system but the population is stabilised and maintained by these functional connections (Orlove 1980). Generally speaking, Neo-Functionalism attempts to demonstrate that there are certain typical aspects in the social organisation and culture of societies which exactly fulfil those functions that make possible the adaptation of a population to their respective environment. Well known examples of this notion are for instance the culture-materialistic answers offered as solutions to so-called ethnographic riddles, such as those of India's sacred cows (Harris 1966) or Aztec cannibalism (Harner 1977).

In analogy to White's organisation of culture into technological, social and ideological components, the culture-materialistic interpretation system also consists of three organisational levels, termed infrastructure, structure and superstructure, at which the cultural functions of a population become effective (Harris 1974; Ross 1980). Modes of production and reproductive conditions belong to the level of the infrastructure, the structure level contains aspects of the domestic and political economy, while religious, aesthetic or philosophical aspects are part of the superstructure. The infrastructural level is of particular significance, since it is here where production and reproduction are causally linked and believed to jointly affect the demographic, technological, economic and ecological links between culture and nature. Both are thus relevant to cultural evolution and selection. Modes of production and modes of reproduction are considered the essential survival tasks of populations, which have to be reconciled in functional terms and brought into accord. Far from a biological explanation and quite in the sense of materialistic views, for example, the meaning of children and the attitude of a society towards children, are seen as a function of their usefulness for the economy, their value as workforce (p. 171ff in Harris and Ross 1987). Reproduction thus becomes a function of production and *vice versa*. The success of such parallel optimisation strategies is evaluated in the sense of adaptation, however in contrast to biology again not from the point of view of natural selection – which would not operate at the population level – but in the sense of a strategy which is generally geared towards the improvement of levels of survival in a population. Implicitly it is assumed that culture traits would fulfil the function of keeping the population density below or at the capacity limit and to achieve homoeostatic stability of the system.

Through its emphasis on subsistence, one of the merits of Neo-Functionalism is certainly to have compiled and provided a multitude of specific descriptions of food procurement strategies. Furthermore, it allowed showing that characteristics of both the environment and the social organisation and culture of populations are systematically connected by the elements of production and reproduction. Also, the pronounced reductionist notion of culture-materialistic explanation is attractive, at least to biological anthropologists. In some important points, however, the theoretical approach received criticism (see Orlove 1980). First, it is a functionalistic fallacy to assume that human populations are generally in homoeostasis, because this overlooks that there are populations which indeed degrade their environment so strongly that the capacity limit is exceeded as a consequence of overexploitation. In addition, the identification of homoeostasis requires a temporal observation depth, which usually cannot be provided in fieldwork situations. Second, it remains unclear to what extent social and cultural characteristics of populations, selectively or as coherent characteristics of a population, influence resource use. And finally, although energy is an important element, it is not always inevitably the central limiting factor in a habitat, where water, animal protein or certain trace elements can also be severely limiting due to low abundance. Borderline energy outputs in a population or a production unit, for example, may be due to the necessity of having to produce for a market in addition to providing for one's own subsistence. Nevertheless, even such external constraints can lead to adaptive strategies of production, without being inevitably limited by energy as a minimum component.

1.1.7 Behavioural Ecology

Theoretical models for the explanation of human/environment relations introduced so far operate at the population level or regard the population as the actual subject of analysis. When during the 1970s the concept of evolutionary Behavioural Ecology (Krebs and Davies 1991), originally developed in animal ecology, began to be applied to humans as well, methodical approaches became available by which a desideratum of Ecological Anthropology finally could be redeemed, namely the question as to how individual behaviour and its effects on the population can be measured empirically. If one, for example, assumes that a certain individual behaviour developed as a result of optimising adjustments to the efficiency of resource use, then the sum of these behaviours would consequently result in an accordingly optimised population behaviour. Observable differences between human groups would then have to be attributed to differently successful behavioural adaptations as a reaction to different environmental conditions (Baker 1988).

Behavioural Ecology can thus be regarded as a sub-discipline of evolutionary biology, which analyses behaviour in an ecological context (Smith 1992a,b; Smith and Winterhalder 1992). It has been beneficially complemented by research into life history strategies (Hill and Kaplan 1999). The approach is based on two basic assumptions. (1) The behaviour and phenotype of organisms are shaped by natural selection. This entails the expectation that individuals behave in such a way that their personal reproductive success and/or their inclusive fitness is maximised. Today's observable phenotypes would then be the expression of those genetic make-ups which, due to higher adaptability, reproduced proportionately more frequently. (2) In any ecosystem time and energy are limited. Therefore, in view of these constraints, individuals have to weigh the costs of certain behaviour against its benefit. What has been invested in one type of behaviour in terms of energy and time is no longer available for another. An example would be the potential conflict in trade-offs between reproduction and longevity.

Based on these assumptions, it is possible to examine why individuals employ time or resources the way they do; and it can be tested whether at any one time this behaviour is adaptive in the sense that it increases personal fitness. One can, for example, ask why under given environmental conditions a female forager would harvest a certain fruit or dig a tuber compared with possible alternatives at her disposal. A typical explanatory hypothesis of Behavioural Ecology would be that from the chosen selection of harvested plants a gatherer would receive the optimum intake of nutrients under the available circumstances. This inquiry is eventually geared towards understanding the ultimate function of behaviour with regard to fitness gain. A frequently used procedure by which such hypotheses are examined is the application of optimisation models for food procurement (e.g. Optimal Foraging Theory, Sect. 2.3.1.3), which exactly raise the question of connections between the fitness-oriented benefit of (nutritional) behaviour and the costs resulting from it (e.g. Smith 1983). With a view to the evolutionary aspect of Behavioural Ecology, covariance between behaviour and ecology is assumed, with the individual being at evolutionary equilibrium. This is to express the fact that organisms are well adapted to the specific constraints of their habitat whenever these have already been familiar to their ancestors in earlier generations, i.e. if there was sufficient time for adapting behaviour to environmental constraints. Enduring limiting conditions of the habitat will in the long term lead to optimal solutions in order to moderate these constraints (Hawkes 1996).

Behavioural Ecology differs from other ecological approaches to the analysis of human societies and populations by the fact that individual cost/benefit considerations are seen in comparison with the conflicts of interest caused by other individuals of the group who show the same behaviour, i.e. who for example use the same food procurement strategy. Emphasis is thus laid on the consideration of inter-individual conflicting aims, from which behavioural variability between individuals can be observed within a population, and it incorporates behavioural variability.

Regarding the genetic basis of the observed variance in behaviour, Behavioural Ecology builds on the working hypothesis that the possible breadth of behavioural responses available to an individual as reactions to certain situations is a result of natural selection. The ultimate goal of having as many of one's own genes represented in future generations creates no fundamental dissent between sociobiologists – the purists among Behavioural Ecologists – and a "non-specific notion of inheritance" in the context of an ecologically aligned Behavioural Ecology (Smith 1992a). In both cases, behavioural expressions are concerned which eventually become recognisable in phenotypic characteristics and are indeed selected through exactly these phenotypes.

1.1.8 Global and Political Ecology

Conscious of and concerned with past and present outcomes of human environmental exploitation, research areas were established which are more strongly geared towards the ecological implications of human action on a supra-regional or global level. The approach of Global Ecology is closely connected with the topics of Environmental Anthropology, which through considering and analysing evolutionary and historical trends of environmental exploitation attempts to identify and point out the resulting likely consequences for future generations. Of particular importance is the reverting to the culture-comparative method of Steward, in order to weigh and process information about causes and effects of local and regional environmental changes and to test to what extent the results of these analyses can be transferred and applied to current problems (p. 383ff in Moran 1996; see also Bennett 1990).

Another more recent spin-off of original Cultural Ecology has to be mentioned which is connected with the term Political Ecology (forerunner e.g. Bennett 1990), a concept where the traditional areas of inter-relations between environment, technology and social organisation meet. It is characterised by an analysis of social conditions against the background of political economy, the unequal distribution of resources, increasing criticism of conventional development aid programs and the threateningly increasing environmental consumption (Netting 1996). Like the concept of Global Ecology, and in view of the complexity of a Political Ecology, the discursive discussion of many human-related natural and social sciences is indispensable for a linkage of these disciplines.

Increasingly and as a result of their exponential consumption of resources and equally exponential reproduction, humans are regarded as the main cause of change on this planet. Stocktaking of resources world-wide and correlated global problems of population growth, food production or refuse disposal, just to mention a few, leads to the formulation of shortage and damage scenarios. Justified appeals for a reflection on the resource-sensitive handling of nature and the necessity for a sustainable economy follow naturally. Emphasising the population problem and the shortage of raw materials, the loss of non-renewable energy sources and the burden of environmental pollution for future generations are frequent topics of relevant publications (e.g. Goudie 2000; Nentwig 1995; Glaeser 1995; Southwick 1996; Diesendorf and Hamilton 1997). Insights from these investigations of human/environment relations feed into outlines of a preventive environmental policy (e.g. Glaeser 1989; Becker and Ostrom 1995; Ostrom 1999; Dietz et al. 2003).

Emphasising such topics is of central importance in order to be able to measure the consequences of current action and the way humans treat nature, to issue warnings and if necessary to counter-steer with suitable measures. Such concepts and notions of human ecology are on the whole not connected by a unified theoretical framework, yet hierarchy theory and modelling to account for the complexity of different encaptic levels of human ecosystem organisation have been proposed as suitable approaches (Gare 2000; Wu and David 2002).

The role of the social sciences is interesting here. It was only in the 1970s that Sociology even recognised the existence and dynamics of human/environment interaction, a conceptual breakthrough termed "The New Ecological Paradigm" (Catton and Dunlap 1980; Dunlap 2002), at a time when all the major anthropological ecologies had long been outlined. It seems, however, that the epistemological gap has been closed and the social sciences have established a firm place in the debate over environmental change and human impact (Scoones 1999).

1.1.9 Environmental History

Also strongly committed to the cross-disciplinary discourse is Environmental History. With long traditions in the UK, it can be considered a relatively recent branch of research in continental Europe. At present this area is still heterogeneous, since it is occupied by different disciplines from the humanities, social and natural sciences, with their respective different methodical apparatus and theoretical foundation. However, it has been claimed that the overcoming of this still-existing disciplinary compartmentalisation is exactly the precondition for a 'method' of Environmental History: the acquisition of the data sources is to remain with the proven competence of individual disciplines, while their evaluation and appreciation would however necessarily require an interdisciplinary discourse (e.g. Herrmann 1996). Environmental history is thus understood as an area of comparative research, which is geared towards a systematisation of phenomena across time and nature. It seeks to describe and categorise relations between humans and their environment in terms of constants or generic principles and to analyse their causes and their direct and indirect consequences. At present, there is still no general consent over the development and use of theoretical foundations of Environmental History, although there is a plea from the science disciplines calling for clear connections to Cultural Ecology and reference to an ecosystemic approach.

Environmental History shares large conceptual subsets with Environmental Anthropology (Moran 1996), an area of comprehensive research which is particularly prominent in America. Here, again, reverting particularly to the areas of Cultural Ecology and Ecological Anthropology and their system-theoretical approach is important for emphasising the interdependence of natural conditions and human actions. The concept not only encompasses individual-centred approaches of behavioural ecology but also includes the time factor by examining the history of resource use in human communities and its effects for forming the environment, i.e. the history of environmental transformation rather than human adaptability.

The main ideas of the theoretical trends presented can thus be summarised as follows. Concepts about the influence of the environment on humans and their cultures were shaped by two very opposite positions at the beginning of the last century. According to the deterministic point of view, it was primarily factors of the natural environment that had a coining and dominant effect on the development of technology, customs, traditions and even political structures of populations, while possibilism claimed that the respective historical and cultural characteristics at disposal under the given environmental conditions were decisive. The environment exerted a determining effect through deficiency, not through variety of supply. In a very similar way in which the environment-centred approach could offer possibilities of overstretched generalisations, they were waived by the culture-centred approach with its stress on situative peculiarity.

As a reaction to this notion of static human/environment relationship, Cultural Ecology put forward an interactive understanding of the mutual influence of the natural environmental condition, subsistence and social structures of a population. Subsistence technology and resource use are considered a reflex on the environment and are mediated by behaviour, which is typical for exactly these utilisation strategies. Thus under comparable environmental conditions, adaptive processes should develop that can be recognised as cross-cultural regularities.

By the same token, the neo-functionalist approach of cultural materialism looks at consequences of production conditions for population dynamics. Social reality is examined in the light of mutual influence of production and reproduction; and adaptive success is measured against effectiveness of resource extraction using cost/benefit calculations.

With the assumption of principles of general ecology and the ecosystem concept, the population is becoming the central unit of investigation, whose interdependence with other organisms and particularly with other human populations can be examined in a given habitat. Adaptability is measured against the ability of a population to aim at dynamic equilibrium in principally open systems with the help of biological and cultural responses. The emphasis of this Ecological Anthropology is on a quantification of correlations between subsistence and social system, e.g. by investigations of flows of energy and matter.

By adopting models from behavioural ecology, the final step is accomplished for analyses at the individual level; and methodical access is available which can empirically seize the adaptive value of individual behaviour. Fitness is the cause-effective unit that allows the success of behaviour to be examined under given ecological conditions and to be evaluated in its effects at the population level. The weighing of costs and benefits of certain behaviours is made possible by a reconstruction of life history strategies or optimisation models.

This brief overview has two important outcomes. First, there is an increasing particularisation of ecological research that can be taken as a sign of ever more strongly specialised questions, approaches and viewpoints, by which human use of the environment is observed. Second, however, its divergence bears the inherent danger of losing internal cohesion. Certainly not just by chance, as the above-mentioned more recent approaches show, increasingly concepts under large comprehensive contextual umbrellas are launched in order to try and unify the multi-layered approaches and numerous disciplines dealing with humans as products and formers of their environment. There is both demand and necessity for an integrative human ecology which is equipped for diachronic analysis and a comparative view committed to a systemic structuring principle and that, both on the level of individuals and populations, makes adaptation ascertainable as a biological and cultural goal. Such a concept, which is fostered in this book, will have to be necessarily eclectic, depending on the research question or the complexity of the topic. But by consequently reverting to the principles of ecology, it can offer a basis for the disciplinary discourse that is uniform and geared towards comparability. In the following section, therefore, first of all basic and subsequently more specific aspects of human ecosystems will be dealt with.

1.2 Humans as Parts of Ecosystems

1.2.1 The General Framework

Organisms interact with their environment in different intensities and at different levels. For a description and interpretation of relations between groups of organisms, and organisms and the environment acting at these different levels of integration, there is a common epistemological category in ecology: the ecosystem (Tansley 1935; for the history of the ecosystem concept, see Golley 1993). It is characterised as any self-contained entity which, in a given area, encompasses all organisms interacting with the physical and chemical environment, so that there is diversity of biological relations and material cycles and energy flow that create clearly defined food chains (see e.g. Odum 1983; Begon et al. 2003).

An ecosystem is composed of a set of components which act in combination within the system and which can be divided into the classes of abiotic and biotic components. Among the abiotic components there is first of all solar energy, which propels the entire system, and furthermore chemical factors in the form of inorganic and organic compounds, as well as physical factors in terms of temperature, light, wind and precipitation. These are contrasted by the biotic components, which create biomass, i.e. life, by utilising the material provided by the abiotic components. The biotic realm is divided into autotrophic producers, which synthesise complex organic compounds by direct utilisation of solar radiation (photosynthesis), and heterotrophic consumers. The latter are not capable of direct conversion of solar energy, but can only exist by utilising the organic compounds produced by autotrophic organisms. Within the group of heterotrophic organisms, one differentiates between macro consumers, e.g. herbivorous, carnivorous or omnivorous organisms, and micro consumers, the so-called decomposers, which break down biomass and thus feed the inorganic and organic building blocks back again into the system.

Abiotic and biotic components are connected through structuring principles, which at the same time denote fundamental characteristics of an ecosystem. Accordingly, ecosystems are characterised by the spatial and temporal distribution patterns of their components, by the transport of material (flow of matter) and utilisation of energy (energy flow), by the exchanging and passing-on of information (information flow) and by the properties of change and evolution. How comprehensively an ecosystem is defined depends on the observer's viewpoint. Ecosystems do not have a given size or structure. They can take the form of a pond, a limestone beech forest, a savannah or a tropical rain forest. Each of these units is in itself an ecosystem in the sense of the above definition and also possesses all characteristic properties mentioned. Because these properties are functionally integrated, however, additional specific characteristics can occur at higher levels of complexity which are not predictable from the structure of the less complex system. In other words, ecosystems are defined as inclusive structures where the whole is more than the sum of its parts.

By applying a systems concept, it follows that the components and elements of the ecosystem are interrelated and that they represent certain conditions and degrees of organisation whose development can be adjusted by feedback mechanisms. Predator/prey relations are a classic example. Interactions between system components thus operate through cybernetic processes. Under ideal circumstances, such control loop relations lead to temporary self-adjusting equilibrium states (homoeostasis).

The units of space, time, matter, information and energy form the key ecological categories (see e.g. Odum 1983; Zwölfer 1991; Begon et al. 2003): The distribution of ecosystem components in *space* forms the framework of factors that enable and limit biological processes. *Time* determines their course of development, the irreversibleness of interactions and the evolution of system components. Together, space and time constitute the major axes of an ecosystem. Within this coordinate system, organisms display distinct distribution patterns and behavioural characteristics, their habitats and niches. To paraphrase Odum's illustrative analogies – habitat denotes the typical location of an organism, its 'address'. At this address, the organism has a certain function or 'profession': it occupies a niche where it has a defined role in the flows of energy and matter and where its behaviour facilitates interaction with other organisms and reactions to given environmental gradients. All processes in an ecosystem are tied to *matter* of some sort, which forms the substrate for the composition, change and decomposition of its structures. The necessary components originate from biogeochemical cycles of matter, in which the chemical elements are recycled between environment and organisms. Slow processes form cycles, which are constantly fed from the predominantly abiotic reservoir of the environment. Exchange between organisms and their immediate environment takes place in accelerated processes, i.e. biotic material cycles. Of some 100 known stable chemical elements only 30–40 are necessary for the maintenance of vital functions of the organisms and form the basic set of building blocks for a variety of very different metabolic performances. The transmission of *information* within an ecosystem is based on genetically coded biological information in the first place. Information cycles thus represent the implementation of genetic information in terms of structural blueprints of organisms, metabolic actions, etc. In this sense, biological evolution can be essentially understood as the advancement of genetic information which forms the basis for blueprints and operational networks of organisms, where 'advancement' implicitly means 'under the conditions of natural selection'. Apart from the material side, however, there is information flow by which learned behaviour and experience which shape the interaction of organisms with their environment are passed on, often in terms of behavioural traditions. This kind of extrasomatic transmission of information contributes crucially to the high plasticity which organisms, particularly humans, show in their reactions to ecosystem conditions. *Energy* is the component necessary to drive cycles within an ecosystem, to structure flows of matter and to build and maintain the different levels of structural organisation. In contrast to matter, which is constantly being reused, energy cannot be recycled. According to the first and second Laws of Thermodynamics, it holds that: (1) Energy can neither be created nor destroyed, but is changed, i.e. it is exchangeable in its manifestations, and (2) the transformation of one manifestation into another is always connected with the loss of energy in the form of heat, i.e. there is a constant degradation of energy (dissipation). Processes of energy conversion run spontaneously only towards the low-energy condition, which owes itself to the physiological principle according to which the maintenance of structures and processes always needs to take place against energy gradients, since all systems are inherently characterised by an increase in entropy, i.e. a decrease in their internal order.

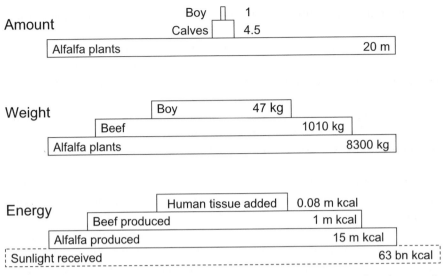

Fig. 1.1. Energy and material conversion for a hypothetical scenario to show how much veal would have to be produced so that a boy feeds on hamburgers for an entire year (modified from Southwick 1996)

A chain of activity initiated by solar energy and leading from autotrophic to heterotrophic organisms determines productivity and the resulting trophic relations in an ecosystem. Food chains constitute the simplest case where food energy produced by autotrophic plants is used in several steps on its way to the highest order consumer. Normally, however, several such food chains are interconnected to form a food web. Both cases are characterised by energetic relations that lead to the formation of several trophic levels, for example in the successive order of producers, herbivores, omnivores/carnivores. Since the transition between trophic levels, i.e. a switch from one stage of energy conversion to the next, is connected with an 80–90 % net loss of energy via a heat sink (see above, second Law), the number of possible trophic levels within a food chain or a food web is limited. Therefore, with regard to productivity, trophic structure and population densities, ecosystems are organised in a pyramidal fashion (Fig. 1.1).

1.2.2 Theoretical Considerations

Wherever human populations are part of ecosystems, they are subjected to the same cycles, regularities and processes of ecological sequences as are other organisms. They show characteristic spatial distribution patterns in their habitats, which are linked to resource density. They participate in material and energy flows and are part of food webs. In other words: just like any biotic component of an ecosystem humans are tied into structural and functional relations with living organisms and the inanimate environment. An ecosystem is thereby regarded as the overall situation within which human adaptation occurs or against which adaptation becomes necessary (p. 8 in Moran 2000). In habitats occupied by humans, their niche is "a specific set of capabilities for extracting resources, for surviving hazards, and for competing, coupled with a corresponding set of needs" (p. 394 in Colinvaux 1982). It is true; and this definition of niche would apply to almost any other organism. However, it is based on the notion that the particular quality of human individuals and populations is their specific ability to intentionally interfere with, steer and change interrelations with their environments through cultural and social systems. Not only do they have to respond to given conditions of the habitat or ecosystem they live in, but they are able to alter these conditions by changing their environment. To refer to Odum's analogy of the niche concept again, it is part of the human profession to change landscapes, to exert influence on species diversity of a habitat and to regulate energy flows in order for resources offered by the habitat to be used effectively and on a long-term basis.

It is obvious that humans do so, but how and to what effect and outcome is likewise part of these very niches. Unlike all other species, human interference with the environment makes use of flows of non-genetic information. It allows available options of environmental transformation to be translated into strategies, rules and arrangements, which become part of their culture, part of their institutions. Culture thus forms a cache of information that, unprecedented in size and scope, can be retrieved, altered and adjusted at short notice. As cultural information is passed on by way of traditions across generations, it helps stretch the time dimension of ecosystems. Humans perpetuate culture into an ecosystemic category that is available on a long-term basis and the very use of cultural information is a characteristic trait of their niche. Humans require culture to fulfil their niche role (Hardesty 1975). However, the term "cultural species" sometimes applied to characterise the distinctiveness and uniqueness of certain societies may actually overshoot the mark, as it suggests a kind of exclusivity and incompatibility between human groups that does not fit. No human community is so isolated in its habitat – maybe with the notable exception of the Palaeolithic – that there are no structural relationships between different populations. The regular and intensive contact and exchange between foraging societies and their neighbouring farming communities is a case in point. Culture is a property of human ecosystems and as such it both expresses the fundamental meaning of cultural characteristics for the functional and structural relations of human populations with other components of the ecosystem and clarifies that the diversity and plasticity of human ecological niches essentially results from exactly such cultural characteristics.

It follows: since humans are subject to the same ecological principles as other components of the ecosystem, they can be examined under system-theoretical aspects (see e.g. Odum 1983). This viewpoint opens up possibilities of theory-led analysis applied to strategies of survival or resource appropriation as part of the functional and interactive processes, by which the components of an ecosystem are intertwined. The relationship between humans and their environment then appears as a sequence of constant mutual influences within a given system. Not only biological characteristics, but in particular cultural structures can be explained against the background of this systemic connection. The human condition is biocultural and thus denotes the situation where biological conditions and outcomes can be explained as a result of the establishment and perpetuation of cultural strategies. A human ecology will therefore have to explore these connections and draw its explanatory value for the emergence of biocultural systems from the principles of ecological integration.

The evolution of concepts that allow viewing humans under an ecosystemic paradigm was outlined earlier. In the following section, aspects of ecological theory formation and methodological development will be revisited and those issues dealt with individually which, due to theoretical considerations and practical experience from field studies, saw change and adjustment and which are important in the context and scope of this book.

The fundamental meaning that is attached to the link between system-theoretical and anthropological investigations is powerfully expressed in the ongoing United Nations (UN) research projects of the International Biological Programme (IBP), the Man and the Biosphere (MAB) scheme and other large-scale research efforts. They enabled two approaches to be linked for mutual benefit that have numerous common grounds and show similarities: on the one hand the systems approach that identifies humans and their cultures as components of ecosystems, and on the other hand the principally holistic approach of anthropology as a science aiming to understand the natural and cultural history of hominids (Moran 1990). Both are based on the idea that, at any one time, a totality is considered whose individual parts cannot be understood without knowledge of the whole. Steward's concept of Cultural Ecology (Steward 1955) has to be seen as an important starting point here. The basic idea of this concept, the fundamental links between modes of social organisation in human communities and their modes of subsistence, can both be studied cross-culturally. Social institutions are regarded as functional units for the solution of recurring subsistence problems. This culture-centred view was later, as it were, balanced by the introduction of a more strongly biologically oriented ecological paradigm, as the application of the ecosystem concept paved the way for an assessment of structures, functions and equilibria in human communities (see Winterhalder 1984). Now human behaviour was no longer primarily located in the sphere of cultural phenomena, but was likewise analysed from the point of view of ecological processes.

Meanwhile no one would question the suitability of an ecosystem-oriented ecology for anthropological questions; and the fundamental discussions about theory may to a large extent be regarded as overcome and terminated (Casimir 1993). In relation to the principles of a general ecology and/or ecosystem theory, however, a set of prerequisites and basic assumptions were specified (Orlove 1980; Moran 1990, 2000) which ought to be discussed in some detail, since they are important for an application of the ecosystem concept in anthropology.

1.2.2.1 Structure and Boundaries of Ecosystems

Ecosystems are generally regarded as open systems which need a constant supply of material and energy in order to sustain their system properties (p. 1ff in Schulze and Zwölfer 1987). As a pragmatic systematisation of space, time and hierarchical levels of the analysed topic and the scientific question, however, it can be helpful to specify boundaries of ecosystems in order to define a clear spatial framework, even though it is acknowledged that this may pose a problem (p. 691 in Begon et al. 2003). An ecosystemic study of human communities would be conducted in a comparable way and local populations be selected by employing a spatially limited territory or effective range as the unit of investigation. Most often in human ecological studies this is equivalent to the analysis of a certain ethnic group or society, for example the Inujjuamiut in Alaska (Smith 1991) or the Turkana in East Africa (Little et al. 1990), but it can as well be the inhabitants of a village, who in this sense would not necessarily be regarded as a distinct ethnic unit, for example the inhabitants of the Swiss village of Törbel (Netting 1981) or a skeletal population.

The choice of local populations has been associated with the potential drawback of implying that, by their sheer limitation in space, such clearly defined groups could be characterised as being self-sufficient and closed units; and it was criticised that this would consequently lead to a simplified, reductionist concept of the ecosystemic integration of human populations (Winterhalder 1984; Moran 1990). The reason is that human ecosystems are characterised by two sets of factors. First, they are defined by the climatic, biological and geomorphological framework, i.e. the natural defaults of the habitat they occupy. These could also be called the primary ecological basic conditions, since they entail a pre-steering of, for example, subsistence activities. This externally defined system is complemented by specific secondary ecological basic conditions defined by the population living in a habitat. The human population can be regarded as a unit tied into one or more ecotopes, e.g. as resources from one or more habitats are utilised. At the same time, however, human communities act as the arranging and changing agencies in their respective habitats and thus themselves define their ecosystem as the area, or space/time structure, that they use. Through the socio-cultural organising of their lives, humans decide on the way they act within their habitat.

Local populations are not self-sufficient, but are connected with other populations outside their habitat by trade and barter of goods, by exogamous marriage relations or by regional and supra-regional networks and alliances. The boundaries of ecosystems shaped by human activity are open and variable, because they are to a substantial part also defined by such functional contexts. For the analysis of systemic links, it is therefore also necessary that spatial delimitations can be explained functionally and analytically. Depending on which point of view is chosen something may thus appear as a closed and self-regulating system on one level, while on another level of analysis, e.g. outside a given spatial framework, it appears open with regard to the first level and would thus correspondingly be no longer self-regulating (Winterhalder 1980; Ellen 1990; Moran 1990). It therefore appears to be useful to follow latest recommendations (e.g. Begon et al. 2003), which suggest that rather than trying to define boundaries it is more in keeping with ecological reality to view communities at certain levels of organisation and hierarchy – a view that certainly fits human societies.

Regardless, a distinction of functional levels is of importance, because it helps avoid the discussion whether ecosystems are self-adjusting entities or not (see Engelberg and Boyarsky 1979; Patten and Odum 1981). Ecosystems are structural and functional units, but they do not possess organismic characteristics, according to which they would follow the rules of self-regulation and homoeostasis. It is true that their components are subject to the laws of biological evolution, but they are neither cybernetic as such, i.e. not by themselves capable of self-regulation through the use of information flows (Engelberg and Boyarsky 1979), nor will ecosystems develop survival strategies that would make their further existence more probable.

It can be observed, however, that within systems there seems to be a teleonomic tendency towards equilibrium, as by way of natural selection a condition of relative stability is reached on a long-term basis with continuous patterns of individual interaction and small fluctuation of populations (Krebs 1978; p. 58ff in Moran 2000). In this sense, ecosystems transform their order as a reaction to external and internal change. This however is only seemingly an active role of the system. The process is rather caused by the fact that the individual components of the system are stabilised in the long-term, or to quote Patten and Odum (1981; p. 888): "... cybernetic attributes emerge passively out of large and complex, decentralized system organization". The permanence of the biosphere, for example, is perceived as a structure formed by self-organisation (Abel and Stepp 2003).

As a consequence of this internal flexibility, there will be no static, but resilient equilibrium states, i.e. continuous adjustment towards stability domains. Within this dynamism, humans on the one hand react with cultural responses to changes in their habitat, e.g. climatic shifts, seasonal resource scarcity or the like; and on the other hand humans themselves transform the environment through organised socio-cultural activity, for example by altering landscape for the purpose of food production. Besides, humans are able to affect ecological dimensions such as population size and resource use through cultural decisions and institutions. Humans do not simply adapt to the conditions and constraints of nature, but actively change the ecological framework through active management and strategies. They intervene in cycles and processes of these systems and are able to adjust them in an attempt to control material and energy flows. In doing so, they at the same time create the prerequisite for the continuation of their own existence in the very system they shape and use. The entire range of human regulative actions defines an ecosystem. In this sense (only) an ecosystem used by humans can also be regarded as self-adjusting (Rappaport 1990).

If later in the book, for example, adaptive strategies of food acquisition are addressed (see Chap. 2), this implies that such observable mechanisms of habitat use are indeed conditions of system stability which have passively adjusted by the repetition of certain interactive operational sequences between humans and other system components. Within a larger time-window or in view of historical processes, these stable conditions then correspond to the re-establishment of stability domains, after e.g. subsistence change (Chap. 3) has temporarily disrupted the stability of the system.

1.2.2.2 Energy Transformation

The description and analysis of mechanisms of energy production, storage and consumption, i.e. energy flows in ecosystems, have been of central importance to ecological anthropology for a long time. Inspired by White's (1943) theoretical model, according to which cultural evolution is determined by the control of increasingly larger amounts of energy, there is ample empirical confirmation for the close relationship of energy extraction and social organisation in both extant and historic societies (e.g. Tainter et al. 2003). For example, studies on strategies of subsistence acquisition and connected energy flows in pastoral societies show to what extent similarities in the reactions of human communities to certain environmental constraints allow for conclusions about similar adjustments of subsistence modes to be drawn (Casimir 1993). Energy is the one currency here which is convertible between societies of the most different organisation and subsistence (p. 83 in Ulijaszek 1995). Meanwhile there is only little debate as to the orders of magnitude in which energy flows through a system – this has been dismissed as "caloric obsession" (Moran 1990) – but increasingly decisions are being considered and analysed that lead to the change and control of energy flows within human communities. In this sense, for example, individual segments of food production processes can be quantified (Winterhalder 1984) and differential efficiencies determined. In other words, the social mediation of energetic relations in an ecosystem can be broken down into their individual components and the relative contribution and importance of each can be considered.

In view of the impossibility to provide ecological total balances in principle (for a historical example, see Herrmann 1997), simply because not all human activities can be transferred into monetary or energy-equivalent values, the use of energy flow analysis on the macro level will be limited to uncovering general regularities (see e.g. Harris 1977). Connecting the aspects of social organisation with those of subsistence modes, however, allows the study of the energetic bases and the efficiency of strategies of resource use in human communities (see Chap. 2). In contrast to this and in view of the social mediation of energetic relations, the application of energy flow analyses on the micro level appears to be particularly suitable, both theoretically and practically, since it is focussed on individuals and refers to a basic production unit, such as a family. It is true that such small units also operate as part of an open system (see above), but by specifying ecosystem boundaries for the purpose of delimiting the range of analysis, the spatial and analytic system can be defined accordingly. Independent of the advantages and explanatory potential of the respective viewpoint, the general problem of investigating energy flows in human communities remains in that these attempts represent static snapshots whose stability is usually not examinable in the course of time and whose enduring adaptive significance remains hidden.

1.2.2.3 The Historic Dimension

Ecosystems have a history. They evolve along a time axis, in the course of which – according to classic concepts – a characteristic ensemble of organismic and abiotic components develops (e.g. Odum 1983). But even without having to refer to this construct of natural succession, developments take place in an ecosystem which can be caused both by external and internal changes. External factors typically fall within the realm of natural phenomena and hazards, while internal changes take place, for example, by human interference with ecosystemic processes. However, humans are equally subject to systemic changes by external factors and thus do not only adopt the role of the disruptive disturbers who alter and transform natural into cultured landscapes. If change and alteration are a system property of ecosystems, then it is the more surprising that the neglect of time factors and historic change in previous studies had to be deplored at all (Moran 1990). The resulting overemphasis of stability and homoeostasis (Jordan 1981) can only be explained by a narrowed punctual view of current conditions in any one population at the point of examination – convenient but insufficient. Studies in archaeology and historical anthropology, which employ a systems approach (e.g. Butzer 1982; Hastorf 1990; Padberg 1996) reveal, however, that only by the inclusion of historic processes in ecosystemic analyses is it possible to gain an empirical approximation to the question of stability or change in the system.

If one considers ecosystems at a given point in time, they seem to be at equilibrium. If one however does so over a period of time, continuous and cumulative change can be observed (Lees and Bates 1990). This change leads to structural transformation in the system, since it is the expression of a process in which the relations of system components with each other are redefined, for example in the form of adaptations of human subsistence strategies to changing natural conditions (see Chap. 3). Rather, systems appear to be constantly in disequilibrium and humans in each case attempt to achieve stability domains enabled by socio-cultural institutions and strategies. This in turn means that the basic condition of equilibrium is subject to oscillations of varying amplitudes, which are an expression of historical change of ecosystems to which human individuals and populations have adapted. A sufficiently small time-frame would create the impression of strong fluctuation and disequilibrium or multiple stable states. The bigger the time-window and thus the lower the resolution, the more the system appears to be stable and at equilibrium overall. There is no necessity to deviate from the concept of steady-state in principle, if one considers that systemic equilibria are not rigid but dynamic, and not constant but punctual or in flux. The underlying mechanism would be largely equivalent to stable limit cycles (p. 374 in Begon et al. 2003).

Including the historic dimension into ecosystemic investigations is of importance for yet another reason. It demonstrates that, for the analysis of human/environment relations in historic times, the same questions apply and can be used that are being pursued in studies of contemporary societies. For example, suitable methods providing, the significance of natural or environmental default factors for the development of adaptive subsistence strategies can be demonstrated for past times, in principally just the same way as adaptations of food acquisition can be observed today – even though perhaps not with the same amount of detail (see Chap. 2). Likewise, again given the appropriate methodical approach, connections between resource use and social inequality within a society can be established and without contradiction be integrated into a general framework of the evolution of social systems (see Chap. 4). By integrating time-depth, historical developments become recognisable which, if regarded as processes parallel or analogous to modern events, also become accessible by ecosystemic interpretation. At the same time, they can sharpen our awareness for and significantly increase our knowledge of possible solutions for ecological problems that are recurring in similar fashion (cp. Janssen and Scheffer 2004).

1.2.2.4 The Decision Level

System-ecological studies run a risk of being taken in by semantic careless-ness, which can then turn into an epistemological problem. What is alluded to here is the widespread notion that cultures, populations or species are capable of adapting to certain systemic basic conditions. This certainly does not apply, because such generic terms only denote some kind of entity or a group of individuals with similar traits or characteristics, respectively. It therefore has to be exactly differentiated, on which level adjustment is being observed, for example on the level of cultural entities (for Egypt, see Butzer 1980), and on which levels cause and effect of adaptation take place. For example, there is good evidence for correlations between subsistence mode and fertility (see Chap. 5), which are observed at a generic level, the cause and emergence of which however occurs at the individual level. Thus, even if certain systemi-cally effective reactions only become recognisable on the level of the popula-tion or within a culture, they are still and foremost based on individual deci-sions (see Smith 1992a, b). If the emphasis is heavily biased towards the population aspect only, the role of individuals as the actually decisive agencies is neglected. Individual decisions, however, are the socially acceptable conver-sions of personal goals. Therefore they are characterised by two ideally con-fluent but sometimes conflicting aspects: the one that, biologically speaking, represents fitness maximisation and the other that stands for the decision being embedded into collective agreement within the system. Both points of view, the neo-Darwinian evolutionary and the culturally systemic, should thus complement each other in the analysis of the ecology of human commu-nities (Moran 1990).

This has concrete bearings on the respective research questions and their interpretation, for example, in which way a particular individual decision taken from a choice of several behavioural options will affect energy flows; or what significance cultural expectations within a community, rank, gender hierarchy, age, etc. will have on individual decisions that will eventually be col-lectively effective and have outcomes at the community level. Only a comple-mentary appreciation of these two very different levels can help avoid falling into the reductionist trap with the dilemma of having to accept deductive statements about the overall system from the individual case (see p. 3 in Schulze and Zwölfer 1987). In this sense, the ecosystem concept is a suitable way to understand both individual decisions and aggregate processes and outcomes. In favourable cases, such ideal requirements can sometimes be met even for historic times (see Chap. 5), while it cannot be taken for granted that both aspects are being considered in the same way in studies of recent com-munities (Smith and Winterhalder 1992).

1.2.2.5 Adaptation

Adaptation has been defined as "... the ability of an organism to adjust to a changing environment such that survival and reproduction are enhanced" (p. 122 in Little 1995), i.e. adaptation denotes any response to physical and chemical factors of the ecosystem, or to interactions with other species and with individuals of the same species, which increases the probability of survival and thus the probability of reproduction. Adaptation is therefore always oriented towards a certain environment or occurs in relation to a certain environment, natural and social, and is subject to natural selection (Little 1995; see also p. 236ff in Ellen 1982; for problems related to choosing the right meaning of adaptation in studies of human behaviour, see Caro and Borgerhoff-Mulder 1987; Irons 1998).

Modes of adaptation can be distinguished by the different reaction velocities they show towards environmental changes (see p. 5ff Moran 2000) or by whether they spread biologically or culturally (Irons 1996). Both aspects are connected by the fact that successful adaptations can be measured by reproductive success in the long run (Alland 1975). For example, long-term adaptations that developed over evolutionary time-spans have led to global distribution patterns in the course of human natural history, e.g. morphotypes of physique or pigmentation. These mechanisms are usually observed or analysed on the population level, although selection, of course, is geared towards and occurs through genetic change in an individual. Adaptation would lead to a shift in the frequency of a certain set of genetic traits, but becomes visible eventually on the population level only through differential individual reproductive success.

In contrast, short-term modifying adjustments occur through immediate morphological and functional reactions of individuals to changes in their environment. The short time-scale at which they occur allows a higher flexibility of responses to changing environmental constraints than would be possible with long-term genetically embodied characteristics. For example, through adjustments during the growth phase, an individual can react to certain given and continuous environmental conditions and develop irreversible morphological characteristics, such as a larger thorax and increased lung volume as a reaction to low oxygen partial pressure in high altitudes of the Andes or the Himalayas. Such structural adjustments cannot be acquired any longer after cessation of growth, yet partial compensation is possible at a functional level through acclimatisation (e.g. Beall et al. 1992). Modifying adjustments are reversible as long as they did not lead to morphological alteration during ontogeny, since they cause temporary change of an organic condition, e.g. an increased erythrocyte titre and capillarisation for an improved oxygen supply, or an increase in skin pigments as a reaction to increased UV radiation, which usually only continues as long as the appropriate environmental stimulus persists. This allows for acclimatisation to take place as a temporary adjustment even without evolutionary adaptations to an environmental condition.

The greatest possible flexibility and plasticity will, however, be achieved by regulatory adjustments which develop in the form of strategies that embrace the cultural, social and behavioural characteristics of individuals and populations. These allow both established reaction patterns to be maintained and new and variable responses to changes in basic conditions to be developed. Through regulatory adjustments, humans marshal a large repertoire of combinable adjustment options which govern their survival in a habitat and relations with other human groups. Examples include the knowledge and skills necessary for building houses, sewing clothes and making tools or technological solutions for different subsistence modes, e.g. hoes, ploughs or harvesters. Just as important, however, are different ways in which individuals and communities organise their social life, their domestic and political economies, or even create and establish their rituals, i.e. those formalised behaviours by which certain concepts are collectively expressed and which have a stabilising, reassuring function within the system. Such cultural achievements are not adaptive for all humans in the same way. They can, for example, be tied to certain environmental conditions, or may have been developed consciously for the advantage of one group and the disadvantage of another. They are not compellingly durably adaptive, or invented for a long shelf life. This would, indeed, be almost counter-productive given the large plasticity by which regulatory adjustments allow the most diverse of cultural responses to the largest variety of conditions.

Regulatory adjustments are also one of several ways of competing for fitness, although very effective ones. To the extent in which cultural achievements and the social tuning of behaviour offer individuals and populations improved possibilities of survival in a habitat as well as under conditions of change, there is an increased probability that adaptation through regulatory adjustment will also be reflected in reproductive success. Fitness, be it on the individual level or as inclusive fitness on the group level, thus becomes a measure of adaptability and adaptive success. At the same time, however, fitness is the crucial unit by which the ultimate goal of organisms, their teleonomic reason for existence as it were, is measured in evolutionary biology, or with particular reference to humans, in socio-ecology, i.e. by its reproductive success (Krebs and Davies 1991). The more effective strategies humans develop that make possible their survival, their reproduction and concomitantly the long-term use of their habitat, the better should their reproductive options be. The variability of cultures and behaviour would thus become explicable as a function of underlying reproductive strategies.

By employing such a more strongly culture-centred understanding of adaptation, it is assumed that differential social reproduction of culture traits takes place in the course of time, i.e. some traits spread faster than others and may be more successful or versatile than others. Cultural adaptation thereby

consists of characteristics or trait complexes, which on a long-term basis promote the reproduction of those individuals who have certain adaptive "culture traits in their heads" (Irons 1996). This way, preference for certain adaptive behaviours would enable humans to acquire culture traits or learn certain
skills, which in turn would promote adaptive behaviour in a given environment, for example, by introducing and establishing a new hunting technology
that would allow individuals or groups better chances of survival and concomitantly eventually improve the reproductive conditions. Moreover, in a
primarily cultural context, adaptation has also been termed as that process by
which populations, societies or cultures are being maintained. Cultures would
be regarded as adaptive systems which themselves consist of a set of adaptive
single cultural traits. According to Ellen (1982; see p. 246), the adaptation of
such larger units to certain environmental conditions is facilitated by the
cumulative effect of similar behaviour traits, which eventually creates the
impression that a population behaves adaptively. The use of a new cultivar, for
example, would often start as an individual invention which subsequently
spreads quickly within a community and in such a way can contribute to the
adaptability of the population. In its effects on genetic success, this can indeed
lead to differential reproductive success for certain individuals, whilst on the
population level this must not necessarily result in population growth. It
might simply be about preserving and stabilising a framework for the entire
population in which reproduction can be optimised, but does not have to be
maximised.

   Bearing this in mind, the examples in the following chapters will employ
the term adaptation as denoting general stabilising mechanisms which are
observed on the population level and which can be understood as common
adjustments to certain environmental conditions. For example, strategies of
food acquisition will be dealt with in this sense (Chap. 2). In other cases, however, where the effects of certain cultural practices on the reproductive success of individuals is examined, adaptation will explicitly be considered
emphasising the aspect of fitness (Chap. 5), e.g. if the question concerns to
which extent culturally specified goals such as achieving status and wealth in
a society are connected with differential reproductive success.

1.2.2.6 Summary

Reviewing the considerations above, it can be said that taking a systemic view
in analysing the ecology of human populations as parts of ecosystems is no
longer questioned. Such ecosystems can be regarded as open in principle,
since humans are also tied into and interchange with other habitats and systems beyond their immediate areas of activity. The boundaries of human-
shaped ecosystems, apart from the defaults of natural space, are largely and in
particular specified by human activity itself. They should be identified and

thus become recognisable on the respective analytic level, in order to avoid the danger of a unilateral spatial definition of the system. For quantification of ecological relations, and given an appropriate research question, energy flow charts can still be regarded as a suitable heuristic model that can also allow cross-cultural analysis. The explanatory value however will depend on the observation level. An ecosystemic approach facilitates the inclusion of historic scenarios and diachronic interpretation of adaptations to given and changing environmental conditions, with the dynamic re-establishment of equilibria being the analytic model on which such an approach is based. Successful stabilisation and its adaptive effects can be considered in different ways. In principle, there are adaptive achievements on generic levels of, for example, populations. It is understood, however, that those are the cumulative effect of individual decisions. These can be assessed by applying the concept of fitness and, in doing so, individual reproductive success can be examined for its effects at the population level. Independent of whether adaptation is seen as a result of biological or cultural selection, reproduction remains a suitable unit to measure adaptive success.

Thus the approach to be employed here is in keeping with a number of recent proposals advocating the inclusive integration of dynamic concepts in ecology. On the one hand, it subscribes to the notion of nature being discontinuous across scales (e.g. Abel and Stepp 2003), resulting in the formation of lumps or wholes in nested hierarchies – which can be seen as an analogy for cultures as ecosystem components – with, at the same time, a shift in focus towards the use of more idiosyncratic explanations – which is exactly the approach that can embrace the specificity of human cultural responses to systemic constraints. On the other hand, it adopts ideas recently put forward under the label of macroecology (p. 272ff in Gaston and Blackburn 2000) that favour the combination of small- and large-scale approaches, bearing in mind that ecology seeks to understand the processes that lead to patterns of distribution and abundances of a species. Humans and their ecological relations can thus be analysed through patterns and processes that define their environmental context and structure their responses at the population level, as well as through exploring interactions that constrain the lives of individuals within their communities. Whilst complementary, a reconciliation of both approaches may not always be possible without residual contradictions and inconsistencies, but this is inherent to multi-level analysis.

1.3 Human Ecology as a Biocultural Approach

Against the background of theoretical considerations for an ecosystemic view and analysis of human communities, this book attempts to explore and systematise biocultural adjustments, with 'biocultural' denoting the biological outcomes of cultural strategies, or more elaborately: all reactions, strategies, factors and processes, humans developed in the form of predominantly socio-cultural responses to habitat- or ecological system-specific conditions, which affect their biology, their survival, their reproduction and their spatial distri-bution. The emphasis will thus not necessarily be on those processes and adaptations which by somatic variability (e.g. Molnar 1983) led to what today is referred to as geographic population differentiation. This aspect will be considered whenever suitable throughout, i.e. if physiological adjustment to limiting factors of the habitat measurably leads to an improvement in survival options (e.g. Chap. 2).

Usually such characteristics are phylogenetically old and the result of per-sisting genetic adaptation. This does not diminish their adaptive value. How-ever, cultural means of coping with certain stressors of an ecosystem are often more effective and more variable than the slower way of genetic change would allow for in a species with a relatively long generation length. Inhabitants of the circumpolar regions, for example, whose body proportions and improved peripheral blood circulation generally provide favourable physical conditions for coping with their inhospitable habitat (So 1980), invented cultural means such as the igloo or the anorak, which alone allow these people to produce a microclimate corresponding to the geographical latitude of Sicily. At the same time, their social institutions and belief systems help establish collateral sup-port structures that secure long-term survival in an environment which is characterised by low biological productivity and fluctuating food resources (e.g. Balikci 1970; Berkes and Jolly 2001). Religious concepts and rituals, for example, emphasise the profane significance and spiritual meaning of ani-mals foraged for food and thus anchor the ubiquitous dependencies from the environmental defaults within a structured system, which is familiar and re-assuring to everyone from everyday life experience. And in terms of social networks, which are organised according to kinship affiliation but, depending on resource availability, allow a flexible definition of criteria for affiliation, principles of co-operation and mutual assistance are specified and organised within the society, which on a long-term basis benefit everyone and thus increase socio-cultural resilience and stability in the habitat.

Cultural solution strategies as part of regulatory adjustments thus appear as those behaviours which facilitate a more pronounced and more varied independence from or adjustment to the basic conditions of a habitat. Such flexibility should then also be reflected in recognisable advantages and adap-tations, which in the long run affect survival in a given habitat and become visible in terms of fitness. However, what can be termed so handily by the word survival is actually the sum of numerous different and complex problem solutions and strategies, by which the components of an ecosystem (material and energy in space and time) are used and interconnected with the help of the system component of information, in order to create what Little and Mor-ren (1976) termed the human life support system.

It is these kinds of biocultural adaptations in human communities that are at the heart of this book. They will be systematised within a functional framework, whose structure takes into consideration central areas of life, or such areas that allow the demonstration of effects, success or failure of biocultural strategies aimed at coping with the environment.

1.3.1 Scope and Aims

The central areas selected here both immediately and indirectly refer to resources; and their main basic assumptions can be outlined as follows.

1.3.1.1 Modes of Production and Food Acquisition

The omnivorous nutrition of humans is a necessary, but not sufficient condition to explain how in the course of their history humans successfully inhabited all biomes of the earth. Dependent on prevailing ecological conditions, humans developed strategies of food acquisition which are responsive to limiting factors of the habitats and which enable them to interfere with material and energy flows. In addition, socio-cultural regulations and mechanisms of resource management are required as adaptive solutions that embody these strategies and support them within a community, so that a population can adapt to the conditions of a habitat in order to achieve long-term survival.

Observable modes of production therefore develop in close affiliation with the conditions of natural space and by coupling ecological and economic structures. They have repercussions on forms of social organisation and thus also constitute a dependence of socio-cultural structures on the condition of the natural base of living. They emerge however not as stereotyped responses, but arise from traditions which in the sense of a probabilistic cultural mediation of environmental conditions are typical of the respective human community.

The further people emancipated themselves from subsistent food acquisition, the more increasingly determined the efficiency of food production became, by integration into complex operational sequences of energy exchange between different populations and habitats. Apart from energetic efficiency, reliability of resource supply is a crucial determinant of subsistence; and it is among those obvious factors, control of which can be accomplished by effective resource management and thoughtful intervention of material and energy flows. Modelling such flows allows observation of the efficiency along the multitude of food chains on which subsistence in human communities is based. Resource management thus becomes comprehensible through energy management.

Variance in the efficiency of resource management has an effect on differential probabilities of survival and fertility within populations. By focussing on the individual level, where operational decisions towards the control of material and energy flows are made, articulation with the environment can be analysed with regard to possible reproductive consequences and can be understood in the sense of energy utilisation as a mechanism of fitness enhancement.

1.3.1.2 Subsistence Change

Subsistence strategies in human communities serve the adjustment and preservation of dynamic equilibria in the use of food resources. The variability of modes of production, however, is the result of evolutionary changes in strategies of food acquisition. If human communities over many generations have accomplished ways of social and cultural mediation that can sustain a mode of production under certain ecological conditions, why is there subsistence change at all? Subsistence strategies change if the ecological or sociocultural basic conditions change. Subsistence change is therefore always connected with social and cultural change.

From a systems ecological point of view, subsistence change represents the temporary disturbance of an equilibrium. Modelling should therefore allow the processual dynamics of subsistence change and its individual components to be described and analysed. The achievement of a new equilibrium is causally connected with the length of a lead phase, in which subsistence change can be prepared. Thus, by changing subsistence modes, a new or improved co-ordination of material and energy flows is accomplished in a given habitat.

1.3.1.3 Resource Use and Social Organisation

General principles of subsistence acquisition and the change in subsistence modes provide the framework of utilisation and regularisation of material and energy flows in a habitat on the population level. The evolution of resource use in human communities shows that strategies are closely linked with both economic management and the complexity of social organisation. Typically, control of access to resources within a community is co-variant with food production and increasing social differentiation.

How then is access to food resources and their use regulated within human communities? Does social affiliation decide on the quality and availability of such resources? Tangible consequences of socially different access to resources can be ascertained not only for recent, but even for historic times without written records, and analysed for their social and social-historical causes. Socially determined resource allocation as the general principle emancipates itself from the natural defaults and becomes observable in a different ecological provenance.

1.3.1.4 Population Development and Regulation

Distribution patterns of human populations in a given area and their size and composition as a biological or ecological unit result from effects that arise from strategies of food acquisition, conditions and sequences of subsistence change and mechanisms of regulating access to resources. Their structures can be analysed in a unified approach as responses to constraints of natural conditions and the emerging socio-cultural strategies.

Likewise, reproductive patterns in human populations are closely related to economic possibilities and strategies, i.e. the modes of production. The choice or practice of a certain subsistence mode affects the expansion options of a population and therefore is connected with conditions and possibilities of reproduction. These, in turn, are informed by resource distribution in the given habitat. Subsistence change thus, beside the new economic mode, also offers the possibility of an optimised reproductive strategy. Differential access to resources is equivalent to unequally distributed resource allocation and can involve differential reproduction. Under certain habitat conditions, even adaptive strategies evolve, which lead to a synchronisation of the reproductive cycles of humans and animals. Reproduction and resource supply thus cause each other. The adjustment of population densities to resource supply, i.e. the regulation of population, largely reveals itself as an economic counterbalance.

Humans neither are nor have ever been in a natural state, however that concept is perceived. Humans are unable to survive without culture: it is their evolutionary lifeline, their key to success and their prime resource for the future. Their biological nature, their bodily shell is the vehicle for genetic continuity, yet it is highly uncompetitive without cultural mediation and overprinting. Human parts and roles in their ecosystems require the consideration of both aspects, but the primacy will always have to be biological outcomes of cultural strategies, not the reverse. Culture as the main driver for our biological existence has accompanied us in the past in the very same way as it will continue to accompany us. The ecology of humans is thus an analysis along a time arrow, but with mutual information of extant and past processes and developments. History is a long-term natural experiment of human/environment interaction and as such is very real and not constrained by controlled approaches. Ecological patterns and processes that characterise and govern human communities have a long past that can ideally inform circumstances and conditions yet to come.

Taking this into account the book pursues three major aims.

1.3.1.5 Understanding The Dual Nature of Human Ecology *I*

Food resources will serve as an example for the direct expression of the conversion of material and energy flows in a habitat. It will be demonstrated that biological and socio-cultural characteristics of human populations are connected, mutually affected and changed through strategies of resource use. The biocultural adjustments following on from this can be systematised and compared for common grounds and differences. The study is based on theories originating both from the sciences and from the humanities. Ecosystem theory will be used as the central structuring principle and extended by the cross-cultural aspect of cultural ecology on the one hand and the evolutionary aspect of behavioural ecology on the other. This will enable a comparison of the adaptability of humans as parts of different ecosystems and an examination of the effects of these adaptations. By including socio-cultural characteristics this approach clearly goes beyond the range of biological/physiological adaptability, but at the same time makes possible to recognise the biological consequences of socio-cultural adjustments, as for example in the case of population dynamics.

With regard to its disciplinary allocation, this approach is deeply rooted in biological anthropology, in its orientation however it is geared towards cultural fields within general anthropology, as numerous references to ethnographic and prehistoric examples throughout the book will demonstrate. Thus, an attempt is being made to unite biological, cultural anthropological and archaeological results in a discursive context, in order to show that a human ecology is hardly conceivable without the close linkage of results and methods from science and the humanities. The application of ecological systems theory forms the formal and contextual bracket. The biocultural nature of anthropology is almost self-explanatory and lends itself to both the mediating and the essentially integrating function it occupies at the interface of most different single disciplines. Human variation is still one of the central topics of anthropology and in multinational projects such as the Human Genome Diversity Project this claim is programmatic today, as it continues to be in the large UN projects (e.g. IBP, MAB) launched during the 1960s and 1970s. Variability, however, does amount to more than just the primarily biological conversions of human adaptability. Even when considering all their physiological adaptations to certain natural constraints of habitats, humans are still genetically very similar. The actual cause of their variability lies in the reciprocal effect of biological and cultural traits in human communities and not only, in fact hardly at all, in somatic differences. To claim meaning for one without understanding the other would mean to neglect integral areas of anthropology. A unification of both areas is consistently derived from ecosystem theory and its application to human-shaped ecosystems and thereby forms the basis of a human ecology.

1.3.1.6 Reaching Out Into The Past

Whenever possible and appropriate, chapters incorporate results obtained from studies of human skeletal remains, e.g. those discussing resource use, subsistence change or social variability of resource availability. The examples will demonstrate that there is a promising and meaningful access to questions of subsistence conditions through the application of element and isotope analyses that in principle comes close to the quality of modern studies. The advantage of such data is that they exploit information encoded in the most direct of all conceivable historical sources that allow unravelling the biology of past human societies. This applies in particular to those times with no or hardly any written sources and for which access through material culture in the absence of archival evidence is limited. The significance of bone-chemical analyses lies in the fact that they provide information of nutritional behaviour and subsistence modes of past times, which are not available by other methods. In this way, testing certain hypotheses about habitat and land-use (Chap. 2), the process of subsistence change (Chap. 3) or status-related control of resources (Chap. 4) becomes altogether possible in the historical context. In these cases, a suitable methical approach facilitates access to ecological questions and their answers in the context of an ecosystemic view.

These examples refer to the resource-related context this book is subscribing to. Other findings from historical skeletons are conceivable, which could likewise be subject to ecosystemic analysis, for example the diachronic change in body height, which can be interpreted as an expression of adaptive technological changes during human evolution (e.g. Frayer 1984), social change, or differential social conditions in general (e.g. Bogin 2001). The crucial aspect, however, is that questions about biocultural conditions and correlations can be successfully applied to historical times in the context of a theoretically founded concept.

A human ecology oriented towards historic events will inquire about the same reactions, strategies, factors and processes as the ecology of recent human populations. It has, however, necessarily and unavoidably material- and source-related limitations in its methical approaches, with the likely consequence that living conditions in historical populations are more difficult to reconstruct. The restrictions do not necessarily lie in the methods available, though, but rather in the fact that possibilities of access cannot be as precise and detailed as with recent studies. It is the research questions that provide the bracket for different time-periods; and these questions can in turn be derived from the principles of ecology. And it is particularly this theoretical framework which permits the inclusion of the historical dimension and thus takes a wider view. Considering the current increasing historical focus of biological anthropology and necessity to employ historic sources, this appears to be significant and, in the interest of the disciplinary profile and safeguarding, even necessary and desirable.

1.3.1.7 Assessing Human Biocultural Agency

All too often the term human ecology is frequently and still used to essentially denote the balancing of world-wide human over-exploitation of nature by dealing with the serious and ever-damaging interference of humans in existing equilibria. Accordingly, human ecology would be occupied with attempts to issue reminders and suggestions for ways out of over-population, growing waste disposal sites, parasite load or climatic degradation (Freye 1978; for a modern approach, see Southwick 1996). This is justified and regularly demands worldwide interest in the form of forest damage status reports and other frightening manuals like the Club of Rome's theses or the Global 2000 study. Additionally, in view of the increasing globalisation of material, energy and information flows, it is not only politically required but also theoretically consistent, if the entire earth is regarded as an ecosystem. The responsibility for the consequences of human actions can probably not be put on anybody else but exactly humans.

Nevertheless, the assignment of a role which sees humans as the eternal cause of disruption and destruction in an ecosystem, is neither new nor original. At the latest, since humans began to produce their food they also irreversibly began to affect and change their environment and to rip it from its postulated harmony. The human fall of permanent breach of rule against ecosystemic equilibrium, as it were, is fiction. Quite the contrary, it is an inherent part of human ecosystems: it is one of its components and thus at least in theory also possesses the possibility to again lead to a new equilibrium. Large parts of what is regarded as an ecosystem component in terms of cultural achievements actually belong within the realm of those solutions and responses which humans developed and invented, including the consideration of ethical constraints (Gare 2000), to be able to live with exactly these constraints of their habitats. It would be unreasonable to deny that these strategies are extraordinarily successful. Exactly this is their value. Strategies for successful survival do have a high adaptive value. Their adaptive potential, the range of variation and plasticity of biocultural adjustments could be regarded as offers for the solution of more up-to-date (ecological) problems which frequently can be attributed essentially to resource conflicts. The fundamental significance of raw material deposits such as oil or other sources of energy becomes a political and ecological argument: the Gulf War was probably essentially about maintaining access possibilities to exactly these raw materials, i.e. a war of resource control. And, as the smouldering conflicts about the water of the rivers Euphrates or Jordan suggest, the power of limiting factors in a habitat must not be called into question. It is not just by chance that the origin of ecological views of human communities developed from the analysis of subsistence strategies and their interrelation with forms of social organisation (Steward 1955). Subsistence strategies however, as will be shown, have to do considerably with the possibilities of resource allocation. Maybe the mercantile relations that are at the basis of natural processes of life are so intuitively comprehensible to us because our economic life functions according to the very same rules of natural operational sequences (for an early yet somewhat vulgar example of political economy, see e.g. Carey 1837). Even if superficially it is about economy, the background is often ecological in the sense that humans operate in ecosystems.

Whatever a human ecology has to offer by systematising biocultural adaptations for survival, it is not about ingenious solutions which are easily transferable. On the contrary: only thorough consideration of regional characteristics and their careful inclusion in and adjustment to existing successful strategies bears a chance for success. But this requires knowledge of both the variability of local and regional strategies and biocultural solutions and the effectiveness of strategies, the reasons for their emergence, their development and their integration into the respective cultural frameworks, in short: the conditions and the ecosystemic context of their success. The case studies presented in this book are only a small selection from the probably inexhaustible reservoir of biocultural adaptations of human communities, be they from the past or the present. They show successful principles and they show the consequences of maladaptation. But they certainly demonstrate the fundamental meaning of anchoring survival strategies in the social realm and their integration into social consent. This, too, is both the chance and the task of a human ecology.

2 Subsistence Modes

In the course of their history, humans successfully settled in such a large number of different habitats that, compared with other species, this ability became their hallmark trait. They were able to find sustainable solutions for the supply of a sufficient nutritional basis and the acquisition of essential food components, which allowed them to spread into a most diverse array of ecological zones. Not least, this was and still is facilitated by the unusually broad niche humans occupy within ecosystems, i.e. by the multiple ways in which they extract nutrients and energy from their environment.

From a physiological point of view, two conditions of long-term survival in a habitat arise. First, a sufficient amount of food energy must be available in order to fulfil the energetic requirements needed to maintain basic metabolic rate, growth or thermoregulation. This amount of necessary energy shows large individual variability and, among other things, depends on human body size or activity patterns. The recommended daily requirement is ca. 3,000 kcal for men and 2,000 kcal for women (WHO/FAO 1973). Whilst this may serve as a suitable basis for industrialised countries and average physical exercise strain, it represents an approximation only that cannot be taken as a standard that would be transferable onto a world-wide scale (Leslie at al. 1984). During pregnancy alone, an additional 80,000–85,000 kcal is needed and subsequent nursing requires at least 500 kcal/day more (Adair 1987; Worthington-Roberts 1989; Dewey 1997). Thus, there is an additional constant energetic requirement independent of specific local, habitual or individual conditions of energy output gained from food resources.

Second, a well balanced nutritional supply is required which provides the necessary quantities of building blocks, such as protein, fat, carbohydrates, minerals, trace elements and vitamins. In other words, those nutrients which are necessary for the maintenance of vital functions of humans as heterotrophic organisms must be present in sufficient quantities. After all, food intake serves the purpose of making energy available to meet vital metabolic demands. Intake of chemical compounds from the environment ensures that organisms can accomplish energy production for both build-up and maintenance of body tissues. Macronutrients are made available for metabolism through digestion by being broken down by enzymatic reactions and transferred into components suitable for intestinal absorption. Again, generic recommended dietary allowances (RDA 1991) of food items are available, but they only provide proxies that may vary depending on the availability of items in the habitat and/or varying nutritional habits.

The supply of minerals, in particular essential trace elements, can be regarded as a food-related limiting factor. Zinc, for example, is required for proper histogenesis of bone, skin and hair and, besides this, is a constituent of numerous metalloenzymes. Daily requirement is indicated as 2–3 mg, but for pregnant women and juveniles an even increased intake is necessary. Zinc is mainly supplied through meat and other animal-derived products and to a limited extent by cereal grains. Higher amounts of cereals in the diet combined with little consumption of meat, however, frequently lead to deficiencies, since grain contains large quantities of phytic acid due to its high fibre content. The subsequent formation of largely insoluble phytate complexes in the digestive tract aggravates the absorption of zinc. This effect can be counterbalanced, e.g. by the consumption of leavened bread (Sandström 1989). In areas where this is traditionally uncommon, for example in the Middle East, a high prevalence of growth disturbances related to zinc deficiency is reported.

Additionally, because humans are high-order consumers, they are frequently end-members of food chains within their habitats and the choice of food acquisition strategies is of importance, as the quantity of usable energy decreases from one trophic level to the next. The general rule of energy conversion states that the shorter the food chain, the more effective is the energetic yield; and it would thus follow that population densities regularly decrease with their increasing position in any food chain. Human populations, however, are unique in that they frequently deviate from this pattern, being able to maintain population densities at high levels. This can be explained by their generalist and opportunistic nutritional behaviour, which permits large flexibility in the extraction of energy from the most diverse food items and/or biotic components.

At this basic level, however, human populations may face difficulties in providing sufficient levels of energy density and food in the long-term. The permanent settlement and use of a habitat then requires mechanisms to be developed that work effectively against resource shortage or which help to buffer shortages in resource availability. Therefore, humans typically employ strategies aimed at maintaining the yield of certain food-relevant plants or animals. Their variability of modes of production, among other things, results from such different strategies of resource management, which enable humans to intervene in and change energy and material flows. Examples would be regular changes in the spatial distribution of a population, e.g. as practised by foraging or pastoral societies, or territorial resource use aimed at a more balanced extraction and subsequent avoidance of local overexploitation. Whilst foragers deliberately interfere with the natural environment, e.g. by burning grassland (Cane 1996), other strategies entail more substantial environmental transformation and change of landscape, for example those connected with horticulture and agricultural farming. Ecosystems used and formed by humans therefore vary substantially in composition and characteristics.

Particularly with a foraging or, more generally, a non-industrial way of food acquisition, modes of production are closely linked with or bound to the respective properties of the habitat. Conditions of geomorphology, climate, seasonal cycles and soil quality characterise the ecological circumstances and, as a function of these factors, the spectrum of naturally occurring or cultivated plant and animal species varies, too. Resource supply and the possibilities of participating in material and energy flows are the result of concurring biotic and abiotic factors and constitute the ecological quality of the habitat. Consequently, the bio-geographical distribution of resources in each habitat, or in terms of larger units in each biome, exhibits a specific composition of biotic and abiotic components, which results in certain limiting factors or stressors, respectively. Arid areas, for example, are characterised by little and unpredictable precipitation, savannah environments by extended periods of drought, high altitudes by reduced biological productivity (Moran 2000). Depending on where human populations want to settle and eventually do so, they are therefore faced with the specific conditions and constraints of their environment. In order to tackle and solve problems entailed with the appropriation and exploitation of existing resources, humans develop a wide range of strategies of food acquisition to provide a sustainable and secure subsistence basis by interacting with other human groups and their environment to facilitate long-term use of the habitat.

To this end, every human population employs techniques to acquire and utilise the resources of their habitats. Such techniques are made up of a combination of material artefacts and the knowledge (= information) needed to use them. Normally, a population will use more than one such technique in parallel, which then join into a mode of subsistence. To stress the potential adaptive value of these techniques and the associated interaction with the environment, each mode of subsistence can be described as a subsistence strategy. Just as essential as the adaptive value of the material–technological side, however, is the one that develops from the social mediation of subsistence strategies. Subsistence is tied into social conventions that regulate and define its application, for example in terms of who is involved in food production, how division of labour and professional specialisation is organised, or which modes of barter or trade relations exist between individuals or groups (p. 128 in Ellen 1982).

Maintenance of a subsistence strategy does not have to be limited to the immediate habitat of a population. If the autochthonous local conditions prove to be insufficient, trade networks or other exchange relations can help balance deficits in local modes of subsistence. Optimised food acquisition strategies can be, but do not necessarily have to be, accompanied by population growth. Rather, it appears to be crucial that, once occupied, a habitat can be used continuously and does not have to be abandoned. The adaptive value of a food acquisition strategy would thus have to be measured against the duration of its use.

With that, the variability of ecosystems shaped or affected by humans can be regarded as a function of the many different subsistence strategies societies employ. Adjustments resulting from this variety should fulfil the following purposes: (1) to secure sufficient caloric production and constant energy flow, (2) to make available a food spectrum which principally enables the supply of essential nutrients, (3) to consider limiting factors of the habitat, dietary constraints and their possible effects, (4) to synchronise subsistence activities with seasonal cycles and (5) to co-ordinate strategies of food acquisition by embedding them into social and/or political institutions and decisions.

Even though each one alone has to be seen as a matter of course, only the conversion of all conditions mentioned make successful settlement and use of a habitat possible. This means that complex cross-linking between subsistence opportunities provided by a habitat and operative decisions of a human population are necessary in order to utilise the productivity of the habitat for securing economic sustainability. Thus, the availability of resources is not only determined by the physical and biological characteristics of a certain habitat alone, but likewise by individual and collective knowledge and decisions, by familial, social, economical and political conditions. Nutritional behaviour and resource utilisation therefore also implies the constantly changing circumstances of human subsistence behaviour, which in turn is affected by the selectivity of human interactions (see p. 8ff in Jochim 1981). Both these biological and ecological 'default values' and their cultural moderation constitute the appearance of human subsistence strategies and emerge from their mutual influence[1].

The following section will first give an overview of different modes of production, which refer to patterns frequently used in economic anthropology as pragmatic functional entities. Afterwards, subsistence strategies from differ-

---

[1] This is true even without having considered yet that the motivation or cause for a certain subsistence strategy may not exclusively lie in perhaps all too obvious energetic or caloric necessities. 'Environment' is always perceived from two sides; the etic (external perception, outside view) and the emic (self-perception, inside view) point of view (cf. Harris 2001). Observations, which are interpreted rationally by a researcher, must not necessarily be explained and understood in the same way by the population under study. Energy flow diagrams would thus have high heuristic explanatory value for the ecologically working anthropologist and are of particular importance with respect to comparative analyses (cf. Sect. 2.3). Yet, the emic view may perceive the efficiency of a subsistence mode or a certain method of cultivation as meaningless in energetic terms, its social or ritual connotation, however, as vitally important (e.g. for Tsembaga Maring, see Rappaport 1968).

ent biomes, thus considering different limiting factors to which human populations adapted, are introduced using examples of extant and historic societies. A decisive criterion for their selection was that the respective societies be ethnographically well documented and, if possible, information about the energetic utilisation of their habitat be available. Optimal Foraging Theory (OFT) will be presented to introduce methods that allow outlining and empirically examining optimisation models on food acquisition. The concluding section attempts a comparison of subsistence strategies, emphasising energy as a system component.

2.1 Modes of Production

There are different approaches to systematise the variety of food acquisition strategies employed by human communities. The most obvious one takes a pragmatic view and follows functional categories of resource utilisation by establishing a set of modes of production (e.g. Sutton and Anderson 2004). Such an approach seeks to establish cross-cultural common traits or similarities in societies that can be assigned to the same adaptive solutions under comparable conditions; and the emphasis is on subsistence patterns, which were developed as coping mechanisms in different ecological settings and their constraints. The structure of such a classification of human communities is informed by an increase in the appropriation of nature and environmental change, or in other words different amounts of technological investment in the environment, and is aimed at the long-term organisation of food acquisition and energetic survival. Two principal categories can be differentiated: the first comprises foraging or hunting/collecting/fishing communities, while the second category covers cultivating, agricultural, pastoral and industrial societies, i.e. the distinction lies in the presence or absence of food production rather than food processing[2].

It is true these categories and their components ought to be regarded as ideal denominations; and such a course grid can hardly account for inherent and existing variety. None occurs in pure terms, but each shares characteristics with the others. In our culture, for example, hunting game and collecting fruits are still traditionally part of food acquisition, yet their contribution is negligibly small and has no essential meaning for securing survival. Pastoral societies, in contrast, are dependent on the seasonal cultivation of arable crops as buffers against incalculable climatic conditions. Although by defini-

---

[2] This pattern implies what some may perceive as a quasi-natural succession of cultural solutions (e.g. Steward 1949). Measured with a rough temporal and rather global yardstick, this may by and large approximate real cultural history, notwithstanding a staggering amount of temporal and geographical variance.

tion more than 50 % of their food basis is provided by animal husbandry, some kind of plant cultivation forms a regular and indispensable part of their subsistence. Finally, foraging societies engage in barter and exchange of goods with neighbouring societies of other subsistence categories.

Differences between groups lumped into the same category can be substantial. Societies like the Tsembaga and the Miyanmin from highland New Guinea would traditionally be assigned to the subsistence category of horticulture, or more specifically slash-and-burn agriculture. While for the Tsembaga crop farming is of paramount importance and accounts for more than 90 % of their diet, the respective value for the Miyanmin is scarcely 60 %, since particularly pig keeping and hunting significantly contribute to their living in terms of 'secondary' subsistence techniques (p. 141 in Ellen 1982).

Despite this overlap, such categories appear to be suitable at least as heuristic models that represent general characteristics of subsistence organisation. They allow for a modification of single characteristics due to local peculiarities and are commonly used today in economic anthropology (e.g. Jensen 1988). None of the extant non-industrial societies, which represent a wide variability of modes of production, can actually be regarded as 'forerunners' of certain adaptive strategies. Rather, they should be taken as examples of certain recent variants of ecological types, which in many cases may exhibit local or regional specialisations that should be explicable from properties of the respective habitat. They can probably be best considered in analogy to adaptive radiation, whereby extant ethnographic cases represent preliminary culminating points of specific developments rather than 'frozen' developmental stages in the organisation of human communities. The advantage of a structure according to subsistence techniques thus lies in the clear demarcation of the large, different categories of modes of production, in particular through the distinction between extracting and producing modes of food procurement.

An alternative approach could be provided through a classification based on institutional categories, such as kinship and descent, residence or belief systems, i.e. patterns of social and political organisation in a human community. The advantage of this more strongly sociology-oriented structure would be the emphasis on institutions of social interaction and, particularly, on the differences within comparable contexts of conduct. In its own kind, however, this is just as limited as a structure based on technological aspects. For example, the organisation of kinship according to the 'Hawaii system' (kin is distinguished between the generations only by using the terms 'related' or 'not related') or the 'Sudan system' (kinship terminology reflects the whole scope of genealogical variation; p. 219ff in Vivelo 1981) as such has no bearing on modes of resource utilisation within a society. These patterns first of all serve as a systematic grid for regulating social interactions and are meaningful as to how people recognise and address each other and how they mutually fit into their social environments. From such distinction alone, an ecological context or for that matter modes of production and food acquisition can only be inferred indirectly, because it is not clear whether and in which way both areas are interconnected.

However, when for instance access to and the use of resources are regulated based on kinship relations (see Chap. 4), i.e. if it is crucial whether possibilities of food acquisition are affected by affiliation to a certain social group, or if it is of importance for certain tasks of food procurement whether labour is divided and specified by gender or age, decisive meaning can be attributed to such patterns. It is these associations that make human subsistence strategies what they really are, a mode of food procurement integrated into a particular social context. Ellen (1982; see p. 175) illustrates this principle by pointing out that, indeed, hunting can be described ecologically and technically as human predatory behaviour directed at a large animal, involving the expenditure of high amounts of energy. However, it is not at all clear from this description whether and in which way food is divided and whether access to resources and land is equally distributed among all the people involved. Strategies of food acquisition presuppose social relations and that the set of traits that is finally characteristic of a population's mode of subsistence develops from the articulation of ecological, technical and social aspects. This, in turn, has bearings on differential individual survival and thus on reproductive options, as will be further explored in Chap. 5.

The following short overview of modes of production will, therefore, include features that denote ways and conditions of food procurement and production, and traits typically connected with socio-political organisation. As a compilation, it considers general categories, bearing in mind that there may be numerous location-typical deviations or modifications (for a general overview, see Sutton and Anderson 2004; for foragers and peripatetic groups, see Rao 1993, Panter-Brick et al. 2001; for pastoral nomads, see Bollig and Casimir 1993; for agricultural societies, see Köhler and Seitz 1993). They are, however, intended to clarify patterns of those different basic conditions, under which biological as well as cultural responses and solutions to subsistence problems were developed. In this sense, the following must not be understood as strictly apodictic information.

2.1.1 Hunting and Gathering

Communities with an acquiring or extracting mode of subsistence are generally referred to as foragers or hunters/gatherers/collectors. Their nutritional basis primarily consists of wild plants and animals and their implements are manufactured from naturally occurring materials. There is generally no cultivation of domesticated species, i.e. the hunting and gathering mode of subsistence implies a lack of direct human control over the reproduction of exploited species (Panter-Brick et al. 2001). Key properties of their economy are ecological in nature (Winterhalder 2001). Foraging communities consist of a few nuclear families that form groups of rarely more than 50 individuals. Size and composition are flexible and characterised by fission and fusion, leading to a changeable membership. This peripatetic way of life is adjusted to the opportunities of exploiting seasonally available food resources and requires high mobility in large home-ranges. It usually varies in connection with habitat-specific requirements, as the group may divide, for example in order to procure food at different places in the habitat. Despite this loose and variable structure, the family is still the central unit of social organisation.

Usually there is no reinvestment in the maintenance of existing resources, so that a temporary or complete exhaustion of resources can occur. Because of the nomadic way of life, material possessions are few: the tool kit shows little specialisation and is universally applicable. Division of labour is neither vocational nor based on a specialisation relating to crafts, but is complementarily organised according to gender and/or age groups. Hunting is traditionally part of the men's sphere, while women are engaged in collecting food and doing the majority of household work. However, this is not a strict pattern and exceptions or reversals do occur and are socially sanctioned (e.g. Noss and Hewlett 2001). In fact, there is considerable overlap in male and female subsistence tasks, with the only notable exception being big game hunting, which appears to be confined to men across cultures (Brightman 1996). In Inuit societies, for example, girls may be trained as hunters while boys can be called upon for tasks within the domestic range, depending on each individual family's sex ratio. As the household forms the basic economic unit, collected food goes to the respective households, whilst bag is usually divided among all group members, securing an even supply of animal protein.

The political organisation is characterised by an absence of formally defined positions of leadership, although an informal guidance of the *primus inter pares* type is common. Such an individual would not enforce orders, but receive status as an esteemed counsellor solely from individual achievement and experience. Decisions, such as those referring to the relocation of camps or a move to new resource-sites, are negotiated and discussed in an equal and general manner, so that information travels fast between group members. The organisation of groups is egalitarian (or acephalic), i.e. there is no social stratification from which auditing rights over resources could be derived. Thus, there is free access to resources for all group members. The political economy does not encompass added value or a dependent clientele. There is no personalised property of land or resources, yet people do have material possessions, also in relation to subsistence. A weir built by one or more families, for example, justifies exclusive use of the caught fish, but other families can be granted joint rights of use. The catchment area has no strict notion of territory attached to it, but is rather regarded as a home-range. Its use is primarily exclusive for the respective resident group, but rights of use for other groups are negotiable. Aggressive conflicts take the form of feuds and are settled individually or through small-scale group encounters, while more comprehensive conflicts against enemies violating the home-range are possible.

Whilst these general characteristics apply to a large number of non-specialised hunter/gatherers, e.g. Australian aborigines or !Kung, it is worth mentioning special forms which owe their existence to specific adaptations of food acquisition caused by peculiarities of certain natural units. They are facilitated by the relatively reliable seasonal occurrence of food resources, which may consequently lead to storage, specialised technologies or even different forms of socio-political organisation. This applies to so-called specialised hunters, for example native Americans of the North American prairies or Inuit, a substantial part of whose base of living depends on hunting for bison or sea mammals and caribou, respectively.

In Inuit societies, the necessity for expert knowledge and experience required for hunting led to the establishment of positions of formal leadership, e.g. as in whaling captains for the co-ordination and lead of whale hunts. Reliable occurrences of anadromous fish allowed native Americans of the northwest coast (e.g. Haida, Kwakiutl, Chinook) to establish permanent settlements that were able to sustain population sizes otherwise only known in agricultural societies. Specialised gatherers (collectors), e.g. Paiute or Shoshone, acquired techniques for using seasonally abundant vegetable food, such as acorns and piñon nuts, which are not edible in a raw state but by preservation through roasting are turned into a reliable food resource. Moreover, collector communities also exhibit transitory developments towards the domestication of plants. Compared with 'classic' foragers, these societies developed a pronounced social differentiation in connection with an elaborate form of storage techniques.

2.1.2 Horticulture

This term indicates an agricultural technique which is marked by the use of simple devices, such as digging sticks or hoes, but no ards, ploughs or draft animals. The diversity of cultivated and domesticated plants, the technologies applied and the subsistence strategies employed is high and varies according to the expenditure of time and energetic investment spent on food production (p. 187ff in Sutton and Anderson 2004). Social organisation covers a wide range of principles, from forager communities to regional farming polities. Because of this high diversity, this mode of production will be described in very general terms only.

Laying out and managing fields and gardens, i.e. investment in the environment aimed at food production, requires a minimum of local constancy. Therefore, there is a tendency towards (usually temporary) permanence of settlement; and the more reliable supply of cultivated food items allows larger group sizes and higher population densities. However, horticulture often involves extensive land use, so that higher demand for land leads to increased spatial dispersion of hamlets or villages. Especially when slash-and-burn agriculture (or shifting cultivation), the most frequent and space-craving mode of horticultural production, is practised, arable land becomes the limiting factor within the system. Without additional influx of matter, e.g. manure or fertiliser to improve soil quality, the land is quickly exhausted and a fallow time of 15–20 years is necessary to allow for complete regeneration of the soil in tropical habitats. Intensive cultivation, in contrast, is characterised by the preservation of soil fertility at the site, e.g. through different ways of fertilisation, allowing the formation of larger economic units with less group mobility and fewer changes in place of residence. Work in the fields and gardens is organised according to a gender-based division of labour.

The necessity to implement a co-operative and durable mode of food production leads to an increasing significance of social organisations beyond the family framework. The structure of the society is based on relatively stable corporate descent groups at local and regional levels. There are vested titles in produced goods and commodities and communal titles in terms of crops and arable land, thus providing the conditions for social stratification. Political organisation varies from incipient formation of institutions, for example in the New Guinean highland, to complex state systems such as in Polynesia. In any case, the concentration of people, land and produce requires more clearly defined executive functions, which are practised by persons of institutionalised authority. Territoriality is pronounced and is related to both the intensity of cultivation and the availability of land.

2.1.3 Agriculture

Food production in agricultural systems is commonly characterised by a substantial investment of time and energy in the environment. Depending on the overall ecological or habitat conditions of environments principally suitable for agriculture, a number of different subsistence modes develop, which are related to certain technological achievements, e.g. agriculture employing ploughs, paddies or terraced fields. This diversity is also reflected in the socio-political systems.

Frequently, agriculture and the presence of other technical developments, such as metalworking, the wheel or larger edifices, are correlated. A crucial factor of agricultural production is a reliable natural or artificial water supply, as well as the use of beasts of burden or draft animals. Local groups generally settle permanently and their composition is fairly stable. An increased requirement of workers and the possibility to supply larger numbers of people with produced food leads to higher population densities and larger populations. Regional and national polities are a regular feature, often associated with central rule where individuals of clearly defined authority serve political institutions. The social organisation is complex and accompanied by differentiated social stratification within a society, which among other things is characterised by differential access to resources and non-equitable possession of property. Claims of ownership and property for land or livestock play an increasing role. There is a pronounced division of labour; and professional specialisation leads to the emergence of vocation-related groups or guilds within the society. Both property and the control of use and access to resources entail fully developed territoriality. Frequently, agriculture is connected with urbanisation, which eventually leads to the establishment of farming communities or estates; and the production of surplus is distributed in markets to supply the non-farming (urban) population.

2.1.4 Pastoralism

The pastoralist mode of production and way of life is shaped by keeping and breeding domesticated livestock. Typically occurring in the Old World, animal husbandry can be found in the New World, albeit often caused by contact with invading European settlers, for example in subsistence systems based on sheep-breeding (Navajo) or horse-keeping (Plains Indians). South America, however, has a long indigenous tradition of camelid husbandry. Pastoralism requires the neighbourhood and complement of agriculture to be sustainable, since vital goods such as salt or vegetable products must be acquired from farmers by means of trade or tribute payments. Pastoralists, despite being primarily nomadic, are often found to complement their food procurement with seasonal subsistence agriculture, especially in savannah environments.

Pastoralism is a special techno-economic type of food production based on animal husbandry. Because there are so many habitat-specific adaptations, it is difficult to find a uniform definition. Vertical nomadism, or transhumance, would be an example for a specialised form, practised in mountainous areas, where during the summer herds and herders stay on meadows at high altitudes, such as alpine pasture, while during the winter the animals are kept in the more protected valleys.

Typically, pastoralism is found in arid or semiarid areas, steppes and savannahs, or in tundra areas. Suitable pasture and water are limiting factors of these habitats, whose accessibility is decisive for the periodic or seasonal mobility of populations. The animals bred and kept vary with the environmental conditions of biomes and endemic species, for example cattle and camels in east Africa, reindeer in Lapland. Whatever animals are kept, the human population is supplied with a variety of raw materials (wool, hides), food (milk, milk products, blood, meat) and even a means of transport (pack and riding animals).

Pastoral communities are relatively small, usually consisting of several households only; and group composition is flexible depending on economic requirement. The division of labour is essentially based on a gender-related separation of tasks for the supply of livestock, but without a strict pattern. Usually, there is no or only little specialisation in labour, since products that cannot be manufactured by the group have to be obtained from neighbouring farmers or travelling traders. Social organisation tends to be egalitarian but may vary with respect to social stratification and the existence of larger corporate groups. Territoriality is understood in general terms and is related to the area used by the entire community, although individual groups may establish rights of access and use with respect to certain areas. But, the necessary extensive nature of land use can lead to martial expansion and/or violent defence of the pasture.

This short overview was not meant to present more than the essential features characteristic of the most frequently found modes of production. They are to serve as background information for the following examples of food-acquisition strategies that were developed as different modes of resource management in response to different habitat conditions.

2.2 Strategies of Food Procurement

The necessity for human groups to develop strategies of securing subsistence can be substantially determined by limiting factors within the respective habitat. Such ecological constraints may be present, for example, as a result of typical climatic conditions in a biome, but they relate just as much to the availability of basic food components, biological productivity, species diversity or geomorphological characteristics of a habitat (e.g. see p. 6 in Moran 2000). This becomes particularly evident in those biomes exhibiting an extreme or strained climatic situation due to their geographical location, i.e. those not located in the relatively favourable areas of the temperate zones. Taking the following adaptive strategies of food acquisition as examples, it will be demonstrated to which extent these allowed humans to mediate serious habitat-specific stressors by biocultural adjustments and to establish conditions for (long-term) settlement in less favoured regions of the earth. At the same time, case studies from different habitats are introduced as examples of certain modes of production (see above). As far as information is accessible from the literature, data are presented on the food spectrum, adaptive nutritional strategies and energetic efficiencies; and behaviour patterns and operative decisions are considered which are of importance within the context of food acquisition.

2.2.1 Foragers

The following section consists of two case studies representing unspecialised and specialised forms of hunting and gathering, respectively. Their introduction will be complemented by a discussion of assumptions developed from behavioural ecology that allow modelling and predicting of foraging behaviour.

2.2.1.1 Unspecialised Foragers in Arid Areas: !Kung

Apart from low biological productivity, arid areas are above all constrained by the limiting factors of little and uncertain precipitation and high rates of evaporation. The availability of water thus becomes a serious problem for safeguarding the survival of human populations in arid areas and forms the crucial component in the context of their subsistence strategies. The significance of water results from its structural task as a medium of transport and disposal. Its function for thermoregulation is just as important. The loss of just 5 % of the body water can lead to death. Even smaller losses can impair body functions, for example during neuronal data processing (e.g. Baker 1988). In arid areas, therefore, the danger of encountering critical situations caused by the availability, or rather the lack, of water is particularly acute.

In arid environments, both freely accessible surface water and liquid bound to food have to be factored into the quantity of water necessary for daily survival. Detailed observations have been recorded for the !Kung San groups of the Kalahari (e.g. Lee 1976, 1979; Tanaka 1976). Surface water crops out in permanent water holes, in earth lenses impermeable to water in dried-out riverbeds, or in *malapo*, water-filled recesses between sand dunes. Also water collected in recesses or holes in trees is used. Depending on the availability of surface water, !Kung cover their water needs from succulent roots as well, for example from plants of the genus *Fockea, Raphionacme,* or *Ipomea*. This, however, is connected with laborious search and digging and is energetically very cost-intensive, considering that the necessary quantity amounts to approximately 20 pieces per person and day. Melons, roots and aloe plants are available in !Kung home-ranges, of which the bitter melon (*Citrullus naudinianus*) is particularly productive, with an average yield of approx. 500 ml liquid per person and day and the added advantage that it can be harnessed [*harvested ?*] almost all year round.

In an acquiring economy, broad knowledge of usable vegetable and animal resources is necessary for the sufficient covering of energetic and nutritional requirements. Frequently, the number of species known and principally suitable for nutrition clearly exceeds the spectrum of those plants and animals which are actually used. !Kung differentiate more than 500 species, of which only [*considerably*] less than half, i.e. 105 plant and 78 animal species, are regularly chosen as food items (Lee 1979). Such a selection reflects several operational decisions of food acquisition which are related to each other: the exploitation of certain food resources depends on their respective seasonal availability and accessibility. As those food sources are preferred that account for the highest calorie yield per unit time, the relation of assigned costs for food search and the resulting benefit is optimised and understood as a function of complex decision sequences (see Optimal Foraging Theory, Sect. 2.2.1.3). The knowledge of species less frequently utilised as food resources can thus be regarded as a knowledge reservoir that allows changing over to less productive or less esteemed nutritional items, if the food of premier choice is not accessible.

Preferred and readily available species are collected in the first place and, in doing so, the area covered by foraging excursions is continuously expanded. Only if these resources are exhausted does one fall back upon less appreciated species. This translates into a foraging pattern by which young adults employ a spatial strategy, accepting longer trips for obtaining highly esteemed food, whilst the elderly rather pursue a hierarchical strategy in closer proximity to the camp and also forage for less valued items. Among the most frequently gleaned plants of the Kalahari, the Mongongo (*Ricinodendron rautanenii*) takes a prominent and fundamental position for safeguarding subsistence (see Lee 1979). Providing 27 % protein on average and an energy yield of approximately 650 kcal/100 g from the edible portion of the nut and 312 kcal/100 g from the flesh, it ranks among the most important items gathered. It is available all year round, with the whole fruit being consumed from April to September and the nuts only from October to March. Besides that, the Baobab fruit (*Adansonia digitata*) is of importance, being rich in Vitamin C, calcium and magnesium. Its energy yield comes to 506 kcal/100 g and contains approximately 34 g protein/100 g fruit. Also, the Marula nut (*Sclerocarya caffa*, 624 kcal, approx. 31 g protein/100 g) and the Bauhinia bean (*Bauhinia esculenta*, 554 kcal, approx. 32 g protein/100 g) are collected in considerable quantities.

Animal protein and fat amount to 20–50 % of the diet. Burrow-dwelling small mammals and birds are most frequently hunted, less frequently so large ungulates. Hunting for small animals is energetically less demanding than pursuing large animals and, besides, means predating on a resource that is regenerating more quickly due the shorter reproductive cycles in small mammals. In general, however, gathering activities provide the more reliable food supply to foragers, less dependent on the imponderables of luck than the hunt and thereby contributing more to subsistence than hunting. The average energy yield obtained from plants and animals exceeds the !Kung's daily average individual energy demand by approximately 380 kcal. At the same time, only 2–3 days per week are necessary on average to manage all work in connection with and resulting from subsistence, repair, maintenance and housework (applies to Dobe !Kung; Lee 1979; see Sect. 2.3). The energy balance of !Kung has been calculated as a gain of 7–11 kcal (Harris 1989) for every single calorie put into subsistence activities. This high-energy output and the relative security of the subsistence economy employed can only be maintained, however, by specific adjustments to seasonal fluctuations in resource supply. Also, it has to be considered that, because of the climate, storing food (e.g. meat) is not possible at all or only to a very limited extent.

Figure 2.1 shows seasonal activity patterns and the conditions of subsistence for !Kung in the annual cycle. The availability of water is the main determining factor that governs the seasonal operational sequence. It entirely depends on sufficient rainfall in the winter as to whether the ponds and water-holes will contain sufficient water during the summer and whether even in the late summer there are permanent water-holes to go to. Accordingly, mobility of groups and their dispersion or fusion into smaller or larger foraging bands varies according to distance from food resources to water-holes. Thus, flexibility of social organisation is essentially determined by ecological conditions within the habitat. This also applies to the relative significance of subsistence techniques, both hunting and gathering. The paramount importance of the Mongongo fruit as a reliable and predictable basis of subsistence is indicated by its all-season availability, in contrast to other plant items such as berries, roots, fruits or leaves that can only be collected seasonally. Hunting is erratic altogether, despite the variety of hunting techniques, such as game drive, trapping or hunting with bow and arrow (see Lee 1968).

As Wilmsen and Durham (1988) demonstrated for the prey species hunted by !Kung in the /ai/ai area of northwestern Botswana, bag is available in this region the whole year round. Its accessibility, however, strongly depends on the time of the year, as local density and distribution of prey varies. Also, vis-

Month	Jan	Feb	Mar	Apr	May	Jun	Jul	Aug	Sep	Oct	Nov	Dec
Water availability	Rainy season (summer)			Dry season (autumn, winter)					Dry season (autumn)		Rainy season (summer)	
	Temporary water holes			Permanent water holes			Few water holes				Many water holes	
Relative mobility	Distance between water and food minimal						Increasing		Maximal		Minimal	
	High dispersion			Concentration			Maximal concentration			Beginning of dispersion		
Hunting	Hunting with bow and arrow											
	Chase hunting				Trap hunting on small game					Chase hunting		
Gathering	Collecting mongongo fruits/nuts											
	Fruits, berries, nuts				Roots, tubers, resins					Roots, leaves		

Fig. 2.1. Seasonal cycle of subsistence activities for !Kung, Kalahari Desert, Botswana. The availability of water is crucial to patterns of spatial dispersal, group size and composition (modified from Lee 1968)

ibility is a factor, largely dependent on local seasonal differences in soil mois-
ture, humidity or vegetation cover. Therefore, food procurement is not only a
function of productivity and hunting technology, but also a function of external factors, including basic ecological conditions and socially negotiated factors of spatial distribution and labour in the human population. Comparable
conditions of varying availability apply to food plants. During the investigation period documented by Wilmsen and Durham (1988), meat yield was
largest from March until October, while between December and February
hardly anything could be hunted. Reasons for this can be found in seemingly
little, yet meaningful details of the subsistence technique: The least hunting
activity is observed during the season with highest precipitation. The snares
made from *Sansiviera* fibres, used for trapping small animals, swell or rot due
to high humidity and become useless. Likewise, the strings of bows lose their
tension and become ineffective. Only during the warmer seasons can the traditional hunting techniques be used again. During the cold-dry period from
May to August, hunting concentrates on large animals with high meat yields,
e.g. giraffe or Eland antelope, even though the more highly esteemed kudu can
be attained more easily.

Comparable figures of highly varying meat yield are also described for the
G/wi bushmen of the central Kalahari (Silberbauer 1972). The fluctuation in
monthly meat consumption observed here, however, shows completely different seasonal variation. The highest bag yields are reported as 5.4–6.8 kg/month
per person between January and June, but in the remaining months the quantity comes to approximately 2.0 kg/month only. Whilst this points to local peculiarities, at the same time it shows that even under conditions of an altogether
positive energy balance, there is a necessity to react to seasonal fluctuations of
resource supply by an adjustment of food acquisition strategies. It is understood that, strictly speaking, the weights of meat portions or collected plant
items have to be related to the physique, body proportions and activity levels of
the populations under study. They must not be taken as an absolute indication
of nutritional requirements or dietary availability, but are mainly to illustrate
variances in seasonal supply.

2.2.1.2 Specialised Hunters of the Arctic: Inuit

The ecological constraints of the circumpolar regions are characterised by
low biological productivity, extreme and continuous low temperatures and
pronounced seasonal light/dark cycles. Against this environmental background, Inuit[3] developed subsistence techniques which in the course of sev-

---

[3]   Inuit is used here as a generic term in accordance with a resolution passed by the
      Inuit Circumpolar Conference, Barrow, Alaska 1977, denominating circumpolar peo-
      ples colloquially referred to as 'Eskimo'.

eral thousands of years led to a close adjustment to the prevailing conditions. The procurement of the main food components is achieved by hunting sea mammals, fish and caribou, while vegetable foodstuffs are essentially missing. Thus, the available menu determines a diet that is rich in protein and fat, but poor in carbohydrates. Based on their traditional pattern of nutrition, Inuit are predominantly carnivorous. Isotope analyses revealed (Bocherens 1997) that, within the arctic food web, Inuit can be identified as the top predator. Unusual for an omnivorous species like humans, this nutritional behaviour not only reflects a highly adaptive plasticity, but also confronts Inuit with special problems of food procurement and resource use regarding the sufficient supply of nutrients. The traditional Inuit subsistence system nevertheless supplies a balanced diet.

Basically, all nutritional requirements are met and the supply of essential food constituents is secured by both the kind of food available and its preparation (Draper 1977; for a detailed account, see Gerste 1992). The high amounts of meat and fish provide a sufficient supply of essential amino acids, vitamins of the B-complex and vitamin K. Also vitamin C is available in adequate quantities, since meat and fish are predominantly consumed raw and/or just sautéed. This prevents the vitamin from being thermally degraded and to be in sufficient supply to buffer against scurvy. In addition, berries collected in the summer and preserved in seal train oil serve as a supply for the winter in the southern and central regions of Alaska. Fat and train oil from fish and sea mammals provide lipids, particularly long-chained unsaturated fatty acids, and the fat-soluble vitamins A and D. A nutrition strongly geared towards meat, however, also involves clear risks. On the one hand, these relate to the possibility of a marginal calcium supply, which leads to increased absorption of the bone mineral, particularly in older individuals (Mazess and Mather 1978). It is known, though, that Inuit are tapping sources of food rich in calcium by consuming animal blood and cut-up or coarsely ground animal bones or fish heads as part of the traditional diet.

On the other hand, with an amount of only about 2 % of carbohydrates in the food, problems can occur during the maintenance of glucose homoeostasis. An insufficient dietary supply of carbohydrates leads to a rapid depletion of glycogen stores in the muscle tissue and in the liver, in the first place. In order to further secure sufficient energy input, either increasing amounts of fat have to be metabolised, or biochemical pathways employing glucoplastic amino acids be utilised by increasing the breakdown of proteins. A pure meat diet supplies 10–20 g glucose/day as glycogen. However, to maintain stable brain functions alone, an adult requires more than 100 g glucose/day. The extraordinarily high protein supply allows Inuit to utilise an energetically costly but effective metabolic pathway, in which proteins are broken down into their amino acid components and used for conversion into glucose (gluconeogenesis), so that adequate blood sugar levels can be stabilised and sustained.

This leaves the sufficient supply of calcium as the only factor remaining that really seriously limits the nutritional basis. Computer simulations based on long-term Inuit energy balances and accomplished using data from a group of 50 Netsilik (Keene 1985) generated the same conclusion, despite the fact that overall the energy balance obtained through hunting and collecting was positive. The vast amounts of dietary protein secured through traditional Inuit foraging thus yield an energetic surplus (for Inujjuamiut, see also Smith 1991).

Even though on the whole this subsistence strategy appears to be optimised, it must not be overlooked that, in order to actually achieve and employ it, a considerable amount of fine-tuning is necessary that has to take into consideration the strong seasonal fluctuations of both food supply and climatic conditions and the fact that the annual cycle of food harvesting activities differs from year to year (for Inuvialuit, see Berkes and Jolly 2001). The Igloolik of northern Canada for example (see Godin 1972) go hunting for sea mammals (seals, walrus) throughout the whole year, whilst caribou are essentially available only during the late summer and autumn months, terrestrial carnivores (above all fox) during the wintertime and birds and their eggs in the short arctic summer. This seasonal dispersion of food supply requires an appropriate sequence of different hunting techniques and activity patterns. At the same time, this is linked to the necessity of temporarily high group and individual mobility, as the adaptive pressure of ecological conditions works against larger groups or settlements. From November until May, this may result in frequent expeditions on dog sleds to control the traps for the hunt of small mammals, as opposed to the summer months from May until September that account for the fishing and whale-hunting season. The annual cycle of subsistence activities depicted in Fig. 2.2 applies to Igloolik only in the first place. However, a general principle of food acquisition strategies can be derived from this operational sequence, as it also refers to local variations in resource availability and to seasonal needs or preferences (see Freeman 1988; Hertz 1995).

In addition to different foraging techniques employed to use the seasonally varying nutritional resources, there is a system of differential habitat use by groups within the population, divided by age and sex. Such behavioural strategies help stabilise the acquisition of food in a non-complex, thus relatively unstable, human ecosystem by allocation of tasks and space. While elderly people and children primarily exploit food sources inland or in the coastal tidal zones, the younger adults additionally exploit the more dangerous locations like offshore islands and cliffs. Food acquisition on the open ocean, e.g. hunting for seal, walrus or whale, is a task usually limited to the young middle-aged men (Laughlin and Harper 1976; cited in Freeman 1988).

There are further characteristics of Inuit social organisation connected with food acquisition which can be characterised as behavioural response in the sense of an ecological adjustment (e.g. Damas 1972). Because of the reduced predictability and thus varying availability of food resources, it has been pointed out that the command of different subsistence techniques and a

Month	Jan	Feb	Mar	Apr	May	Jun	Jul	Aug	Sep	Oct	Nov	Dec
Activity			Sled hunting								Sled hunting	
		Trap hunting									Trap hunting	
							Whaling					
						Fishing			Tent camps			
Prey					Seals							
			Walrus				Fish				Walrus	
		Caribou		Caribou						Caribou		
	Bear		Bear		Rabbit			Rabbit			Bear	
		Fox						Wolf			Wolf, Fox	
						Birds, eggs						
Environment		Ice increasing				Ice at maximum				Ice increasing		
			Longer dark phases				Light phase				Longer dark phases	

Fig. 2.2. Seasonal cycle of subsistence activities for Igloolik, Melville Peninsula, Canada. Hunting and settlement activities follow the abundance of major prey species (after Godin 1972)

detailed environmental knowledge provide Inuit with a relatively high level of personal autarky. The most successful head of a household would be the one who has the most comprehensive set of alternatives of food procurement at his disposal (Berkes and Jolly 2001). This results in a flexible composition of economically independent households within the community which, as additional security, are interconnected by social networks and exchange relations. Such formalised relations between small subsistence groups provide mutual support and assistance and permit Inuit to even enter home-ranges of neighbouring groups for food acquisition. This is part of a general principle of reciprocal bonds between families, which at the same time allows the establishment of alliances necessary for survival. It may include food-sharing, often beyond the immediate (family) group (Sabo 1991).

Even though the traditional pattern of seasonally changing modes of hunting and settlement still play a role today, a change in the way of life by forced adoption of a western habit affects the nutritional basis and thus the general state of health of Inuit populations. The changed nutritional situation has led to an increased occurrence of diseases typically associated with today's lifestyle. This, for example, includes vitamin C deficiency due to the cooking of meat, increased blood cholesterol because of an increased consumption of fats with a high proportion of saturated fatty acids, or an increase of periodontal disease and caries, because more refined sugars and flour are consumed. Also, dysenteries due to lactose incompatibilities from increased milk consumption are frequently observed.

Taken together, though, the consequences of this partial dietary conversion are not regarded as a serious degradation of the nutritional status (Gerste 1992). Even under conditions of dual economy, i.e. where traditional and contemporary ways of nutrition are practised in combination, neither a reduction of subsistence procurement nor a weakening of the associated social actions shows up. On the contrary, communal hunting, the distribution of meat and communal dining are kept and preserved as a meaningful constituent of daily life, because the way of how to live like an Inuit is considered important and is retained within the traditional social networks. Thus, adaptations which were developed over many generations in order to cope with the specific subsistence conditions of the circumpolar region and which led to an extensive minimisation of ecological constraints are maintained and serve as a buffer against the inevitable changes of external living conditions. Coping mechanisms thus facilitate the enhancement of resilience.

Having introduced subsistence strategies of forager societies now allows taking a closer look at considerations that aim to analyse and predict patterns of food procurement. They take into account behavioural–ecological categories which govern the choice of food items under given circumstances of resource availability and environmental constraints and combine them in the formulation of optimisation models.

2.2.1.3 Optimal Foraging Theory

Societies with an acquiring mode of subsistence have extensive and detailed knowledge of usable food resources occurring in their habitat. However, as can be seen from the !Kung example, usually only a small part of all known edible plants and animals is selected to cover the nutritional requirements, even if other, less preferred, resources are encountered during hunting or gathering trips. This can mean two different things: either this reflects a pattern of differential appreciation, which leads to selective food procurement, and/or the selection is associated with the efficiency with which a certain quantity of energy, protein or other basic food item can be procured per given unit of time. To explain the implied variability in nutritional behaviour of individuals and groups, a model was developed based on considerations from evolutionary ecology entitled "Optimal Foraging Theory" (for further details, see e.g. Smith 1983, 1991; Mithen 1989, 1990; Kaplan and Hill 1992; Winterhalder 2001). It allows putting forward hypotheses in order to examine and test links between subsistence strategies and fitness gain.

The vast majority of such modelling was accomplished using foraging societies from different ecological settings and this informs the selection of the following examples. Only a few attempts have been made so far to apply optimisation models to food-producing societies, e.g. horticulturists (e.g. Gage 1980; Keegan 1986), perhaps because of the difficulties in dealing with increasing complexity.

OFT is based on the assumption that all organisms will optimise their food procurement behaviour in such a way that, in relation to the costs incurred, they will obtain maximum benefit in food returns, normally expressed as caloric and energetic yields. Optimised food acquisition should, therefore, be adaptive in the sense of natural selection, assuming that a certain nutritional behaviour will allow flexible reactions to changing resource situations, so that in the end the greatest possible individual benefit is achieved in terms of survival and reproductive success (Darwinian fitness).

Since costs and benefits of food procurement strategies are difficult to measure as an enhancement of individual fitness, proxies are introduced as presumed correlates of fitness. Accordingly, it can be predicted that during their quest for food humans behave in such a way that the net yield of energy or nutrients per time unit, i.e. foraging efficiency, is maximised. Thereby, however, fitness is limited by the respective influx of energy and nutrients, by the time available for activities not related to nutrition and by a possibly increased risk while foraging compared with other activities. Therefore, even though people would attempt to maximise the respective primary goal, e.g. energy yield, the overall outcome is the result of an optimised strategy. The evaluation of the optimisation strategy is based on the introduction of a currency (e.g. energy), a behavioural goal (e.g. efficiency maximisation of foraging), a set of constraints (e.g. delimitation of available foraging options) and a set of options (e.g. possibilities of individual choice or decision during food acquisition). The emphasis, therefore, is on general strategies of nutritional behaviour, which are subdivided into separate categories of decision-making. A distinction is made between the spectrum of foodstuffs in the diet, the rank order in which certain resources are selected, the resource distribution in the habitat, the duration of foraging trips, how many individuals make up a foraging party and, finally, where the most favourable settlement place is laid out as a function of resource distribution in the habitat[4].

With the help of these categories, individuals (or groups) are thought to make decisions for foraging strategies considered optimal under the given circumstances and which are assumed to lead to adaptive results, as they prove preferable under the conditions of natural selection. Thus, OFT allows a compilation of general decision rules for strategies of food acquisition that are based on cost–benefit considerations and that can be derived from principles of behavioural adjustment by natural selection. Of course, the individual foragers, or a group of people in the quest for food, do not themselves perform these kinds of cost–benefit considerations in reality. Rather, they as it were behave as if they do; and behavioural–ecological options can be put forward

---

[4] It is acknowledged that non-human primates, in fact many other organisms, follow the same underlying principles of feeding behaviour. Humans therefore are no exception to the general rule of optimised resource input for maximum resource yield.

OFi developed for non-human animals

as hypotheses and tested on the basis of empirical data. Such hypotheses can be derived from a number of key items or categories.

2.2.1.4 Basic assumptions and categories

Item 1. Diet breadth, prey choice. The composition of the dietary spectrum results from several food resources of different importance. The rank order in which these resources are included in the menu varies as a function of energetic net yield (e.g. Charnov 1976). Accordingly, a specific source of food would only be selected if the energetic balance is more positive than the average yield for food components receiving higher appreciation, after subtracting all necessary costs incurred with procurement (search, transport, processing). Based on a high correlation between encounters and food choice, energetic return rates have been identified as being the best predictors of foraging patterns (Hill et al. 1986). A high supply of preferred food should result in a more specialised diet, while there would be a shift to less preferred items, thereby expanding the food spectrum, should highly esteemed foodstuffs be scarce. Furthermore, if encountered, a productive and preferred resource will always be used, whereas all other available foodstuffs would enrich the dietary spectrum by demand and practicability. This means that the choice of a certain food resource will only depend on the availability of components of higher appreciation, not on its own availability. Thus, the appreciation of a certain type of food does not reveal anything as to how frequently it appears in the food spectrum, but only whether this source of food is used, if encountered.

Item 2. Patch choice, time allocation. In habitats where different sources of food are not evenly distributed, a problem may arise as to location and time. Would foragers be supposed to call at every patch where the resource crops out and exploit it; and how much time should be spent in each case? Under the primacy of optimising the itinerary of a foraging trip, only so many locations will be successively visited that keep the total time expenditure per resource yield minimal. Also, in combination with item 1, the relative amount of time allocated to different tasks, i.e. foraging versus processing, will impact on ultimate fitness gain.

Item 3. Group formation, optimal group size. These variables particularly refer to effects of group formation and co-operative foraging, which aim at an increase in efficiency during food acquisition. The following situations are considered advantageous: (a) the average per capita yield is increased, because the location of resources is discovered more quickly as an overlap of visited locations can be avoided, or because food acquisition can be organised by division of labour, (b) co-operative foraging levels out the variance of individual success, i.e. the differences in individual return rates, (c) the exchange of information allows the pursuit of food acquisition from a central settlement place and also facilitates the search for resources less well predictable in their occurrence. More recently, the aspect of sexual division of labour has gained importance by considering gender-related foraging abilities and patterns that lead to differential return rates (Hill et al. 1986).

2.2.1.5 Optimal Foraging Among the Aché

Of those studies that applied OFT to human communities, the most comprehensive was carried out for the Aché of the Amazonian rain forest in Eastern Paraguay (e.g. Hawkes et al. 1982; Hill et al. 1984). In a foraging economy, hunting is regarded as a strategy entailing high risk and small yield, while gathering represents a strategy with small risk but high and constant yields (Lee 1968). Among the Aché, however, men are considered particularly effective hunters who are able to accomplish a calorie yield from bag about four times higher, even on an average day, than could be returned from collected vegetable and animal food. Since empirical studies could not find a correlation between the total energy from hunted and collected resources altogether, Hawkes et al. (1982) considered this a suitable starting point for a test of OFT in general, and for a test of why Aché women would gather altogether, in particular[5].

According to the assumptions derived from the optimal dietary breadth model (see above), every resource would be selected whose energetic benefit would increase the total yield after all costs incurred with search and processing had been subtracted. In contrast, every resource would be ignored that would not lead to an improvement of the energy balance. Taking the 13 food components appreciated most by the Aché, from white-collared peccary and armadillo to oranges, honey, larvae or palm hearts, it was examined how the ratio of calorie yield per time spent for obtaining each of these components would be affected in comparison with the average yield and in relation to food acquisition in general, if a further component were added in each case (Fig. 2.3). It could be shown that, despite generally good hunting success, collected plants and insects on the whole improved the ratio of energy to time, so that they are considered worth collecting during food procurement. However, it was noticeable that not all options available were actually used to collect food. Palms, of which fruits, hearts and fibres can be eaten, were used less than for instance oranges, honey or larvae. This was explained as having to do with the spatial concentration of the respective resources and the time needed to

---

5   These considerations are solely based on energetic return and deliberately do not include aspects of differential nutritional values of foodstuffs. As the results of the study by Hawkes et al. (1982) implicitly reveal, female foraging success indeed contributes items of important nutritional value.

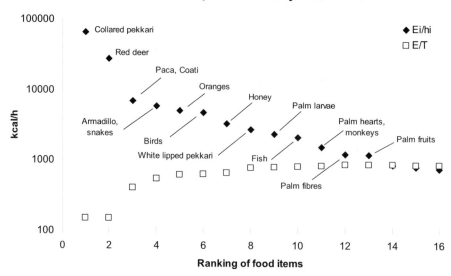

Fig. 2.3. Energetic yield of various food items procured by Aché, northern Paraguay. *E* Energy return from prey, *T* time, *Ei* energy return from new prey, *hi* handling time for new prey, *Ei/hi* average profitability of new prey. Every additional item justifies the energetic expenditure incurred for its procurement until, from item 14 onwards, the balance becomes negative. The 13 food items Aché appreciate most are also those that provide the best caloric gain in relation to the energy invested (After Hawkes et al. 1982)

harvest them once discovered (cp. patch choice). Whilst it is worthwhile to exploit palm locations only if the hunting yield stays below a certain value, it is beneficial at any time to even interrupt the hunt in order to use the high energetic yield and nutritional value per time provided by the three other highly productive resources, i.e. oranges, honey and larvae.

Another remarkable case in point can be made for an application of OFT by addressing the question whether children in forager societies would necessarily contribute so relatively little to subsistence, as was suggested and later generally accepted following studies with !Kung. Hawkes et al. (1995) were able to show for the Hadza from Tanzania that conditions found to apply with !Kung are not generally transferable. Hadza children, indeed, are very active and efficient members of food-procuring parties. There are even some resources they can exploit almost as effectively as adults. It can be more favourable, therefore, for a woman to move to a distant location together with her children, for example to pick berries, since despite the larger move the energetic benefit derived from this teamwork is higher than if she collected e.g. tubers at a closer location on her own. The advantage resulting from this pattern of food procurement is determined by age-specific yield rates of the Hadza children. In contrast to Mongongo nuts, the most important vegetable resource for !Kung, berries do not need to be processed any further and thus supply high and immediately available energetic and nutritional yield per unit of time. By not accompanying their mothers to the relatively distant Mongongo groves, but staying in the camp to crack the collected nuts, !Kung children actually practise the most productive and energetically most favourable activity possible, given the ecological conditions of !Kung habitats.

The results of such investigations thus confirm predictions from OFT models, according to which the choice of certain food components does clearly follow cost–benefit considerations; in all likelihood not in the sense of deliberate decisions, but rather in such a way that behavioural patterns emerge that can be observed and interpreted as result and consequence of optimisation strategies, i.e. as if people would deliberately entertain cost–benefit calculations before and while engaged in foraging trips. These strategies are accessible to empirical testing and can show that humans are subject to fitness-enhancing mechanisms (even) in their nutritional behaviour and thus follow a universal adaptive behavioural pattern. By achieving that, under given circumstances, food procurement supplies the highest possible energetic yields or foods of high dietary quality per time unit, humans maintain or improve their nutritional status. Consequently, this will have a knock-on effect on food procurement and thus on the chances of their offspring's survival. Such interpretations consistently fit in with considerations from life-history models – exactly because of the assumption that individuals who have more resources at their disposal have higher chances of survival and, eventually, higher fertility because they have better means of buffering negative energetic trade-offs. This, of course, would likewise provide better probabilities of survival for their offspring. For those individuals who can rely on fewer resources only, there may be a higher price to pay and trade-offs resulting from reproductive decisions may become more costly (see e.g. Hill and Hurtado 1996). But things appear to be more complicated. Whilst it has been a long-held view that the sexual division of labour would result in co-operative parental provisioning, males and females are known to pursue different provisioning strategies, which do not always lead to maximising the average long-term rate of food intake. Rather, co-operation or conflict seems to vary with reproductive goals involving childcare trade-offs for women and mating investment trade-offs for men, depending on which of the two has a higher fitness gain under given circumstances (Bird 1999).

2.2.2 Cultivators in the New Guinea Rain Forest: Tsembaga

Natural tropical rain forest displays an enormous gross productivity, due to high solar radiation, long growing seasons and ample humidity. In comparison with other biomes, it exhibits the highest density of plant biomass, characterised by a large diversity of species, albeit small numbers of species per unit area. The fauna is rich in species as well, though it predominantly consists of small animals and microfauna, resulting in a low animal biomass per unit area. Therefore, increased predator pressure in order to meet the protein requirement associated with rising human population density inevitably leads to rapid depletion of the animal stock. In addition, because of the small humus stratum, soils are altogether poor in nutrients and can, moreover, be easily washed away by heavy precipitation.

Despite a broad variety of potential food resources, natural tropical rain forest can only provide the nutritional basis for less than one human per square kilometre. Agriculture and thus the possibility of relative population concentration, can only be practised with the help of techno-cultural measures that make the limited nutrient supply of the soil lastingly available. Slash-and-burn cultivation (also called shifting cultivation) has proven a particularly suitable agricultural strategy in tropical rain forests and is practised world-wide by approximately 250 million people (p. 280 in Moran 2000).

In the rain forests of New Guinea, this type of agriculture was obviously developed independently approximately 9,000 years ago and has remained essentially unchanged up to the present time. Shifting cultivation is based on the felling and clearing of a plot of forest land with subsequent burning of the dried timber, followed by the planting and harvesting of cultivated crops. Before the soils are exhausted, the area is abandoned to allow succession of secondary rain forest and a new clearing is laid out at another location. Burning of the felled timber leads to a clear improvement in soil quality. Not only are possible crop parasites killed, but also the soil is supplied with nutrients in terms of mineral fertiliser from the wood ash. Whilst the burning transforms nitrogen or sulphur into the gaseous phase, the soil is enriched with calcium, phosphorus, magnesium and potassium and so its improved nutrient supply favours the growth conditions of cultivated plants. A further benefit is achieved in reducing soil acidity due to the ash cover. These ameliorating effects, however, will last for approximately two years only, at the end of which the supplied nutrients have been either washed out or essentially used up by the cultivated plants. Continuous cultivation on the same plot would lead to a clear reduction in crop yields. By shifting the gardens to a new clearing plot, however, the previously used area can regenerate through natural succession.

The gardens of the Tsembaga, a group of horticultural farmers in the highlands of New Guinea belonging to the Maring language family, are mainly in rain forest areas at altitudes between 700 m and 1,500 m; and in areas further uphill only marginal agriculture is possible for topographic reasons. Even on those plots preferentially used for cultivation, the soil requires fallow periods of 15–25 years to re-establish its original fertility and at higher altitudes it can take up to 35 years. As the garden soils are exhausted after two years at most, there is substantial demand for arable land and the Tsembaga thus inevitably employ extensive strategies of land use. More than 90 % of their entire territory lies fallow (Rappaport 1971) and thus forms the regenerating and/or not yet utilised reassurance for the crucial resource of land and its continued fertility.

Several plants are cultivated in the gardens at the same time. This allows for great diversity of species produced on the cultivated plot and leads to structural similarity with the characteristics of the surrounding rain forest. Plants of different heights efficiently use the incident sunlight and protect the soil against sun and rain; and the risk of harvest loss by pests is spread. By skilfully utilising different locations in the habitat to lay out the gardens, it is ensured that the cultivated plants can grow under optimum conditions. The sweet potato, for example, is tolerant to lower temperatures and can be planted at higher locations, while taro prefers warm, damp conditions in only slightly converted rain forest areas. In gardens situated on slopes, cut timber is left lying as protection against erosion. In the course of weeding, care is taken that tree seedlings are spared in order to accelerate regeneration of the plot during the subsequent fallow period.

Cultivated crops account for more than 90 % of the Tsembaga's daily food. Among the most important are taro (*Colocasia esculenta*), sweet potatoes (*Ipomoea batata*), yams (*Dioscorea* spp) and manioc (*Manihot dulcis*), with in addition bananas (*Musa sapientum*), sugar cane (*Saccharum officinarum*) and further vegetables. Particularly roots and tubers, the bulk staples, contain only little protein. About a quarter of the arable crops, especially the less appreciated sweet potatoes and cassava, are invested into rearing and keeping domesticated pigs, to produce at least a small protein supplement. Pork comes to about 8 % of the diet, but is very irregularly consumed (Bayliss-Smith 1982), since pigs play an important role in cyclically recurrent rituals between Maring groups rather than serve as mere calorie supply (Rappaport 1968). Even though the Tsembaga subsistence system is able to produce an amount of 34 g of animal protein per day and person (Little and Morren 1976), this represents a computational figure only, considering the actual conditions of consumption. In addition, an energetically negligible part of the food spectrum (approx. 1 %) is procured by hunting. However, it can hardly serve as anything but culinary enrichment of the menu.

Energy flow in a Tsembaga community is thus essentially based on three fundamental food chains: (1) roots and tubers → humans, (2) roots and tubers → pigs → humans and (3) other garden produce → humans. In addition, there is a slight supplement from wild fauna and flora and a food chain based on the exchange of food with other communities in the habitat (Little and Morren 1976). The energetic efficiency achieved by the Tsembaga system of food production provides 14.2 kcal on average for each energy unit invested into the system by human labour. With an output/input relationship of 20.9, the roots and tubers food chain, i.e. carbohydrate production, is managed most effectively, while the hunting and gathering route produces a ratio of 5.4 only. Pig keeping is energetically most demanding with 3.2 kcal energy yield per invested unit. This can be explained by the relatively substantial amount of produce from the gardens that has to be spent on the raising of pigs alone and entails an extension of the food chain. From an energetic point of view, the high costs incurred in this food chain clearly show why it contributes only very little to the total Tsembaga diet (Bayliss-Smith 1982).

By performing a mixed subsistence strategy consisting of horticulture and animal husbandry, the Tsembaga are able to maintain a self-stabilising system of food production, on the condition however that a long-term balance between resources and population size can be established. This is accomplished on the one hand by land-extensive use of the habitat employing shifting cultivation and long fallow intervals. On the other hand, it has been postulated that the cyclical capping of pig numbers during ritual events, in which several societies are involved, equally contributes to stabilising the system. In doing so this would avoid that, at a given point in time, the amount of energy invested in pigs would exceed a level eventually causing destabilisation of the energy balance. Moreover, if successful and effective enough, human population densities would always remain below the carrying capacity of the habitat, even if the total number of humans is not referred to the total area of potentially arable land, but only to those areas that have to date or ever been cultivated[6]. Projections based on observations with Tsembaga in the mid-1960s showed that the habitat could have sustained a sufficient subsistence basis for approximately 50 % more people than the 204 individuals who actually lived there at the time of the investigation (Rappaport 1968).

2.2.3 Cultivation and Livestock Farming in the Andean Highlands: Nuñoa Quechua

Life in mountainous ranges requires the development of adaptive strategies which counteract the limiting factors of generally low biological productivity due to low temperatures and relative dryness, night-time cold stress and, at high altitudes, low oxygen partial pressure (cp. p. 155ff in Moran 2000). For the purpose of this chapter, i.e. with regard to modes of production and food acquisition, the question of habitat productivity is primarily important and will be introduced taking examples from populations of the Andean highlands.

Human populations of the Andes are provided with a diet that is rich in carbohydrates, but generally poor in animal protein and fat. Despite the slight protein supply, however, at least adults seem to have an even nitrogen balance.

---

[6] Computer simulations (e.g. Foin and Davis 1984) indicate, however, that homoeostatic regulation of population density by means of ritualised celebrations is too idealistic an approach (see Chap. 5).

Whilst a low fat and rich carbohydrate diet is basically considered suitable to living in high altitudes with low oxygen pressure, it is not yet clear whether this has to be regarded as an adaptive response to existing environmental factors, or whether the small amount of animal-derived food products only reflects the more unfavourable socio-economic situation of populations of the Andean highlands (Picón-Reátegui 1978). When food energy is mainly provided by carbohydrates and arduous manual labour must be carried out under less favourable external conditions, hypoglycaemia is likely to develop as a consequence of reduced carbohydrate absorption. The widespread habit of chewing coca (*Erythroxylon coca*) proves an effective adjustment here to conditions of potential nutritional deficiency.

The most effective way for an organism to produce energy is by constantly supplying it with carbohydrates. Their absorption rate in the intestines depends on the length of time carbohydrates spend at the intestinal surface and on the necessary supply of enzymes responsible for the oxidative cleavage of complex carbohydrates. The sugar molecules resulting from cleavage (glucose, galactose, fructose) are absorbed and transported into the bloodstream. When supplied in excess, glucose is stored in the form of glycogen and can be mobilised if necessary. One of the key constituents of coca, the alkaloid atropine, is responsible for the positive effects of coca-chewing. Since atropine extends the time of contact between carbohydrates and the intestinal mucosa, it considerably increases the availability of glycogen storage and, as a consequence, enables a more efficient utilisation of food rich in carbohydrates. Blood glucose levels were found to increase up to 100 % above fasting level ca. 2 h after chewing 30–40 g of coca leaves (p. 162 in Moran 2000).

Besides, the pharmacological effect of coca can even be increased if it is chewed together with substances providing calcium supplements (Wilson 1985). The daily requirement in calcium comes to several hundreds of milligrams and can be efficiently covered, in particular by milk and milk products and by green leafy vegetables. However, those populations who are strongly dependent on plant food are at risk of suffering from marginal calcium status, since the majority of edible plants are relatively low in calcium. This condition applies to the Andean highlands, since tubers and cereal grains, the main groups of plant staples of the region, do not rank among the calcium-rich vegetable foodstuffs (Souci et al. 1981). Two substances are commonly used as calcium supplement: *llipta* is made from the ash of the stems of two *Chenopodium* species (*C. quinoa*, *C. pallidicaule*), while *cal* consists of pulverised limestone. Both are chewed with coca and used as an admixture in grain mush. The effect of such culinary practice is substantial: it provides the regular food with a calcium supplement in the order of 300–1,200 mg (*cal*) or 200–500 mg (*llipta*) per day (Baker and Mazess 1963). Thus, even with a nutritional basis potentially deficient in calcium, the daily allowance of approximately 800 mg calcium recommended by the FAO (1980) is met. Apart from the nutritional and physiological advantage, chewing coca is an important social activity and meets fundamental needs for the establishment of cultural identity among the Quechua-speaking people of the Andean highlands; and both the handling and the use of coca are subject to defined rules (Allen 1981).

Rather than being an outcome of purposeful invention, the emergence of such an optimisation mechanism has to be regarded as an outcome of trial and error that eventually became adaptively effective (see Rozin 1987). Regardless, it results in alleviating a limiting factor and thereby an increase in the survival options in a specific habitat of the Andes. A similar striking culturally adaptive mechanism, although less frequent in the Andean highlands, but particularly spread throughout Central and South America, can be observed with those populations whose nutrition is substantially based on maize. Maize has an altogether unbalanced composition of essential amino acids, being deficient in lysine and tryptophan, among others. Furthermore, it has a reduced content of niacin, a vitamin of the B complex. Given these conditions, humans relying on a pronounced maize diet run a high risk of developing pellagra, a niacin-deficiency disease with symptoms such as dermatitis, diarrhoea as well as psychotic conditions and mental confusion (p. 908ff in Marieb 1998). But pellagra can also be induced by enhanced levels of dietary leucine, also a characteristic of maize. The three components, tryptophan, niacin and leucine, form a synergistic complex, with niacin being a derivative of tryptophan, whereas high leucine contents inhibit this transformation (Katz et al. 1974). These unfavourable effects can be alleviated by adding calcium hydroxide when a maize dish is prepared, as it alters the leucine/isoleucine ratio in favour of the latter and thus diminishes the inhibiting effect of leucine. As the solubility of certain protein components (the zein portion) of the maize grain is being lowered, the relative content of essential amino acids in the food thus made bio-available from the glutelin fraction is increased and the nutritional quality of maize enhanced. In a cross-cultural analysis, Katz et al. (1974) were able to show that the custom of preparing maize in water enriched with lime, i.e. calcium compounds, is widespread among those societies whose subsistence is based on this staple.

The adaptive value of this practice is obvious, but its propagation and establishment most likely was less so. Dough is prepared from cooked maize, which is then used to make tortillas. The addition of lime to the water makes the dough more workable than if the maize grains are cooked in plain water; and therefore it considerably facilitates the production of this basic foodstuff. And interestingly, but not surprisingly, when women were asked why they prepared tortillas from maize they first cooked in limewater, they indicated it was for this practical reason (Rozin 1982). Certainly, one might not have expected an answer referring to the complex biochemical foundations of preparing maize for better dough, but the example suitably shows that a physiologically effective adaptive nutritional strategy requires cultural mediation to be adopted and permanently used as culinary practice. On the back of such mediation, the nutritional advantage and thus the survival advantage in a certain habitat are co-selected. The 'invention' of more workable dough leads to a life-practical improvement, which is adopted within the community and maintained as a tradition.

These kinds of biocultural strategies, like coca-chewing or the use of lime-water counteract restrictions that result from limiting factors of the habitat. The ecological constraints of the Altiplano, moreover, also evoke special adjustments in the sphere of economic relations.

The ecological basic conditions of the Andean highlands cause a delimitation in plant productivity, which involves a decrease in total biomass production with increasing altitude. In historic times, utilising the biotic zonation of the highland habitat compensated for this restriction of natural habitat options. As groups of people colonised the different ecological zones of the highlands, pre-Incan populations established different 'resource islands' and the produce gained was distributed through interzonal trade, thus achieving an optimised use of the region's vertical ecological zonation (p. 170 in Moran 2000). Recently, strategies of both horizontal and vertical resource utilisation have also been demonstrated for different ecological settings along the Osmore Valley of southern Peru, using isotopic data from human skeletal remains (Tomczak 2003). Such exchange networks between locations at different altitudes are well known also from extant indigenous populations, e.g. the Nuñoa Quechua of Peru. They are based on a fine-tuned co-ordinated system of plant cultivation and livestock production at the respective altitude levels.

Of the domesticated livestock used in the Andes, llamas and alpacas stock at the highest locations within the vertical zoning of the Andes region, while sheep are kept on mid-range altitudes and cattle and horses in the lowest zones. Among these, the camelids have a central function in supplying vital products. Not only do they serve as beasts of burden and transport, but they also supply hides, wool, food and dung. As autochthonous inhabitants of the Andes, they are well adapted to high altitude and the bunch grass vegetation cropping out here. Due to its high caloric value, dung is suitable as fuel and, besides, can be collected relatively easily, since the faeces are preferentially deposited at certain places. As part of the exchange network, camel dung fuel and sheep muck are traded to the lower regions, while in return coca gets to higher altitudes. Sheep muck is enriched in nitrate, magnesium, calcium and potassium and is therefore suitable as fertiliser, which is put on fields at lower locations and thus supplies the crucial nutrient input there in order to ensure the cultivation of potatoes. In turn, coca facilitates improved metabolism of the high carbohydrate portion in the food of inhabitants at the highest locations (see above). Thus, dung and coca are central to an optimised utilisation of the ecological zonation in the Andean highlands (Winterhalder et al. 1974).

Within the respective vertical zones, micro-locations of very different ecological properties can be found, which allow the Nuñoa to grow old-cultivated cereals (Chenopodiaceae) and tubers (*Oxalis, Ollucus, Solanum*) even at altitudes of up to 4,000 m above sea level. Among the tubers alone, there are three different varieties of potatoes, two of which are frost-resistant. Plants domesticated in more recent times usually do not grow above 1,500 m and their use is thus limited to lower locations.

Figure 2.4 displays how such a combined subsistence basis consisting of herding and agriculture is co-ordinated in the yearly cycle. The management of herds is an all-season task, structured by shearing the sheep and selling their wool in the winter, slaughtering animals and the trade or barter with zones at lower elevations. In contrast, the planting cycle is adjusted to the course of the rainy season. Late summer is the time to plant tubers and cereals, which are harvested towards the end of winter and at the beginning of spring. This sub-division of the year into a pre- and post-harvest interval bears some consequences for the energy budget and the allocation of resources within Nuñoa families, as much as it has led to the development of complementary adaptive strategies to interfere in energy flows (see Sect. 2.3).

Three trophic levels are represented in ecosystems managed by the Nuñoa or other Quechua-speaking groups and define relatively straightforward food chains. Primary production from the natural grass vegetation of the *puna* provides pasture for the domesticated herbivores (llama, alpaca, sheep, cattle, horse); and all cultivated plants are used for human consumption. Taking into account produced food energy, energy consumption and energy costs from labour, the energy input/output balance has been calculated for a Nuñoa family of six (Little and Morren 1976; Thomas 1976). According to these figures, from cultivated crops alone about 11.5 times more food energy is obtained than that invested in terms of labour. The energetic efficiency of animal husbandry including the goods exchanged for animal-derived products amounts to 7.5 kcal per expended energy unit.

Month	Jan	Feb	Mar	Apr	May	Jun	Jul	Aug	Sep	Oct	Nov	Dec
Herd-related activities		Butchering; Preparation of dried meat						Exchange and barter			Shearing; Selling wool	
Farming activities		Harvesting and threshing of Quinoa							Planting Quinoa			
		Harvesting and threshing of Cañihua							Planting Cañihua			
		Harvesting potatoes			Drying potatoes					Planting potatoes		
Environment					Rainy season							

Fig. 2.4. Seasonal cycle of subsistence activities for Nuñoa, Altiplano, Peru. Whilst the farming cycle depends on the duration of the rainy season, herding activities occur year round (modified after Little and Morren 1976)

Since agriculture alone cannot cover the entire food spectrum necessary, the mixed economy of the Andean highland serves as an adjustment of mutual benefit under the existing conditions, because livestock supplies nutritional supplement, the necessary fertiliser for the fields and valuable exchange goods. Moreover, the way division of labour is practised positively adds to the total energy balance of the population. As boys or young adolescents are allocated the task of herding the llama, which is not physically demanding, the energy consumption of the human population is reduced by 30% per day on average. This saving in net energy of approximately 100,000 kcal/year per family considerably minimises the danger of hypocaloric stress in the population (Thomas 1976).

2.2.4 Pastoralists in Semiarid Zones: Karimojong

The survival of human populations in the expanded savannahs and grasslands of the earth is closely connected with the natural or managed occurrence of large herds of animals. They can be exploited for seasonal hunts with no or little[7] active interference being necessary to maintain stock, since the large herds migrate across specific areas in seasonal cycles and thus essentially represent reliable and calculable resources (e.g. bison, Plains Indians of North America). In order to hunt them, human groups are forced, however, to co-ordinate their settlement patterns with the (seasonal) occurrence of the herds.

There is a different attitude towards herds whose size and composition are steered by human interference. The East African savannah is among the traditional areas where pastoral nomadism has been and still is being operated. Well known groups of livestock breeders of this region are e.g. the Karimojong of Uganda, or the Maasai, Turkana and Samburu of Kenya (e.g. Fratkin et al. 1994). With its expanded grassland vegetation, the savannah offers an ideal biome for animal husbandry and pastoral farming. Herds of cattle, sheep and goats convert the grass and shrub vegetation, which cannot be directly utilised by humans, into 'living food' in terms of meat, milk and blood, i.e. into a form of energy readily accessible to humans. In the sense of habitat-specific resource utilisation, this transformation is the subsistence task that lies at the heart of pastoral societies.

Among the common characteristics of traditional African herding communities is the lack of ranching or dairy farming, i.e. animal husbandry for market production, but there is subsistence pastoralism, which entails taking the flocks to pasture and water. Their economic goal is the daily supply of

---

[7] Plains Indians, for example, regularly burnt the perennial grass cover of the prairies to stimulate growth in the new vegetation period.

basic food, leaving little room for producer surplus, rather than the conversion of green fodder into marketable products, since keeping as many animals as possible is necessary to regularly supply as many people as possible (Dyson-Hudson 1980; Casimir 1991). To keep a herd of livestock maximally large under given conditions requires operational decisions of herd management that have implications beyond the basic and immediately vital meaning of herds as living subsistence capital and that extend into many areas of life. The respective herd stock constitutes personal property and indicates personal wealth, social status and influence, it forms the object of inheritance from father to sons and it is subject to exchange in formal friendship treaties between pastoral groups as much as it is the material basis of a legally binding marriage (Dyson-Hudson 1972). Therefore, safeguarding a herd as populous as possible is the strategic goal of pastoralists, as this is the only way to facilitate the long-term utilisation of this resource and maintain social obligations.

The savannah, being a typical biome of subsistence herding, is a highly differentiated ecosystem with discernable sub-systems, which essentially differ by the amount of precipitation and thus by the proportion of characteristic vegetation of trees, shrubs and grasses. Such climatic conditions require extended periods of time without rainfall and cyclic droughts to be taken into account. Therefore, often subsistence agriculture is practised besides animal husbandry, wherever facilitated by the natural habitat conditions, including the cultivation of staple crops such as sorghum or millet. This secures energy supply by alternative sources of food to avoid being solely dependent on one type of resource, livestock, due to the uncertainties of the habitat. Pastoralists pursuing this strategy have been termed 'avoiders', as they operate agriculture in dry regions only in the season of heaviest rainfall to ensure a distribution of subsistence risk by utilising the whole diversity of the habitat. Other options include an enduring strategy, pursued by pastoralists in the most arid areas only, or escaping by laying out irrigated fields and thus creating an artificial environment (Agnew and Anderson 1992).

The Karimojong make a good example of the 'avoider' group. They are a community of approximately 60,000 people, who inhabit an area of approximately 6,500 km² in the Northeast of Uganda – at the time the studies referred to were conducted. This area is made up of a central range of good quality soil; sufficient water supply and suitable vegetation cover as well as less favourable peripheral areas. Therefore, long-term settlements are to be found in the central areas, where crop farming on alluvial soils allows ample yields. From September until March, the time of increased precipitation, only the cattle herds are put out to pasture here, while sheep and goats are kept to graze elsewhere throughout the whole year (Dyson-Hudson and Dyson-Hudson 1970).

It is a key feature of pastoralism, though, that animals rather than the environment form the crucial resource. Pastoralist farming leads to local depletion of the local environment and, despite substantial indigenous knowledge about habitat properties (e.g. Bollig and Schulte 1999), generally no sustainability or protection of the environment is intended. Any exhaustion of pastures is met with a change of location. Interfering with the environment in order to improve pasture merely occurs by burning the dry grass in some places to allow fodder rich in minerals and vitamins after germination to be available for the cattle. Since the goal of the subsistence strategy is to preserve the cattle resource, animals are slaughtered rarely and only on special occasions (Dyson-Hudson and Dyson-Hudson 1969). Yet, animals are of course being used for the supply of essential foodstuffs, not only by milking but also by bleeding. Both practises allow tapping the resource without having to kill the animals. Bleeding cattle is commonly performed and involves puncturing the jugular vein for 2–4 l of blood, which is processed for human nutrition, often by mixing it with milk to produce a nourishing coagulate high in protein. Such a quantity does not harm the animal when applied two to three times a year. Bulls, lactating cows and calves are excluded from bleeding, thus protecting the most precious animals. Other societies, for example the Ariaal, bleed camels, which can provide up to 35 l of blood per animal per annum for human consumption (p. 85 in Fratkin 1998).

Milk, besides being instantly consumed, is processed into clarified butter (*ghee*), the only product that can be kept well under given temperature conditions over a long period of time. It is also true that meat from slaughtered cattle can be dried and then stored. However, the owner's family would hardly retain more than a quarter of the meat, since all remaining flesh would have to be distributed among relatives, friends or neighbours as part of strategies to reinforce alliances and social security. Therefore, extracting the predominant part of the nutrition from living animals is more favourable, because only meat must be distributed. Generally, the sale of animals is also avoided, because no other food could be bought from the attainable proceeds that would be equivalent and correspond to the value of a living animal. This even applies to cows milking or reproducing poorly, since they can still be bled or also used within the social network of mutual social reinsurance, e.g. to pay bride wealth, exchange goods or serve formal friendship treaties. If slaughtering of an animal is required, for example in connection with religious rituals, bulls or infertile cows are selected. The meat is then distributed among the participants of the ceremony, whereby the largest portions frequently go to older men. Such ceremonies usually take place in times of low rainfall or after harvest failures. In doing so, a way is found to somewhat reduce the herd size and to distribute meat within the community, when other food is scarce.

Apart from safeguarding the human nutritional basis, the management of large herds of cattle requires further strategic decisions necessary to maintain the resource despite the seasonal change of ecological conditions in the habitat (Fig. 2.5). The task is to guarantee that there is always sufficient water within reach and that suitable fodder is available with the necessary protein/fibre ratio. Furthermore, pasture suffering from high infection risk by

Month	Jan	Feb	Mar	Apr	May	Jun	Jul	Aug	Sep	Oct	Nov	Dec
Pastoral Activity				Relocation of herding camps								
	Herd movement during …									… dry season		
				Herd movement during rainy season								
Farming activity			Relocation of …					… settlements				
		Preparation of fields										
			Planting		Weeding			Harvest				
								Drying millet and maize				
									Drying peanuts and squash			
Environment				Highest probability of rainfall								

Fig. 2.5. Seasonal cycle of subsistence activities for Karimojong, Uganda. Probability of rainfall is the single most important factor influencing the mobility of herds and humans, which requires careful temporal and spatial co-ordination (modified after Dyson-Hudson 1972)

pathogens, e.g. the tsetse fly, is unsuitable in principle. Finally, the occupation of suitable pasture areas has to be co-ordinated and adjusted among those pastoral groups competing for comparable locations (Little and Morren 1976). The unstable and often unpredictable climatic conditions of the savannah further add to the difficulties already affecting these decisions. An extension of the dry season or the non-appearance of seasonal rainfalls subsequently entailing a drought period requires short-term operational decisions, in order to maintain the herds as the foundation of human existence. Therefore, the shifting of pastures and camps has to be carried out the whole year through, as a measure of adjusting to habitat conditions. Mobility and the option of residential change within the entire territory are of vital importance, since it is the herds that must be led to the water and the pasture grounds, as the reverse is usually not possible.

If the herd owner's family is large enough and has sufficient numbers of boys and young men, labour tasks can be divided and the herd split up and pastured at different locations (Coppolillo 2000). At least in the case of boys, it may be assumed that this, in addition, would lead to a similar effect of energy savings as those known for the Nuñoa (see Sect. 2.2.5). Furthermore, with this kind of partial, seasonal residential mobility, it has to be considered that during the rainy season women and young girls will stay in the settlement to till the fields and bring in the harvest. The extent to which the respective produce of a combined pastoral and agricultural subsistence contribute to the diet is dependent on the habitat setting and seasonal and/or yearly fluctuations of climatic conditions. Thus, the probability of rainfall and the possibility of using fresh pasture grounds are of importance. Projection figures calculated for the Karimojong revealed that approximately 34 % of the food energy can be obtained from herding. On annual average, cattle supplied 1.61 l milk/day (Dyson-Hudson and Dyson-Hudson 1970; data from 1957–1958). If all herd animals are taken together, i.e. including sheep, goats and donkeys, then a value of approximately 50 % is attained (Little and Morren 1976). For other pastoral groups of East Africa, the proportion of milk alone can come to more than 60 % of the daily food energy, temporarily even more than 90 % in good years (Little 1989; Galvin et al. 1994; Fratkin 1998). A universal characteristic of the nutrition, however, is the extraction of obviously only relatively little energy, which varies between approximately 1,000 kcal/day and 1,400 kcal/day per person, depending on climatic conditions (Galvin et al. 1994). Whenever possible, pastoral populations will put the emphasis of their nutrition on milk and other animal-derived food products. Thus, being forced to change over to cereals in bad years does not mean a change to the general subsistence strategy, since there is a cultural preference for food produced the pastoral way. More than that, pastoralism is perceived as a distinct way of life that permeates these societies' world views, belief systems and social organisation, to the extent that foodstuffs become important parts of idealised categorisation that constitute food-related cosmologies (e.g. for Maasai, see Böhmer-Bauer 1990)[8].

The unusually high amount of milk sometimes found in the diet of pastoralists points to an adaptation which links genetic and cultural adjustments in an amazing way. Lactose tolerance (for a detailed overview, see p. 226ff in Durham 1991), the ability to digest milk in large quantities even at adult age, is subject to genetic control – the gene for this polymorphic trait is located on chromosome 2 and is expressed by an autosomal-recessive inheritance mode (p. 614 in Vogel and Motulsky 1997).

Humans, like all other mammals, produce the enzyme lactase in early childhood, which enables them to digest the milk sugar contained in their mother's milk. After weaning, the enzyme activity rapidly drops to less than 10 % of its activity in new-borns. Children have meanwhile become acquainted with and used to solid foodstuffs and milk is no longer the exclusive or main constituent of their nutrition. As a consequence of the drop in enzymatic activity, the juvenile organism develops intolerance against lactose, which may lead to clinical symptoms such as diarrhoea and colic attacks when milk is consumed in large quantities, usually the equivalent of the lactose contained in more than 0.5 l in one portion, for some individuals even less. However, some human populations have retained a high capacity for enzymatic cleavage of lactose into glucose and galactose, even at adult age.

---

[8]   Interestingly, quite similar patterns can be observed at our doorstep, as it were, for example in a Breton fishing community (Chapman 1980).

eurofocus

The geographical distribution of this trait shows a pattern of three foci which also gave rise to different hypotheses put forward to explain the spread of this trait (see Holden and Mace 1997).

1. A high prevalence of adult lactose malabsorbers is found in those areas where traditionally milk is not used for food, whereas animal husbandry and dairy farming occur where the frequency of the absorber trait is high. The emergence of this pattern would have to be explained in such a way that adult pastoralists, who were carriers for the absorber trait, had a selective advantage that particularly provided them with a nutritional advantage in protein supply. The development of pastoralism as a subsistence strategy and the establishment of the 'milking habit' (Simoons 1969, 1970) would have had a positive feedback effect on the selection of the genetic trait (Simoons 2001) in the sense of co-evolution.

2. A high frequency of lactose tolerance is found in Central and North Europeans and in their descendants in America and Australia. This observation fostered the hypothesis (Flatz and Rotthauwe 1973) that, in geographical latitudes with reduced solar (ultraviolet) radiation and thus an increased risk of vitamin D deficiency, each physiological achievement promoting the intake of calcium – as applies to lactose – would have an adaptive advantage. The deficiency results in a decrease in both calcium absorption from the gastro-intestinal tract and exchange between the blood and the skeletal system. This triggers a counter-regulation of the suprarenal cortex, which causes intensified mobilisation of calcium from the mineral deposit in the bone. A high supply of calcium-rich food would accordingly alleviate the situation.

3. Lactose tolerance is frequently found in arid areas, e.g. North Africa or the Near and Middle East. With water being a limiting factor, the digestibility of milk would provide additional liquid and electrolyte supply and therefore be of selective advantage (e.g. Cook 1978).

Based on 62 societies from the Ethnographic Atlas (Murdock 1967), with information on the proportion of pastoral and dairy farming contributing to subsistence, on solar radiation and on the duration of the dry season in the respective habitats, these three hypotheses put forward to explain the persistence of the lactose absorber trait were examined (Holden and Mace 1997). Pasture and dairy farming were found to be significantly correlated with frequency of the absorber trait, not however with solar radiation or aridity. It was concluded that selection towards a pronounced capacity of lactose compatibility in adult individuals can be regarded as an adaptation to dairying, while any selective influence of latitude or habitat conditions appears to have been insignificant.

The selective advantage is obvious. Pastoralists seek to maximise cattle, their indispensable resource, and thus each cultural or genetic trait that contributes to preserve resources is of advantage. As milk can be used for food, not only does the nutritional situation improve by the supply of protein and food rich in minerals, but the animals can also be exploited over a long period of time. This, of course, in a similar fashion also applies to animal husbandry as part of a mixed mode of production (see below, Sect. 2.2.7). The co-evolutionary spread of the milking attitude and lactose tolerance can be regarded as a positive-feedback system. It is likely that the development of dairy farming, i.e. the cultural trait, preceded the evolution of a genetic trait of lactose tolerance. The development of dairy farming and thus the acquisition of a new source of food provided a selective advantage to each individual who carried the genetic trait of lactose tolerance. The high selective pressure in favour of the absorber trait would have eventually paid off in terms of enhanced chances of survival and thus reproduction. In turn, the more the characteristic frequency rises within the population, the higher will be the amount of pasture and dairy farming, or at least its significance and acceptance will rise, because a new source of food can be used by more and more individuals.

How this process evolved in detail is not yet fully understood. It is acknowledged, however, that the spread of pastoral and dairy farming took place relatively quickly after the invention of pastoralism in the Near East; and just a few thousand years later it was established in Central and Middle Europe. In an attempt to model the spread of the genetic trait, it was found that, given an initial prevalence of 0.05 % for the absorber allele, a selection coefficient of more than 10 % would have been necessary in order to achieve within 6,000 years the allele frequency of 0.7 estimated today in Northern Europe. Other computations may differ somewhat, but they support the view that the co-evolution of a biocultural trait of high adaptive value took place within the – on an evolutionary scale – short time of approximately 200 generations (see p. 13 in Cavalli-Sforza et al. 1994). This estimation would roughly coincide with the onset of dairying in Europe (Simoons 2001).

Admittedly, there are societies, for example Mongols, Herero or Nuer, who also practice pastoral farming, but nevertheless are predominantly lactose-intolerant. Their strategy to circumvent or avoid adverse physiological reactions is to consume milk as fermented milk products, which are strongly reduced in their lactose content. In these cases, the gene–culture coupling was waived and replaced completely by a culturally steered behavioural strategy. Not only is the adaptive strategy of dairy farming being adopted, but moreover milk is processed into a generally tolerable food item that maintains all the advantages of a milk-supported nutrition. Yet the full selective advantage of subsistence based on milk remains.

2.2.5 Pre-Industrial Agriculture in European Temperate Climates

In contrast to the majority of biomes dealt with so far, the temperate zones can be regarded as altogether favourable areas. With no climatic extremes, a sufficient precipitation all year round and a wide range of different ecotypes, humans are offered numerous possibilities for agricultural production, which in historic times predominantly led to intensified forms of land-use. Pronounced seasonality and temporally limited vegetation periods typical of the temperate zones nevertheless entail ecological constraints that have to be overcome by the choice of suitable modes of production and possibilities of storage. The adjustment of food acquisition strategies to the respective ecological potential of natural units is equally important.

The following section draws from examples of both the more distant and recent past in order to demonstrate that there is an element of timelessness in the actualistic interpretive approach employed in ecology. Populations of the past were subjected to the very same ecological constraints that can be identified for sub-recent societies and their responses can be analysed and assessed in a quite similar fashion, even when the available evidence is chemical rather than archival, as in the following case study. Conventional archival or archaeological sources usually permit an indirect derivation of structural connections between the economy and ecological factors of the habitat at best. In contrast, chemical analyses of archaeological human skeletal remains can provide direct access to topics in economic and social history, even for pre-and proto-historic periods, through deciphering information encoded and stored in the primary source material itself (e.g. Ambrose and Katzenberg 2000).

2.2.5.1 Early Mediaeval Alamanni

In pre-industrial times, the provision of foodstuffs in order to meet nutritional needs is closely linked to the capacity and properties of a given natural unit and is thus dependent in particular on site-typical factors of geomorphology, climate and soil quality. Agricultural production is small-scale and geared towards the immediate supply of produce to the community (Abel 1967). It is true, natural landscapes had to be transformed even in those times. However, this usually never reached today's dimensions characterised by large-scale alterations and the typically extensive consumption of land; and an abolition of original natural settings is even less likely. Against this background, the development of modes of production typical of certain localities, or rather habitats, is anticipated. This may be obvious for large geographical areas such as coastal or inland regions where the acquisition of food is often largely dominated by either marine or terrestrial foodstuffs. However, it can be equally expected for spatially delimited sites in lowland, riverine, mountainous or valley locations. Thus the optimal use of a location would aim at adaptive strategies of food acquisition in small-scale ecological niches. By combining ecological default values of the habitat with empirical data on the palaeodiet, it is possible to trace such past strategies of land-use, i.e. the relative contribution of dairy farming and/or crop farming to subsistence. Populations analysed here represent examples of subsistence conditions in variably favoured lowland and mountainous settings with a geographical dispersion reaching from the eastern edge of the Swabian Alb to the upper Rhine fluvial plain and the foothills of Lake Constance in southwest Germany, which represent basic patterns of location-specific subsistence strategies (Schutkowski and Herrmann 1996).

Concentrations of skeletal trace elements revealed a general pattern that allows distinguishing between locations in the first place, as they reflect the mineral supply typical of the respective area (Burton et al. 2003). These signatures of locality, however, have dietary signals superimposed and indicate that all populations analysed relied on a mixed subsistence derived from livestock and agricultural farming. Overall, a higher amount of vegetable foodstuffs in the nutrition correlates with decreasing altitude of the location, thus with an increase in arable land. As can be expected, this amount is highest in populations originating from the Rhine valley and the expanded plateau areas off Lake Constance. Accordingly, the relative proportion of livestock and dairy farming increases in mountainous locations and rolling scenery such as the Swabian Alb where, due to the geomorphology and soil conditions, crop-farming and subsistence agriculture is only possible to a limited extent. It can

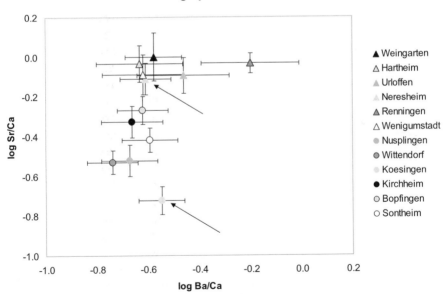

Geographical variation

Weingarten
Hartheim
Urloffen
Neresheim
Renningen
Wenigumstadt
Nusplingen
Wittendorf
Koesingen
Kirchheim
Bopfingen
Sontheim

Fig. 2.6. Variability of element ratios (means and standard deviations) from sites in lowland locations (*triangles*) and mountainous locations (*circles*). While the values primarily represent signatures of localities, they are superimposed by dietary signals which clearly indicate that different subsistence strategies are influenced by local ecological conditions. *Arrows* denote the sites of Neresheim (top) and Kösingen (bottom) (see Fig. 2.7)

therefore be demonstrated that, while pursuing a mixed subsistence strategy, the respective mode of production was still closely adjusted to the prevailing environmental condition (Fig. 2.6). Trace element levels differ significantly ($P<0.001$) between the populations in moderate altitudes and those in valley and lowland settings and/or on plateaus. The distinctiveness of trace element profiles and their representation of certain subsistence patterns clearly correspond with proxies of geomorphology, climate and soil quality, an interesting example being two neighbouring locations in entirely different habitats.

The two sites of Neresheim and Kösingen are situated at the eastern fringes of the Swabian Alb. Only 5 km apart, they belong to distinctly different ecotopes or natural units (Dongus 1961). Neresheim is at ca. 500 m above sea level amidst laminated limestone with weathered lime loams and silt loams as top soils. The autochthonous vegetation is characterised by lime beech forests and lime arid swards. Mean annual temperatures and average rainfall contribute to the overall favourable climatic conditions of this locality. Kösingen is about 70 m higher in altitude, exhibiting a lower annual mean temperature and increased precipitation. Soils are dry and karstic, interspersed with expanded clayish sections. For both ecotopes, there is evidence of settlement activity from the Iron Age onwards, yet in medieval times there appears to have been only the one village on the high plains, while Neresheim shared the favourable lowland location with other settlements (Knaut 1993).

Based on element analyses of human and animal remains, Neresheim most likely represents a typical mixed subsistence mode, consisting of produce from horticulture, agricultural and livestock farming. Diet at Kösingen, by contrast, appears to have been dominated by foodstuffs rich in dietary calcium, most likely derived from dairy farming and some horticulture. These two cases suggest subsistence strategies during the sixth to eighth centuries AD which were geared towards local ecological conditions. Despite the relative altitude of the location at Neresheim, fertile sections of the ecotope were used for agriculture, supplemented by animal husbandry in less favoured areas, whereas the overall unfavourable conditions of climate, geomorphology and soil quality at Kösingen practically excluded large-scale agricultural farming. The scarcity of settlements during the early Middle Ages may be seen as a direct consequence of the relative ecological disfavour that would not support more villages. Each site thus reflects a distinct strategy of land use and food acquisition. While trade or other exchange relations between the two neighbouring locations, even in terms of transhumant relations, are likely or cannot be excluded, the very existence of two separate burial grounds indicates independent entities and communities. Interestingly enough, even today the different modes of subsistence are obvious and thus suggest a continuity of food procurement strategies closely adapted to the environmental conditions (Fig. 2.7).

This successful adjustment, however, was obviously not met with comparable biological success. Neresheim seems to have been better off both in terms

Fig. 2.7. Neresheim (top) and Kösingen (bottom) are close neighbours in very different habitats. The differences in subsistence strategies established during early medieval times basically continue to prevail right through to the present day (see text)

of childhood mortality (19.3% vs 28.6%) and life expectancy at birth (29 years vs 22 years; all data from Hahn 1993). In other words, in historical times under conditions of subsistence economy, the variance of ecological basic conditions comes at a different 'biological cost', whose effects could only be moderated by the development of adaptive cultural strategies – to the extent that element analyses on archaeological bone are actually able to show this. Safeguarding survival for several hundred years seems to have been possible only with appropriate concessions to the effects of the relative ecological favour or disfavour of the habitat.

Another pair of localities involves the sites of Renningen, in a mid-altitude valley location, and Kirchheim unter Teck, situated at the northern rim of the Swabian Alb. At the former, due to topography, a mode of production emphasising livestock and dairy farming would have been expected, yet the slightly more favourable environmental conditions in relation to the surrounding countryside have resulted in increased production and consumption of crops. In ecological terms, such a strategy would be preferred and adaptive from an energetic point of view since, in comparison with animal husbandry, the production of plant food requires a shorter food chain. Energy input into the system is thus reduced and the ratio of invested and produced energy for maintaining subsistence economy optimised. Obviously the choice of habitat management appears to be based on a fairly accurate, yet nevertheless likely unconscious, energetic cost–benefit calculation.

The subsistence pattern reconstructed for the population of Kirchheim unter Teck, however, shows that modes of production may not at all be solely affected by energetic calculations. Temperature and rainfall data (Ellenberg 1955) for the location, set in a climatically favoured open valley, would suggest a mode of production with an emphasis on agriculture. However, reconstructing the nutritional basis of the population actually reveals a substantial consumption of animal protein (see Chap. 4). Since for the early Middle Ages subsistence agriculture can be assumed, this would be in line with intensified livestock and dairy farming. Archaeological findings and historic sources (Däcke 2005) suggest that Kirchheim was a prosperous community with numerous examples of great individual wealth. Additionally, Kirchheim is considered to have been a kind of local centre that profited from the proximity of old Roman trade routes and a regional market, visibly expressed by a stone church already erected at around 600 AD. From an ecological point of view, material and energy flows merge in a concentrated manner in Kirchheim. It therefore appears quite plausible that the population could pursue a subsistence strategy which, due to high levels of prosperity, involved greater energetic expenditure than would have been necessary given the conditions of the natural environment.

Thus, socio-cultural conditions made possible and maintained the partial detachment of food production from the ecological defaults of the habitat. An altogether sophisticated lifestyle and elevated prosperity, and a good supply of protein-rich high-quality food, led to improved living conditions at Kirchheim, as indicated by a higher adult mean age-at-death in relation to sites in comparable natural settings, such as Renningen, where the population fed on a diet less rich in animal protein (45 years vs 37 years; Schutkowski 2000). Thus, a better-off socio-economic standing plus associated beneficial nutritional conditions had a direct positive effect on the biological characteristics of the population. It can only be assumed, yet plausibly, that the relative emancipation from and the effective exploitation of natural conditions of the habitat at the same time created possibilities and conditions for improved economic exploitation strategies.

2.2.5.2 Nineteenth-Century Wiltshire

As shown in the preceding example, adaptive strategies of land use can result from even partial utilisation of the ecological potential in a given natural setting and have measurable effects on the living conditions of the population. The case of Wiltshire (p. 37ff in Bayliss-Smith 1982) shows quite the contrary. It can serve as a case in point to demonstrate that a particular subsistence mode, even though energetically efficient and very well adapted to the existing natural conditions, was sustained without recognisable advantages for the producing population, but rather with disadvantages for their quality of life. People were forced to abandon their pre-existing subsistence agriculture and to produce for a market or a landlord, instead. Here, agriculture loses its original feature of supplying a population with self-generated food resources from their immediate environment by being subjected to the constraints of a capital-orientated market.

During the eighteenth century, Wiltshire experiences profound structural change, which for most farmers is connected with a change from a largely self-determined self-sustainable supply from their own property and the common land, embedded into the social security of village life, to a market economy. The farmer transforms into a dependent agricultural worker, effectively as a result of the abolition of hitherto guaranteed rights of land use and the right of pasture on common land. Being smallholders now, yet still on their own property, most peasants are no longer able to keep up with the requirements of the market. Consequently, an amalgamation of small plots into large production units under the ownership of a wealthy farmer takes place, which for the so far independent farmers means the virtual lack of possessions. What looks like an overall enormously efficient mode of agricultural production at face value proves detrimental to the single farmer.

The natural habitat situation in the south of Wiltshire displays an arrangement of hilly high plateaus separated by broad riverbanks. Apart from a sufficient water supply, the habitat provides meadows and areas of arable land on the banks, with pasture and forest on the hills. Based on this pattern of land use, a functional linkage of meadows, fields and pastures had developed towards the end of the eighteenth century, which supplied the resources for a combined sheep/grain economy. At the beginning of the nineteenth century, presumably only small areas of the arable land lay fallow (Bayliss-Smith 1982). Wheat and barley are mentioned as main crops. Because fallow periods were practically missing, it was necessary to maintain soil fertility by continuous fertilisation. This demand for manure was the principal reason for keeping sheep. By letting the sheep graze on the hills during the day and penning them at night on the areas of arable land, sheep muck was evenly distributed on the fields with minimum human work input. As a supporting and flanking measure, turnips were cultivated as fodder for the winter. Add to that the already harvested hay stocks and the pasture options on irrigated meadows in the spring, and an all-season supply for the sheep was made possible. At the same time, this allowed more animals to be brought over the winter, which then in turn provided more manure for the fields. Likewise, larger areas of more favourable soils and topographic locations for the cultivation of grain became available, because the sheep pasture could be shifted predominantly to the hills (see Fig. 2.8).

This system sustained an extremely effective land use with high yields. At least on the rich and fertile soils, cereal grains could be cultivated twice consecutively as part of a tripartite rotation of crops, typically followed by clover or tubers as the third cultivar. The energy balance of this combined strategy yielded an extraordinarily high surplus. The ratio of energy output (87 % in the form of cereals) to energy intake (almost exclusively human work) amounted to a staggering 40.3. It has to be taken into account, though, that such a huge energy yield was only possible by the use of horses as traction animals. If one includes the total energy required for the horses, the output/input ratio is reduced to 14.2. This, however, shows that a substantial part of the efficiency of this subsistence strategy was actually based on the employment of horses as 'energy slaves' (Bayliss-Smith 1982).

As nearly the entire agricultural production had to be delivered to institutions or persons, the question remains as to what benefit arising from the maximisation of energy was actually left to their producers. By changing the point of view from the population level and taking into account the economic

Month	Jan	Feb	Mar	Apr	May	Jun	Jul	Aug	Sep	Oct	Nov	Dec
Sheep pasture and fodder provision	Hay; Pasture on water meadows	Grass	Pasture on village green (only lambing ewes)		Early summer grass; Pasture on water meadows		Pasture on hillside		Wheat and barley stubble	Pasture on hillside		Hay
Rotation of crops	Wheat				Barley					Wheat		
	Grass									Grass		
	Swedes				Swedes							

Fig. 2.8. Seasonal cycle and connection of sheep husbandry and crop farming under the conditions of pre-industrial agriculture in Wiltshire, southern England (modified after Bayliss-Smith 1982)

situation of an individual agricultural worker and his family, the real conditions of subsistence acquisition as they may have likely prevailed for the majority of the population begin to emerge. The total revenue from agricultural production was only some £30 per year for a family. Produce from the domestic garden, such as potatoes and other vegetables were the only direct contribution to the family's own subsistence efforts. This, however, hardly totalled 20 % of the amount of food needed, so that the remainders, i.e. essentially bread, ham, cheese and tea, had to be bought. What little money was left could only provide for a few non-food items of daily requirement such as fabrics, leather or lamp oil.

If one relates the consumption of food energy required by a household to the energy intake necessary for its production, the ratio is 12.6. While in other economies, e.g. the Tsembaga Maring, such an energy surplus would directly flow back into the subsistence cycle (e.g. as pig fodder), the energy gain of an eighteenth/nineteenth century agricultural worker family from Wiltshire only rarely suffices to buy barley for at least one pig in order to improve the domestic economy (Bayliss-Smith 1982).

Thus, as the result of a forced abolition of subsistence economy, the generated matter and energy flows are almost completely withdrawn from the producing system and external flows of energy are required for the family's own life support. It is true, on the level of the habitat, that a very successful adaptation of food production could be established, which was only given up because of general economic and technological changes as late as the 1920s. At the same time, however, this led to a lasting degradation of living conditions at the level of food producers, due to fundamental changes to their socio-cultural situation. This degradation can be certainly rated maladaptive for the individual.

2.3 Conclusion: Biocultural Adaptations to Food Procurement

Within their ecosystems, humans occupy a very broad niche, which compared with other organisms allows them to occupy an unusually large number of different habitats and which impacts on the formation of subsistence strategies. On the one hand, the survival of human populations is influenced by natural conditions of the habitat, as climate, geomorphology and soil quality affect the spatial distribution and nutritional value of food resources. In this regard humans, like other biotic components of the system, are exposed to comparable problems of extracting energy from material flows. On the other hand, however, they are able to intentionally change their environment to adjust flows of matter and energy specifically to their needs and to manipulate them accordingly if necessary. They can achieve this either directly in the course of those processes that constitute their subsistence activities, for example by applying certain techniques of food procurement, or by behaviours which, although not directly connected with the use of resources, are likely to affect their use and availability, as in the case of differential resource allocation as a result of social inequality within a population.

It is true; humans do not possess steering possibilities to influence the general climatic and geomorphologic impact of the habitat or seasonal cycles of temperature and precipitation. However, connections do exist between these conditions and certain subsistence options which human communities developed as adaptive strategies of energy conversion from natural resources in their habitats (Ulijaszek 1995). These inform the classification of food acquisition in terms of modes of production, e.g. the categories of foraging, horticulture, pastoralism and agriculture (see Sect. 2.1).

Yet this common and pragmatic classification only provides a relatively coarse framework and obscures the fact that human populations may encounter the same kind of environment and thus similar or comparable conditions of e.g. temperature or precipitation, but may nevertheless pursue quite different modes of subsistence. It follows somewhat trivially, therefore, that different human-shaped habitats or ecosystems exist next to each other and are in mutual exchange; even more so because most societies themselves pursue mixed subsistence strategies rather than one subsistence mode exclusively. Many pastoral nomads, for example, although primarily living of their herds, employ horticulture or agriculture of some sort as an alternative food procurement strategy to bypass times of drought that are often difficult to predict, while for other groups trade may be the preferred option.

Human populations can be expected to build their subsistence basis on resources that are reliable, available on a long-term and energetically efficient, as the examples of !Kung, Tsembaga or Karimojong in this chapter demonstrate. If natural conditions allow, however, alternative sources of food are explored and become part of the cultural kit that can serve as supplements, for example in times of continuing drought. With the exception of extreme habitats, such as the circumpolar region, which practically permit no alternative to a certain nutritional regime, human populations are in a position to adapt their trophic level position within the food web to the respective ecological conditions of the habitat. It is not the exclusiveness of a certain mode of production that is characteristic of human subsistence, but the opportunistic combination of several subsistence techniques into a strategy.

Subsistence cannot be reduced, however, to the bare techniques or activities of food procurement. On the contrary, it can only be understood by considering how these patterns and techniques articulate with social relations within a society. Human-shaped ecosystems, therefore, differ substantially from each other not only because populations live in different habitats, but also because humans solve the problem of biological production and the reproduction of social structures in very different ways. For example, in habitats characterised by seasonally varying or durably extreme climatic conditions, strategies accepted and sanctioned by the community have developed which secure an even distribution of essential food resources in the population and thus spread subsistence risk. This behaviour is either restricted to times of distress, or embodied as a general principle. It is almost characteristic of the social organisation of foraging societies (e.g. !Kung, Inuit) that resources from different regions of the habitat become available to several households or groups for reciprocal use in more or less formalised ways. Karimojong distribute surplus from the millet harvest either on the occasion of celebrations, by way of exchange with cattle or even through begging, a behaviour socially completely acceptable and common to the society (Dyson-Hudson and Dyson-Hudson 1970). The social consensus is based on the principle of mutuality and thus secures advantages on a long-term basis that benefit the individual. Also physiological mechanisms of optimisation, e.g. chewing coca in the Andean highlands, are culturally safeguarded and maintained through exchange networks.

On a general level, modes of production can be related to the use of certain resources and the level of interferences necessary (or unnecessary) in terms of the management and control of material and energy flows in a given ecosystem. A simple stage model illustrates the extent to which human populations manipulate their ecosystem, i.e. the extent to which subsistence strategies determine the choice of resources, the cultivation and domestication of plants and animals, land development, land use and landscape change (Table 2.1). A basic pattern emerges, showing that foragers, but also specialised collectors dependent on seasonally occurring resources, are usually to be found in pristine ecosystems, in which resources are generally allowed to regenerate naturally. Intensive use of non-domesticated resources and extensive use of domesticated species, e.g. in slash-and-burn agriculture, is found in partially changed ecosystems, while intensive use of domesticated animals

Table 2.1. Correlations between human ecosystem interference and mode of subsistence (modified from Ellen 1982)   *Simplistic - FIRE*

State of the ecosystem	Extent of manipulation	Type of resource	Modes of production
Pristine	Little	Not domesticated	Hunter/gatherer/fisher
Partly modified	Regular	Partially domesticated	Horticulture, slash-and-burn agriculture, basic animal husbandry
Largely artificial	Only stable by human intervention	Almost entirely domesticated	Agriculture, pastoralism, industrialism

and plants for food production can only be sustained in systems that have to be artificially maintained to a large extent.

Even though within these categories there is large variability, the model shows that the more extensive the manipulation of the environment, the more specialised and significantly changed the ecosystems. The general course of development leads from systems with a high diversity of species and small numbers of individuals to few species with high numbers of individuals as a sign of the increasing specialisation and concentration of food production on few highly controlled species. This development takes place at the expense of an initially higher bio-diversity. Flows of matter become increasingly channelled and have to be maintained in order to actually keep a system going, which is stable almost only by human interference. Accordingly, ever-larger quantities of energy are required that have to be procured from ever more distant parts of the ecosystem to compensate for increasing dissipation (Ellen 1982). Subsistence strategies in human communities thus reflect different depths of interference in their habitats to manage the co-ordination of matter and energy flows.

The inter-relations of food procurement and cognitive appropriation of the habitat, i.e. the exact knowledge of distribution and availability of food resources, have been addressed in the section on optimal foraging behaviour (see Sect. 2.2.1.3)[9]. With the help of information, another fundamental ecosystemic component, humans are able to optimise the use of matter and energy under given conditions in such a way that an optimised solution is achieved regarding both food quality and energetic benefit. Information becomes a crucial determinant that not only allows processing the knowledge of locations and occurrence of food, but also its seasonal supply. At the same time, it can be estimated what group size is best suited for a certain type of food procurement or how much yield is possible per unit of time. Besides, situations of shortage or crises are stored in the collective memory through cross-generational flows of information, allowing experiences that go beyond the lifetime of an individual to be used.

In investigations of subsistence behaviour, particularly in foraging communities, the question of time budget has played some part. Based on Lee's studies with !Kung (see Sect. 2.2.1.1) it had been argued that foragers generally spend significantly less time on the procurement and processing of food compared to societies with producing modes of procurement, even though they rely on more complex subsistence techniques. Meanwhile, this idealistic notion of 'affluent foragers' (Sahlins 1968), capable of satisfying their needs easily and quickly, can hardly be upheld any longer because clear differences can be recognised already between foraging communities. Meticulous

---

[9] It is acknowledged that this is also a behavioural feature of non-human primates, yet this issue will not be pursued further in this context.

recordings of the lengths of time, which for instance Aché men need for work that is related to food procurement and other activities, showed that on average approximately 7 h/day are spent on food acquisition (Hill et al. 1985). This contrasts with data from !Kung (Lee 1979) of only 4 h/day, which is little even when compared with other foraging societies. Other investigations with !Kung, however, arrive at far higher figures, if the expenditure of time is referred to days per week (Table 2.2). Comparisons of time expenditure in forager and horticulturalist communities (e.g. Yanomamö, 5.3 h/day) result in figures of comparable scale. This shows that quite obviously no direct and simple connections can be established between mode of production and time spent on food acquisition. Rather it has to be assumed that observable differences in time budgets result from the special local conditions of ecotopes, which are needed for the procurement of a comparable quantity of energy.

The societies specified in Table 2.2 live in environments as different as the tropical rain forests of South America (Aché) and Africa (Mbuti) and different arid and/or semiarid habitats (!Kung, G/wi, Australian aborigines). This would explain, why a food acquisition strategy that works to the primacy of fitness maximisation, can lead to different time requirements under different ecological conditions, accordingly. If the ultimate goal is to achieve an increase in fitness, it can be expected that optimal subsistence strategies also vary with regard to time costs and caloric or energetic benefits, even with the same mode of production, in order to achieve optimised acquisition of food under given conditions. Just like time budget, therefore, the organisation of subsistence based on the division of labour by age and gender groups and by patterns of spatial distribution of foraging individuals or groups have to be subsumed (see e.g.!Kung, Inuit, Nuñoa or Karimojong). In the end even here it is about optimisation strategies, since the allocation is used in such a way that either the food supply is maximised or energy consumption is minimised.

Table 2.2. Differences in time budgets for subsistence activities in foragers. Listed are average times used for hunting and concomitant activities (data from Hill et al. 1985)

Society	Hours/day	Days/week
Ache	7.0	5.9
!Kung	4.1	4.8 (dry season) 6.3 (rainy season)
G/wi bushmen	6.3	5.3
Aborigines	4.5	No data
Mbuti	5.4	4.6

One of the reasons why modelling is employed in human ecological investigations is the possibility of being able to relate the adaptability of different subsistence modes of human communities to one another. As the comparisons showed, aspects of time budget or the use of certain subsistence techniques are certainly suited to measure the ecological success of such resource use. They can be combined into widespread fundamental strategies. A problem is posed, however, in that such variables are frequently not exchangeable across different ranges of human subsistence behaviour and therefore comparisons are possible only on a quite general broad scale.

Because of the specificity in which environmental conditions and human culture interact in a certain area at a certain time, several solutions of comparable problems can be found even under similar natural conditions, which correspond to different behavioural strategies or chains of action. Cultural reactions do not just develop in a deterministic sense as stereotypical responses to given environmental factors, but evolve from traditions which are typical for the respective human community. Yet, such a probabilistic view of cultural mediation of environmental conditions does not mean that the adaptive value of certain culture traits is constant and arbitrarily transferable in space and time and across human cultures, but has to be understood in the sense of adaptive snapshots, which nevertheless can repeat themselves in their basic patterns (see Casimir 1990).

More important than the cultural characteristics themselves seem to be the goals pursued in each case, respectively. In ecological studies, the acquisition and use of energy are stated as one such goal to which all organisms are subject. One does not have to carry things as far as Leslie White (1943) and regard the cultural development of humankind deterministically as a history of an incessant increase in energy output to realise that energy, as one of the central ecosystemic components, is subjected to more and more complex exploitation strategies. Besides, energy is a currency that allows the comparison of the subsistence activities of different populations in standardised measurable units.

There are, however, epistemological and practical problems associated with such an approach, which were pointed out by Ellen (1982; see p. 117ff). Aspects of social life can hardly or not at all be reduced to calories or efficiency calculations and this raises the question as to what extent energy balances can reflect cultural adaptation at all. Power as a sign of social and cultural influence is just not identical to the flow rate of sensibly usable energy, just as monetary and economic transactions are hardly the equivalent of flows of matter. The simplified approach of reviewing economic operational sequences, legal norms or religious conceptions on the level of calories is misleading, as long as they are not carefully converted into consumer and exchange goods. Meanwhile, however, there is consensus over these issues (see Sect. 1.2.2) and there is agreement on the assessment that statements regarding comparative views of energetic yields can be made on the level of analogies and/or basic patterns. Energy measurements provide quantifiable information on subsistence behaviour, the development of and change in technological conditions and the meaning of different food and activity patterns. Power is certainly not identical to energy flow but, to the extent of how e.g. energy flows are partitioned through social mechanisms and thus have an effect on the energy budgets of groups within a human community, energy becomes the unit which allows such a comparison to be made, provided it can be measured at all.

By viewing energy as an exchangeable currency for culture-comparative purposes, energetic/caloric data provide a standard unit with practicable comparability. They are suitable when comparing the ecological efficiency or the efficacy of nutritional strategies to show the relationship between the energetic expenditure of certain subsistence activities and the benefit introduced into the system, i.e. to reflect the expenditure invested in a certain subsistence system and the conversion of habitat resources into usable energy. Food acquisition strategies aim at a maximisation of energy yields in relation to the energy input. With increasing intensification of food production, however, costs of invested labour rise. Accordingly, the energetic efficiency of different modes of production should be related to different factors in order to allow the energy balance to be kept even or positive. Foragers will thus pursue strategies that aim at optimising their yield either per unit of time or as a function of the spatial distribution of resources (see Sect. 2.2.3). Agriculturalists from the outset invest more strongly in the production of food, as evidenced by the work necessary to prepare and till the fields. They will therefore increase the efficiency per area of arable land. In the case of horticulture or slash-and-burn agriculture, this is achieved on the one hand through a suitable combination of arable crops which are cultivated on the same strip of land; and, on the other hand, the productive capacity of the habitat is at the same time maintained by the regular shifting of fields and an adherence to long times of fallow. But also other strategies have to be taken into account, such as diversifying agricultural production, especially for non-subsistence crops (Herhahn and Hill 1998). Finally, pastoralists will optimise the efficiency of their livestock, i.e. the resource on which their life support system is crucially dependent. In particular, those peripatetics living in marginal habitats with variable productivity pursue strategies where the transformation of solar energy into different food chains is optimised in such a way that a relatively high biomass of both humans and animals can be sustained, while degradation of the habitat is avoided. Adaptive use of energy therefore aims at strategies that maintain ecological stability (Ellis et al. 1979; Coughenour et al. 1985).

Quantification of energetic efficiencies of different subsistence strategies can be achieved by contrasting two commonly used variables: (1) the actual energetic efficiency, which results from the ratio of produced food energy p and the energy expenditure r that corresponds to the human energetic input, in order to gain $p(p/r)$, (2) the total efficiency, which also considers the energy utilised in addition to human energy expenditure (energy p/r plus imported energy). Table 2.3 presents some of these values for populations displaying different modes of production, from which several observations can be derived. In societies assumed to generally have small energetic expenditure, the quantity of energy consumed comes close to the total amount of energy produced. The example of !Kung shows that for the procurement of food no additional energy input is needed, since the energetic efficiency is equal to the total efficiency. The indicated figure however does not consider to which extent food obtained through barter was available. For Inuit no data can be derived from older investigations of energy budgets that would be directly comparable (Kemp 1971), but balances obtained by Smith (1991) for Inu-jjuamiut can provide auxiliary data. Even though they do not permit direct translation with regard to the mode of calculation, they are still comparable in such a way that, for energetic efficiency, the net yield/hour can be related to the energy expenditure/hour; and for total efficiency, all labour costs for the gained amount of energy can be considered. Accordingly, it becomes evident from the clear discrepancy of the two values that (today) the relatively high energetic efficiency can only be accomplished by supporting the traditional subsistence activities of hunting with substantial external energetic input, i.e. by motor sledges and motorboats.

Whilst for the Tsembaga of New Guinea, being an example of horticultural farmers, the values for energetic and total efficiency are still very close, because the amount of imported energy towards subsistence provision is small, a more differentiated picture can be drawn for the Nuñoa of the South American Altiplano. Here, the data permit differentiation of energetic balances for the two subsistence techniques used in tandem: horticultural farming and animal husbandry. The larger efficiency of plant production thereby reflects differences in work effort, as only 12 % of the subsistence labour relate

Table 2.3. Comparison of ecological efficiency of different modes of production (see text). Compiled and calculated from sources mentioned in the text

Society	Energetic efficiency	Total efficiency	Source
!Kung	9.5	9.5	Ellen (1982)
Inujjuamiut	12.9	1.7	Smith (1991)
Tsembaga	16.0	13.4	Ellen (1982)
Nuñoa Quechua	11.5 (crops) 7.5 (herds)	9.5	Thomas (1976)
Wiltshire 1820	40.3	5.8	Bayliss-Smith (1982)
Wiltshire 1970	1,266.7	2.1	Bayliss-Smith (1982)

to this activity, while 88 % are invested into animal husbandry. For Nuñoa, the relatively high total efficiency results from possibilities of exchanging animal-derived products for food of higher caloric value, because they are merged into a complex system of trade and barter relations between the vegetation zones of the Andean highlands.

Wiltshire, as an example of pre-industrial agriculture in temperate zones, shows very high energetic efficiency, when human labour for food production is considered exclusively. Realistically, though, the work force of horses has to be included, since these literally carry the principal workload. In this way, the gross value of 40.3 is reduced to a net 14.2 and thus comes to lie in a region of efficiency comparable with that calculated for the Tsembaga. The total efficiency is even more reduced, since the Wiltshire farmers could not even sustain subsistence economy, but had to produce for a market. Only to a small amount (approx. 20 %) were they able to cover their nutritional requirements from private horticulture, while all remaining food had to be bought. Here, the efficiency of food production in a given habitat is not tied to subsistence economy: instead, its impact is largely blotted out by socio-political conditions. This leads to a decoupling of production from a subsistence economy. The opposite is possible, as in the case of Kirchheim (see Sect. 2.2.7), where conditions of general prosperity have likely led to a partial decoupling of food production from the default conditions of the habitat.

Comparative figures are specified for the county of Wiltshire 150 years later on, in the era of industrial agriculture. An unusually high figure for energetic efficiency is in contrast to a total efficiency here with altogether one of the lowest marks. This example demonstrates that data on energetic efficiency in more complex modes of production with high technological input may give a distorted or wrong impression, since they suggest an efficiency which could not be accomplished without the substantial employment of machines and fossil energy.

In the long run, however, a trend continues – only in a particularly salient form – that could already be observed in the examples of mixed modes of production on the Altiplano and in pre-industrial agriculture in England. With increasing distance from subsistence food acquisition, the efficiency of food production is determined to an ever smaller amount by direct human input. Rather, it is determined through integration into complex operational sequences of energy exchange between populations and habitats. As human populations keep the borders of their ecosystems increasingly flexible and open, the flows of matter and energy become determining factors for the efficiency of local subsistence strategies beyond these borders. At the same time, it becomes clear that, with allegedly 'simple' modes of production, and even more so in systems of complex socio-political entwinements with integration into supra-regional markets, it becomes increasingly difficult to reconstruct energy flows or, at least, the nature of this reconstruction becomes rather mechanical. A multiplicity of human activities really cannot just be converted into energy-equivalent units, even if they act as determinants of energetic conversion and utilisation of energy within the system. The provision of global energy balances, or even those of national economies, is rendered non-feasible exactly because of the impossibility of the comprehensive energetic accounting that would have to consider even the smallest detail.

If the reconstruction of energy flows on a supra- or multi-regional level appears to be impossible, does this imply the end of modelling energy flows in human communities altogether? Does Remmert's (1988) conclusion that already ascertaining the energetic balance of a village, an oasis or a trade settlement has to fail because of the imponderables of external energy inputs necessarily entail having to break with energetic balances? Energy flow models after all represent a starting point from which different subsistence systems and different populations can be compared since, at a very basic level, energy can serve as the common reference unit (see p. 83 in Ulijaszek 1995). Certainly the balancing of total efficiencies of whole populations and systems to a large extent also tends to obscure the sociological variability in the efficiency of certain subsistence activities, as well as the variability brought about by time and groups within populations. Efficiency, however, is but one of several crucial determinants of subsistence. Reliability of resources is another and can be much more important in its significance than the average yield at a certain point in time. Reliability of resource use, however, is exactly one of those factors that can be achieved and contributed to by humans through exerting control in their habitats. Such management of resources allows matter and energy flows to be steered or stabilised. As energy flows are modelled, one reveals a concept of energy use, or rather its efficiency, along the different food chains contributing to the subsistence basis of a human community. Resource management thus becomes comprehensible through energy management.

In contrast to energy balances at the population level, a better possibility of differentiation seems to be available at the micro-level represented by a household or a family. It is true; also a family does not operate within a closed system and is therefore connected in various ways with the outside world by exchange of material, energy and information. However, by selecting this level of observation, one seizes both the core unit of production, as it were the germ cell of subsistence, and at the same time the core unit of reproduction. Both cases allow an individual-centred view of the social mediation of energy flows. The way models of optimal food procurement (see Sect. 2.2.1.3) were discussed under the aspect of fitness maximisation and the reconstruction of life-history strategies exactly provides the link to treating energetic efficiency in the sense of an adaptive strategy. The perspective is thus to obtain a proxy datum for (potential) increase in fitness by descriptive modelling of energy or resource management. Thus, variance in the efficiency of resource management at the level of households or families can provide the basis for viewing possible effects on differential probabilities of survival and fertility.

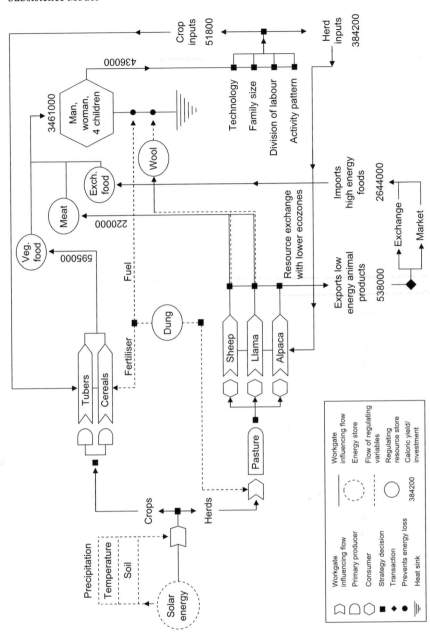

Fig. 2.9. Energy flow diagram for a Nuñoa family (see text; modified from Thomas 1976)

The principle of this consideration – although it has to be said that the data available were not raised under this aspect at that time – can be demonstrated using the Nuñoa example (see Sect. 2.2.5; Thomas 1976; Fig. 2.9). Under conditions of low biological productivity and the effect of rigid environmental stressors in the Andean highlands, it is necessary to have in place a particularly efficient energy management in order to pursue a fitness-enhancing strategy of food acquisition. The energy flow for a Nuñoa family, consisting of parents and four children between 2 years and 17 years of age (cp. Thomas 1976), proceeds along two major rails with different food chains. On the one hand, there is horticultural farming constituting a cereal grains → humans food chain and a tuber → humans food chain; and on the other hand, there is a food chain that employs pastoral farming and leads from sheep, llamas and alpacas to humans. From the first chain, the family gains biomass representing the equivalent value of 595,000 kcal, while an energetic input of 51,800 kcal is estimated for the planting, weeding, harvesting, threshing and processing of the arable crops. In contrast to this, 88 % of the total energy expenditure of subsistence alone (384,200 kcal) are needed for animal husbandry and gain a yield of 760,000 kcal. Of this, the equivalent of 222,000 kcal are supplied to the family as meat, while the remaining amount of animal-derived products, being predominantly low-energy goods (wool, hide), are fed into barter or trade with ecozones at altitudes below the Altiplano.

This strategy provides the actual profit of subsistence economy for a Nuñoa family. The net balance of animal husbandry closes with an input/output ratio of 2.0. However, because animal products such as wool and hides with a relatively small energy value are converted on the markets into food items of substantially higher energetic value, such as wheat or maize flour with an estimated equivalent of 2,644,000 kcal, almost five times the energy flows back again into the system. Instead of a net energy production from agriculture and animal husbandry of altogether 1,335,000 kcal, the family now has at its disposal food energy worth a total of 3,461,000 kcal.

When this amount is compared with the average energetic requirement for a family of this composition per year (data from Table 8 in Leslie et al. 1984), the result is a surplus yield of approximately 260,000 kcal. What is the value of this additional energy? Taking into account that the figures indicating the energetic requirement of the family already include differences resulting from age and activity patterns, Nuñoa to a large extent try to delimitate physically exerting work to save energy under the given harsh climatic conditions. Activities that do not necessarily have to be accomplished by adults are carried out by children, for example herding, a pattern also widespread among pastoralists (e.g. for Karimojong, see Dyson-Hudson and Dyson-Hudson 1970). Just by assigning to children the task of guarding the herds can save approximately one-third of the energy an adult would have to spend on the same activity (Thomas 1976). When this is viewed against the average cumulative energetic requirement for a pregnancy plus a subsequent period of some

2 years of breastfeeding (ca. 445,000 kcal; see Adair 1987, Worthington-Roberts 1989), it becomes apparent that with 260,000 kcal of surplus per year, additional energy can be produced at least in an order of magnitude, which covers just this requirement resulting from pregnancy and lactation. This kind of governing matter and energy flows can to an extent be regarded as the basis for maintaining the possibilities of reproduction for a family, as the surplus can be invested into further offspring and thus into increased fertility. But also in the further course of childhood, adaptive strategies become effective among the Nuñoa that can be interpreted in the sense of fitness enhancement.

The latent hypocaloric stress characteristic of living conditions in the Andean highlands is even aggravated by seasonal fluctuations in food supply. The average per capita input of food energy amounts to 1,150 kcal/day in the pre-harvest period, while after the harvest it rises to 1,519 kcal/day (Leonard and Thomas 1989). Considering the energy requirement estimated by Leslie et al. (1984), a marginal caloric status results for Nuñoa, particularly before the new harvest. Interestingly enough, however, seasonal differences in food supply for children under 12 years, in particular for infants up to an age of 3 years, are small compared to adult individuals. Two mechanisms contribute to this. First, households reduce their size, as at least one young person or adult man put themselves into service as a wage worker outside the habitat in the months before the harvest; and, where this is not possible, care is taken that children get a fair share of the food during meals. At the same time, particularly adults reduce their activity levels in the lean season, while children carry on and further contribute to their share of the production by herding or collecting fuel material.

With this double strategy, the energetic load of the family is moderated by seasonal out-migration of household members and calorie consumption of adults is reduced by the simultaneous maintenance of central household functions through child labour. This results in a complementary adaptive response that even under difficult ecological conditions allows a reduction in the impact of seasonal fluctuations in food energy and improves the children's nutritional status (Leonard and Thomas 1989). In this way, the allocation of scarce resources benefits the children and thereby increases the probability of survival of one's own offspring. By focusing on the micro-level of households or families, future investigations into energy flows in human communities appear worthwhile and meaningful, as studies elsewhere suggest (e.g. Dugdale and Payne 1987; Ulijaszek and Strickland 1993). By addressing the level at which the articulation of individuals and their environment becomes the focus, a perspective emerges to seize operational decisions about interfering with matter and energy flows by groups and individuals and to analyse them with regard to possible reproductive consequences in order to understand energy use as a mechanism of fitness enhancement.

3 Subsistence Change

In the preceding chapter, it was argued that human populations select, establish and refine modes of production which allow a flexible co-ordination of material and energy flows and consequently long-term occupation or settlement in a given habitat, depending on the specific ecological conditions. This usually results in temporary equilibrium states between a human population and the abiotic and biotic components of an ecosystem, including other human communities. Complex strategies are developed which embrace and combine technical solutions of resource use with behavioural adaptations and activity patterns. Since these factors mutually affect each other, strategies geared towards the maintenance of equilibrium states can thus be expected to result in optimised, continuous adjustments typical of a given location and time, and add to the local resistance and resilience capacity.

Human populations focus their subsistence basis on long-term reliable and energetically efficient resources. They develop strategies which facilitate an even distribution of essential food resources within a population and thus spread the subsistence risk. Frequently also alternative food sources are explored, which serve as supplements or an alimentary alternative, for example under conditions of pronounced seasonal availability of resources. To an amazing extent, humans are thus capable of adjusting within the food web their trophic level to the respective ecological conditions of the habitat and to develop adaptive solutions of utilising the environment.

Slight and regional fluctuations which may lead to variations in resource supply are usually known to a population as limiting habitat factors and are likely to be checked by the system once established as part of its resilience, for example the consequences of irregular and unpredictable precipitation. Profound or supra-regional changes, however, which are difficult to predict and which may concern a main component of the system, such as climate, key resources or technology, have a high probability of leading to a destabilisation of the equilibrium. Which behavioural options are at the disposal of a population then as solution to this problem, if serious changes become an endangerment of the adapted life support system? In principle, there would probably always be the possibility of out-migration and subsequent settlement in a new habitat. If, however, this is not an option and a once successfully occupied habitat is not going to be abandoned, one of the possibilities to react in the sense of re-establishing ecological balance is to adjust the subsistence mode to the changed basic conditions.

Since subsistence systems are integrated into a framework of ecological and social relations that at any one time result in certain modes of production, the option of subsistence change is reckoned not only to entail a change in conditions of food production but likewise to affect the socio-cultural foundations of a community. Because these two areas are so closely entwined (see Sect. 2.1) this may lead, for example, to newly defined structures of ownership and dependence and the possibilities of influencing or exerting control over resource use, which would clearly differ from those in the existing system. Subsistence change is thus obviously connected with culture change.

At the same time, subsistence change is eventually responsible for the variability of modes of production and their typical local appearances, as we know them today. Different depths of interference in human ecosystems and changes in the socio-political organisation can be systematically related to subsistence modes. Likewise, human subsistence strategies are geared towards optimising the energetic efficiency of resource use under given ecological basic conditions. The results of such optimisation strategies are temporary equilibria of varying duration in stability domains, which develop from a continuous adjustment of material and energy flows to fluctuations in resource availability. Why then is a strategy given up that has been established as an adaptive solution of habitat use? What triggers subsistence change? Is the increase of energetic efficiency a culturally defined goal and, in the end, the actual driving force behind subsistence change?

At first glance this, indeed, seems to be the case. White's (1943) somewhat reductionist attempt to systematise the relationship between technological progress and the harnessing of energy is a classic example. He proposed that human/environment relations are based on the following five factors: the human organism, the habitat, the quantity of controlled and used energy, the mode of energy use and the product that results from the utilisation of energy for the satisfaction of one's needs. Assuming that culture could be considered on a global scale and that consequently both the total human population and the habitat factor can be considered constant, the degree of cultural development would eventually vary directly with the energy used per capita and the efficiency of its technological conversion, respectively.

Necessarily, such a model remains relatively static. But there is some merit in considering, as it were, discrete energy levels, because empirical data reveal that different modes of subsistence do indeed exhibit a pattern of increasing and a more and more complex distribution of energy consumption. It can be shown, for example, that generally the relative per capita energy consumption is rising with increasing use of technology, i.e. from an extracting economy to the production conditions of the post-industrial society. Likewise, there is an increase in the amount of energy which is not directly converted into food, but which is required to satisfy rising added energy expenditure in order to maintain the respective mode of production – and thus the system properties (see Cook 1971). Whilst such a view is useful in ascertaining the *status quo* of energy requirements and their technological conversion, it cannot provide information about mechanisms, circumstances or causes leading to a change in subsistence activities, i.e. the dynamics of subsistence change.

But if existing subsistence strategies are to be considered as optimised solutions of resource use, and there are good reasons to do so, one has to inquire about the causes for change. Why and under which conditions will a certain mode of subsistence be given up? Which mechanisms or developments lead to a change or a new arrangement of existing modes of resource use? Is it possible (and justified) to inquire about common grounds as to how human communities manage the co-ordination involved in changing given ecological conditions, technological possibilities and energetic yields, if their existing subsistence system is to be successfully transferred into a new one? In other words, there is an issue about the determinants of subsistence change. This, however, is equivalent to a historic reflection of ecological equilibrium states, i.e. the temporal course of the co-ordination of matter and energy.

In a recent systematic and cross-cultural study (Bradley et al. 1990), embracing a wide range of societies (Murdock and White 1969), the historical aspect of the topic was taken up by examining which were the factors pointing towards a change in existing subsistence patterns that arose from the contact these societies experienced with expanding world-wide economic interests and forces during the nineteenth and twentieth centuries. The analysed indicators refer to issues of both food procurement (e.g. intensification of food acquisition strategies, introduction of new food resources, resource loss) and socio-political and socio-economic factors (e.g. change of settlement patterns, introduction and enforcement of wage labour and trade). The study revealed that 84 % of the examined societies entailed partial or complete change of subsistence systems as a direct result of contact with, or rather, coercion into confrontation with supra-regional or multinational economic entwinements – a major impact of political force on traditional ways of living. The most frequent causes were found to be expansion of trade relations, agricultural intensification, introduction of wage labour, loss of key resources and the introduction of new cultivars. Certain combinations of indicators for subsistence change corresponded clearly with geographical distributions. For example, subsistence change and the intensification of non-agrarian modes of subsistence appeared to be a typical outcome for societies of northeast Asia and North America, such as Plains Indians, who expanded the hunt for bison as a consequence of commercial relations with European traders in the eighteenth and nineteenth centuries, clearly beyond a stage that would have been necessary in keeping with their own traditional notions of subsistence hunting (White 1983). In contrast, African and Eurasian societies experienced partial subsistence change due to measures of agrarian intensification and an increase in wage labour. The Nuba, for example, as a consequence of pacification attempts imposed by British colonial rule in the Sudan, were able to shift cultivable land from hill sites into valleys. Even though this provided a possibility to expand agricultural production, it created a requirement for additional workers and consequently led to the introduction of wage labour (Nadel 1947). The temporal agreement is just as remarkable: a complete change of subsistence is predominantly found in nineteenth century accounts, whereas partial subsistence change seems to be a phenomenon of the twentieth century. For the earlier time-period, this pattern corresponds with the propagation of a world-wide commercial system into predominantly non-agrarian areas, whereas in the course of the twentieth century agrarian intensification is the determining cause for subsistence change.

Whilst the systematic correlations found in this study are intriguing and meaningful, it should not be overlooked that subsistence change is not a phenomenon that is necessarily tied to the expanding economic and political power of industrial or colonial nations, as the study by Bradley and colleagues could suggest. Colonial exploitation strategies with the consequence of forced change do not clarify the question about possible evolutionary patterns of subsistence change.

Rather, from a human ecological point of view, the question of ecological determinants of subsistence change has to be addressed. Subsistence change would be expected to occur if components of ecosystems change or are altered in such a way that this would also result in modifications of the subsistence basis of the human population – or in the first place, the necessity to react to changes in basic conditions by restructuring subsistence. Whilst strictly speaking this applies to the above example of change through colonial economic power, the window of time and opportunity represented in the study by Bradley et al. (1990) is quite narrow. The spectrum of indicators and criteria for subsistence change as such are thus not readily transferable and compatible with the examples to be dealt with later, which extend into a pre-historic time-scale and which attempt an explanation of subsistence change at turning-points in human biocultural evolution. Therefore in the following a set of criteria is introduced to serve as a working platform for an identification of causes of subsistence change, and which also allows pre-industrial conditions to be considered. Subsistence change can thus be triggered by:

(a) Climatic change. A lasting shift in ambient climatic conditions, be they caused by natural events or high-impact human agency, would compulsorily require and entail the development of new adaptive subsistence strategies, due to the mostly profound changes to landscape and species diversity in a habitat, if an occupied area is not to be abandoned[1].

(b) Disruption of a practiced subsistence strategy. Brought about by either extrinsic or intrinsic causes, the disruption often results in compulsory abandonment of key resources or a main component of the existing nutritional basis. Whilst, for example, an epidemic such as rinderpest and the concomitant loss of livestock has no human involvement, anthropogenic interference can have comparable devastating effects, as in the case of extended irrigation leading to soil salination in ancient Mesopotamia.

(c) Optimisation strategies. These are related to the intensification of extracting or producing strategies of food procurement or the improvement in the managing of already existing and used resources. Likewise, this category comprises the introduction and use of new food resources (e.g. maize in North America) or new technologies (e.g. the plough).

(d) Spatial expansion. Often a change in settlement patterns entails the expansion of a given area used for food acquisition – or, in fact, vice versa – and may include the exploration of new ecozones. Associated population growth is a common pattern. Frequently, this kind of activities is connected with:

(e) Socio-political changes. Here, two contrasting sequences can be envisaged. On the one hand, subsistence change may occur as a consequence of social differentiation and change within a community; and on the other hand, superimposition of an existing population by a dominant élite may take place that goes with the introduction of a new subsistence system.

Mostly, a change in mode of subsistence will be the result of several of the factors mentioned above, acting in combination as a complex, multi-layered incidence which renders impossible a mono-causal explanation. Whilst the causes of such events are not necessarily biological, the effects of subsistence change certainly are, since they will most likely lead to a changes in the carrying capacity of a habitat and thus, in turn, will affect the population dynamics of human communities. Likewise, there will be pertinent changes of biotic components in the system, since new relations of resource extraction have to be established. If existing subsistence systems were developed in the course of an adaptive process based on the mutual formative influence of both natural

---

[1] Whilst the psychological and emotional linkage of humans to the topography of a landscape may well be a motivation for locality-related subsistence change, this falls within the realm of the formation of traditions (see Schutkowski and Herrmann 1999) and can hardly be substantiated. As a conservative resource-related alternative, however, there is always the option of out-migrating from the traditional habitat. In this respect, the colonisation of temperate zones outside Europe is the most obvious example of the successful translocation of an existing subsistence system (see Crosby 1986).

habitat properties and human culture, then a change of subsistence basis is also connected with modifications of the social organisation and the concomitant effects on political and economic structures. At the same time, this is associated with a change in traditions, customs and attitudes, which is not conceivable without broad acceptance and 'mental preparation' of the new mode of subsistence within a community or society.

The potato provides a case in point for the biocultural nature of such changing circumstances (von Gundlach 1986). The potato came by sea and was well known in the large commercial centres of Europe, at the latest since 1580. Despite this widespread acquaintance, it took more than a hundred years before it was generally accepted and widely distributed as a crop. After its first introduction, the potato was particularly appreciated because it was attributed medicinal properties. Be it for these or nutritional reasons, although the potato was generally known, its propagation was limited to a large extent to the Mediterranean, since there were no suitable growing conditions north of the Alps for the potato varieties available at the time, where attempts at cultivation failed again and again. It was not until the beginning of the eighteenth century that cultivable varieties were at hand that could cope with the harsher climate of the middle latitudes. From a cultivation point of view, this eventually meant the break-through of the potato as a widespread arable crop, although its success actually depended on other decisive conditions.

In the first half of the seventeenth century and particularly during the agrarian depression after the Thirty Years War, there was already a tuber that was commonly used in Europe, the Jerusalem artichoke, which both with regard to appreciation and value for money even competed with cereal grains. Cultivation expanded with increasing demand. After the Jerusalem artichoke became feral however, it lost its exclusivity to the middle classes and the market price plummeted. The acceptance of the crop experienced a 'social decline' and became poor man's food. In addition, the Jerusalem artichoke was now also cultivated on a large scale for its use as fodder, fostered by the fact that the tuber is capable of storing certain carbohydrates that animals, but not humans, are able to metabolise. Concomitantly, there was extensive consumption of land, which left too little for the cultivation of other foodstuffs. As a reaction, the cultivation of the Jerusalem artichoke was declining again after 1700 and was replaced by the potato. With that, a plant was available which was equally suitable as food for both humans and animals, the cultivation characteristics of which as a tuber were already known in principle from the Jerusalem artichoke and which was broadly accepted by the farmers.

By testing hoe-farming as a new subsistence technique for a sufficiently long time and, at the same time, preparing the acceptance of a certain crop, the technical, biological and socio-cultural conditions for a change in subsistence were met. It was the right time, the right place and the right mindset for the potato. The coincidence and mutual influence of such different factors, i.e.

their common evolution, laid the foundation for a new subsistence strategy to be carried through within the community[2].

From the obvious close correlation between modes of production and social structure, it follows that property conditions, consumption patterns, resource allocation and food production, but equally aspects of collective mentality or those related to the social acceptance of certain modes of subsistence, are interlocked. Next, this complexity of linking biological and cultural adaptations to changes in the mode of subsistence are demonstrated and its consequences evaluated. Subsequently, a transition model is presented that allows course and tempo of subsistence change to be described and basic theoretical assumptions to be considered. The chapter concludes by comparing the cases introduced as examples of major subsistence change and by examining the correspondence of empirical evidence and theoretical expectation derived from the model.

3.1 Strategies of Subsistence Change

The following case studies were selected so that the set of complex causes for subsistence change specified above, e.g. climate and social organization or technology, can be considered. Furthermore, there must be sufficient evidence of the temporal course of subsistence change. Examples of successful (adaptive) change and those with negative (maladaptive) consequences for the respective population are dealt with. The emphasis is on proto- or pre-historic cases of subsistence change by referring to classic transitions of food acquisition in human history.

3.1.1 Upper Pleistocene Foragers in Europe

The spreading of anatomically modern humans in middle Europe is connected with the occurrence of a characteristic cultural stage or lithic industry, the Aurignacian [starting approximately 40,000–36,000 before present (BP), depending on the interpretation of radiocarbon dates; Churchill and Smith 2000]. During the ensuing time-period, covering several tens of thousands of

---

2   Notwithstanding this success story, however, the potato may also serve as a suitable example to demonstrate the political economy of decreed subsistence change and its maladaptive consequences. Devastating famine and mass emigration in nineteenth-century Ireland can largely be attributed to the facts that the potato was virtually the only food available to farmers and that the almost non-existent alternative subsistence options after the outbreak of late blight led to substantial social distortions (e.g. Ross 1986).

years up to the beginning of the Holocene, a homogenising cultural development takes place that is accompanied by an adjustment to the ecological conditions of the glacial environment (Bosinski 1989; Bar-Yosef 2002). Of particular importance is the formation of a relatively uniform lithic assemblage
characteristic of the Magdalénian and Epigravettian cultures, which developed after the cold maximum of the last ice age starting from approximately
17,000–18,000 BP.

The material artefact evidence suggests that this homogeneity is lost due to
rapid climatic degradations during the older Dryas (around 12,000 BP) within
a time-span of decades, a process supported and well documented by recent
studies into climate change at the end of the Pleistocene. Artefact assemblages
are now characterised by distinct regional cultures. They are replaced by new
developments during the Allerød interstadial period (ca. 11,700–11,000 BP),
whose much more humid and warmer climate (e.g. COHMAP 1988) leads to
profound changes in the environment.

The colder climate favoured the formation of steppe-like grassland extending over vast areas of late Pleistocene Eurasia, roamed by large herds of ungulates. They provided prey to foragers either in one area, for example horses or
Saiga antelopes, or in the course of seasonal migrations, such as reindeer or
European bison (wisent); and the archaeological record provides ample evidence for the economic significance of these herds of herbivores as key sup-
pliers of food and raw materials for dwellings, tools and clothes (see p. 530ff in
Klein 1999). Yet, other more opportunistic food procurement strategies contributed to a relatively broad food basis as well, for example involving freshwater resources.

Settlement structures exhibit close spatial links to available resources. Base
camps, evidenced by post-holes as remnants of wooden constructions (Bosinski 1981), are believed to have been built at places that allowed for convenient
access to stationary resources. At the same time, base camps are thought to
provide places of long-term settlement in the sense of temporary and serial
sedentariness, which in a comparable way also applies to rock shelters and
caves. They are even described as "village- like settlements with substantial
houses" (p. 382 in Bosinski 1988). Base camps were complemented by hunting
camps set up at places with a high abundance of seasonal local resources, for
example, along the known migration routes of reindeer or wisent, but these
were only established and occupied for the duration of the hunt.

The richness of lithic assemblages and the high variability of tools produced are generally regarded as an indication for the enormous innovative
capacity of late Pleistocene humans, regardless of whether observable differences are dependent on function or reflect stylistic variations that can be
regarded as different cultural solutions in comparable functional contexts
(p. 524ff in Klein 1999). At the end of the late Pleistocene, the entire technological assemblage has been invented which is considered characteristic of the
tool kit of historic forager populations. For the first time, larger areas of east

Europe and north Asia were colonised, which were more demanding climatically for the development of life-support systems than the already previously populated areas of west and central Europe. 'Luxury goods' such as amber or shells were traded over distances of up to several hundred kilometres, just like particularly desirable types of flint (Schild 1984). Furthermore, there is even evidence of incipient 'ceramic' production towards end of the Upper Pleistocene.

This efficient acquisition and use of natural resources developed under varying geographical and climatic conditions can be regarded as a set of highly adaptive strategies, which in the long run also laid the foundation for population expansion at the end of the late Palaeolithic (Henke 1989). Certainly, evidence of the first larger-scale artistic activities in the Upper Pleistocene may be also taken as an indication that, by gaining increased control of the environment, time and energy was released and became available for a hitherto non-existing transcended perception and assimilation of (daily) life in the form of representational and realistic images. Following White (1943; see above), this would be seen as an expression of high technological and energetic efficiency of resource use that generates freedom in other areas of human life, e.g. 'down-time' in hunting activities. In view of the assumed magic–ritual meaning of ice-age cave art with its direct connection to subsistence, however, a purely energetic explanation would fall short. Rather, the artistic conversion of naturally available resources as a constituent of belief systems would have to be considered an integral element of the subsistence strategy.

The climatic change from dry/cold to a more humid and warmer climate at the beginning of the Allerød period, twelfth millennium BP, represents a drastic change in the ecological basic conditions. The prevailing mammalian mega-fauna becomes extinct and the landscape changes due to increasing forestation. The extinction of the mammalian mega-fauna only describes the effect which marks the transition from a glacial to an interglacial period. By putting forward the hypothesis of a Pleistocene overkill, Martin (1967) has instead argued that Upper Pleistocene hunters themselves may have contributed to a large extent to the decimation and finally extinction of the mega-fauna, due to their optimised technology (see also Martin and Klein 1984), a view that has recently been confirmed for North America (Haynes 2002). *too strong)*

Independent of this, however, the climatic shift remains an important determinant of environmental change. There is a loss of numerous food resources used successfully so far. Also, due to increasing vegetation cover and sedimentation, the raw material for the production of stone artefacts may have become less accessible. As a consequence, the variety of large-sized lithic implements characteristic of the late Palaeolithic decreases and there is a gradual change in lithic assemblages towards small and very small artefacts. While this may have been connected with increasing scarcity of raw material [e.g. Azilian of the Allerød period (ca. 11,750–9,800 BP), 'Federmesser' assemblages, see Bosinski 1989; Baales 2001] improved tool-making capabilities certainly allowed a more efficient use of resources. Frequently, several such artefacts are combined to form composite tools. This development is being prepared by an increase in small-sized projectiles and other artefacts already found during the cold phase of the Older Dryas, which precedes the Allerød period. The bow and arrow start to become established as the most important hunting weapon and gradually displace the spear-thrower, in particular with increasing forestation of the landscape. This change in material culture co-evolves in line with the gradual extinction of the mega-fauna and the spread of smaller prey and hunting game that no longer require the use of large stone tools. Table 3.1 shows an overview and comparison of the most important changes between the last cold phase and the succeeding temperate period.

The material culture of the Allerød period exhibits a new quality of spatial homogeneity on a broad scale across the whole of Europe. This has to be viewed in connection with the drastic climatic changes, which become a central determinant for a change in the ecosystemic relations of humans in the late Pleistocene. The loss of food resources represents an external factor beyond human influence that leads to an inevitable change in the subsistence base. The artefacts, however, provide good reason to believe that both a cultural change and an adaptive subsistence change take place as a response to changing environmental conditions. The technological side of necessary new problem solutions and adjustments has already been prepared in principle by the spectrum of tool kit assemblages available in the late upper Pleistocene. Even more so, as the process of climatic and environmental change proceeded and allowed the flow of extrasomatic information to be passed on and used as acquired knowledge. This not only refers to the technological sphere, but can also be shown through connections of artistic traditions between the late Magdalénian and the Azilian (e.g. Bosinski 1990). The changes to the techno-

Table 3.1. Environmental differences and consequences for technology and lifestyle for the transition from colder to warmer climates

Factor	Stadial	Interstadial
Landscape	Open, few trees, steppe-like	Closed, increasing forestation
Fauna	Herds of large mammals	Smaller species
Flora	Grasses, mosses, lichens, berries	Trees, shrubs, nuts, berries, seeds
Technology	Adjusted to large game hunting	Trend towards reduction in size, composite tools
Hunting strategy	Adjusted to larger groups	Adjusted to smaller groups
Settlement structure	Large camps	Small camps

logical side of resource use at the same time lead to conversions in hunting strategies and transformations of settlement patterns.

The development of the bow and arrow as a new distance weapon and the gradual conversion of lithic assemblages coincide with a time period of climatic transition and a change in the natural basic conditions. The technological conversion, which developed as an adjustment to the emerging new environmental conditions, later functions as a pre-adaptive measure to facilitate the exploitation and extraction of new food resources that are now accessible through changes in ecological conditions. The end of the late Pleistocene finally sees a fluid transition into the Mesolithic, which is characterised by a broad-spectrum economy and a new diversity in the use of ecozones and food resources hitherto not available. The foundations for an adaptation to subsistence change were laid by a long preparatory phase, practically since the beginning of the Magdalénian. The change in subsistence is soft and not abrupt, since it was technologically and culturally buffered.

3.1.2 The Transition to Food Production

In analogy to the radical technological changes during the second half of the nineteenth century, which are terminologically established as the 'Industrial Revolution', the term 'Neolithic Revolution' (Childe 1928) has frequently been used to describe the introduction of sustainable agriculture and livestock farming. Nowadays, however, there is general agreement that for this complex process, which not only comprises the change of an existing subsistence strategy but also has considerable implications for population development, settlement and social structure, the term revolution is a misnomer. Rather, it became clear that the transition from an acquiring to a producing mode of food procurement was not a sudden event but a gradual evolution, often accompanied by keeping traditional strategies of food acquisition or by developing a form of coexistence of extracting and producing subsistence activities (e.g. p. 206 in Ellen 1994).[3] The following two examples look at the introduction of agriculture from the joint viewpoint of changes to the management of a previously known resource (Levant) and the adoption of a previously known subsistence technology (Southern Scandinavia) as a result of ecological and social necessities.

---

[3] Whilst the common notion among cultural anthropologists is to distinguish between acquiring and producing modes of food procurement, some researchers argue (e.g. p. 153ff in Bargatzky 1997) that foraging would have to be considered a mode of production (*sensu stricto*) as well, since hunter/gatherers would only extract resources that are 'pre-produced' by nature, which they would then process or even store, if necessary.

Present knowledge suggests that the transition from a foraging mode of food procurement to agricultural production takes place within some 4,000 years only, which is short by all standards, but especially when measured against evolutionary time-spans. This process occurs in different regions of the earth, obviously independently and several times, and it spreads comparatively quickly once initiated. The circumstances which lead to a change of subsistence are closely linked with the respective ecological conditions and the local supply of food resources in the different areas of origin.

With regard to the cultivation and domestication of plants, two groups of crops are of importance: (1) roots and tubers, typically prevailing in semi-humid areas with a short dry season, and (2) varieties of cereal grains, commonly spread across semi-arid areas with a longer dry season (Harris 1973; Hole 1994). It is this interaction of seasonality and concomitant possibilities for the respective plants to propagate which constitutes their potential use in different ecological zones (p. 168ff in Rindos 1984). The domestication of animals, with the exception of the dog, seems to have been developed from the context and as a consequence of agriculture in most cases (p. 77ff in Benecke 1994; p. 210 in Ellen 1994).

3.1.2.1 The Levant

For one of the classic centres of Neolithic transformation, the so-called Fertile Crescent in the Near East, the interaction of climatic change and adaptation of subsistence strategies is well documented. In the following, essentially socio-cultural pre-adaptations will be pointed out that in conjunction with a change of ecological basic conditions have led to a change in modes of production in the area of the southern Levant (cp. Bar-Yosef and Belfer-Cohen 1989a, b, 1992; Bar-Yosef and Meadows 1995; Smith 1995a; Bar-Yosef 1998).

In the first half of the Epipalaeolithic, during the Late Glacial Maximum (ca. 20,000–14,500 BP), the area of the eastern Mediterranean is characterised by a cold/dry climate with temperatures approximately 6–8 °C lower on average than today, but hilly areas receive winter precipitation and are covered with forests. The resident forager population of the Kebaran culture is semi-sedentary and, besides some hunting, supplies their subsistence from collecting vegetable food, above all seeds of wild grain. The archaeological record allows two different settlement types to be distinguished: larger agglomerations of dwellings, featuring stone architecture ('residential sites'; Hovers 1989), which are situated in the lowlands and used during the winter, and smaller, more ephemeral summer camps ('logistical sites') in the highlands. In the Mediterranean vegetation belt, this settlement pattern is kept during the following period of the Geometric Kebaran (ca. 14,500–13,000 BP). Climatic amelioration, including increasing precipitation, makes the use of new, more arid areas and more elevated locations possible. The archaeological record already contains artefacts indicative of vegetal food processing; and carbonised plant remains from a water-logged site date back as far as 19,000 BP (Kislev et al. 1992), providing strong evidence for increased gathering of varied plant resources.

It is assumed that during the following Natufian period, starting around 13,000–12,800 BP, population growth occurs due to an increase in settlement density, particularly in the ecologically stable park-like areas of the Mediterranean vegetation zone (Bar-Yosef and Belfer-Cohen 1989b). The climate is getting increasingly warmer and more humid, only interrupted by a temporary setback during the younger Dryas in the eleventh millennium BP with decreased rainfall. The Natufian is considered the major turning-point in the history of the Near East (Bar-Yosef 1998) and the onset of agriculture in this region.

The warmer and more humid climate favours the propagation of wild forms of cereals and legumes. In fact, Natufian base camps are located in the woodland belt, whose undergrowth comprises grasses with large amounts of cereals. Definitive evidence for the domestication of emmer (*Triticum turgidum* subsp. *dicoccoides*) and barley (*Hordeum vulgare* subsp. *spontaneum*) is available from the end of the eleventh millennium BC onwards (Kislev 1989; also on the morphological distinction of wild and domesticated forms). In the course of the tenth millennium, five further 'founder crops' (Harris 1996b) can be identified: einkorn (*T. monococcum*), pea (*Pisum* cf. *sativum*), chickpea (*Cicer* sp.), lentil (*Lens culinaris*) and flax (*Linum usitatissimum*). However, the available archaeological record suggests that several thousands of years prior to its regular cultivation and domestication wild grains are already intensively used as a food resource and their occurrence and habitat requirements are commonly known. Not least, this can be demonstrated through numerous artefacts of the Natufian, for example harvesting tools such as sickles and food processing tools like grinding stones, mortars and pistils, as well as some kind of storage facilities. They are clearly connected with the use of plants, in particular grasses, and are significantly older than the definitive evidence for domesticated plants (e.g. Kraybill 1977; Dubreuil 2004). While the Natufians still keep characteristics of a foraging lifestyle, e.g. game hunting, they have been described as practicing intensive and extensive harvesting of wild cereals as part of their seasonal mobility (Bar-Yosef 1998).

In addition to these archaeological findings, there is supporting evidence from the study of human skeletal remains. Based on bone strontium analyses[4], Schoeninger (1981) plausibly argued that already during the Late

---

[4] It is acknowledged that trace element analyses on archaeological bone have received substantial and justified criticism since and need to be treated with caution. This notwithstanding, the results referred to provide at least conclusions that are not contradictory to archaeological and other findings.

Kebaran wild grain had formed a substantial part of the nutritional basis in the context of an acquiring mode of subsistence (see also Sillen et al. 1989). Combined strontium/calcium and stable carbon isotope data for human remains from Kebaran and Natufian sites suggest a diet dominated by C3 plants (Sillen and Lee-Thorp 1991). Moreover, Smith (1989b) was able to demonstrate a reduction in lower jaw dimensions and an increase in dental disease as a diachronic trend in skeletons of the Natufian period. This is explained by a change in nutritional habits towards the increased consumption of vegetable food rich in carbohydrates.

The climatic deterioration of the Younger Dryas (ca. 11,000–10,300 BP) is regarded as a major driver for economic and social adaptations, which built on existing knowledge and experience acquired during the Natufian (Bar-Yosef 1998; but for a different view based on zooarchaeological findings, see Munro 2004). The decrease in the natural production of C3 plants and a reduction in the geographic distribution of wild cereals to the western regions of the Fertile Crescent triggered different responses. While people in the Negev and the northern Sinai improved their hunting technology, elsewhere people started to experiment with cultivating wild cereals, a development that would later facilitate the introduction of agriculture in the Levantine Corridor and which was the result of changes in material culture, social organisation and lifestyle.

While many local populations continued a foraging way of life, early Neolithic cultures emerge during later tenth to ninth millennium BP (Pre-Pottery Neolithic A): first the Khiamian, later the Sultanian. The return of Pluvial climatic conditions allowed the formation of cultivating and/or harvesting communities whose populations, beside the cultivation of non-domesticated grains, intensively continued with hunting, fishing and the collecting of vegetable food (Bar-Yosef and Belfer-Cohen 1989b). As a consequence, the success of such a broad spectrum economy (Binford 1968) leads to population growth (not the reverse!), which can be gauged from a gradual increase in settlement sites, the respective size of which can be explained by varying degrees of subsistence dependence on the cultivation of vegetable food. The more suitable the natural potential is for a stable subsistence based on food production, the larger are the settlement sites and thus the more dependent a population is on cultivation as the primary supply of foodstuffs. Artefactual evidence of pounding tools, slabs with cupholes, hand-stones and grinding bowls demonstrates the shift from Natufian tool kits towards increased and more sophisticated cereal processing.

A pre-requisite for this development is that humans are able to develop mechanisms of social organisation that allow them to cope with the delayed return of resource yields from crop farming (McCorriston and Hole 1991). With rising reliance on cultivation, in this case wild cereals, control increases over a certain source of food, both pre-emptive and retrospective. This means that techniques are being used which maintain or expand the growth and

reproduction of resources and resources are being preserved by the storage
and provision of stocks.]

In this context, [cultivation is regarded as an ecological link between
humans and certain food components in the mutual dependence of growing,
attending and using crops in the first place and has to be separated from
domestication as a purposeful genetic alteration of species in the course of
selection] (Ellen 1994; Harris 1996a). This distinction is important since,
according to all available evidence, the onset of sedentary residence in the
Levant, be it temporarily or permanently, takes place in the absence of animal
or plant domestication. Sedentariness is thus a cultural pre-adaptation for
domestication, not however for cultivation. Sedentariness facilitates and
probably necessitates higher birth rates and thus, potentially, population
growth; and only through this would it amplify the trend towards clear
domestication strategies. It also allows acquiring control over territories and
thus resource scheduling and control are prerequisites as, due to changes in
climatic conditions, there is a relative concentration of new usable resources
whose cultivation and exploitation benefits from and necessitates temporary
or long-lasting local residence. Once resources become more constantly available, a sedentary way of life is likely to gradually develop from this.]

One of the reasons why sedentariness has been connected with higher
birth rates is that cereals prepared as a mash provide a suitable food supplement that allows suckling infants to be weaned at an earlier age which, in turn,
facilitates shorter birth intervals (see Chap. 5). The frequently introduced construct of population pressure finally leading to more intensive resource use
and sedentariness (see Boserup 1965) falls short here. The archaeological
findings do not furnish support for this hypothesis. On the contrary, population growth always seems to develop in parallel with increasing intensification of food production or in parallel with subsistence change. Thus, the
adaptive advantage of sedentariness at the time of its formation is the better
and more efficient use of cultivable plants that occur naturally and in high
abundance. Increased and improved knowledge of the species used is a pre-
requisite for their domestication; and sedentariness as the compulsory necessary element is already an established constituent of a subsistence strategy
that proves pre-adaptive and preparatory to domestication. It allows and facilitates a shift in settlement patterns and increasing site size during the early
Neolithic. Under the circumstances of broad-spectrum economics, featuring a
combination of proportionate amounts of cereal cultivation and foraging subsistence, such an economy may still have been strongly adjusted to an optimal
input/output ratio of yield per unit time. This condition would have gradually
changed to optimise the yield per unit area ratio in the course of increasing
domestication and stationary food production, regardless of necessary alterations in the time budget. An increased investment in time is the price that has
to be paid for stable and predictable resources (see Table 2.1, p. 50 in Harris
and Ross 1987).

Within an ecosystem, the interaction of increasing control over resources and population dynamics leads to a gradual increase in the carrying capacity of the habitat. The crucial condition, i.e. the sustainability of resources (here wild cereals), is secured by the more favourable climate. In addition, it is known (Harlan and Zohary 1966) that cereals grow just as densely under natural conditions in the Levant as they do under conditions of cultivation. This allows the cultivation of wild grains to be much more widely accepted which, under conditions of increasing resource control, not only leads to their cultivation even outside their natural habitats (Flannery 1973) but also finally to their domestication. Moreover, increased sedentariness amplifies the necessity to store food, which in turn allows getting over short-term or seasonal food shortages.

In combination, the elements of favourable habitat conditions in natural units, the onset of sedentariness and the acceptance of plant cultivation are reinforced through positive feedback and have a lasting effect on the development of socio-cultural concepts within and between the Neolithic populations (Bar-Yosef 1989). We see the development of complex ownership structures and social mechanisms of barter. Changes in house architecture from spherical to angular ground plans are being explained by an extension of social units from nuclear to extended families. The evidence of plastered skulls and human statuettes in sizeable settlements such as Ain Ghazal, Jericho or Nahal Hemar is seen in connection with a change in the hierarchy of values and the practice of family-bound rituals in Neolithic village communities (e.g. Butler 1989; p. 161 in Parker Pearson 1993).

In the end, all these phenomena have to be regarded as late consequences of the change in food acquisition strategies which commenced several thousand years before as a response to environmental change. They are characteristics of agricultural communities and are founded on pre-adaptive developments which existed a long time before agriculture developed as the main mode of production. The application of planned agriculture as a mode of production did not develop in order to make a new source of food available. Rather, agriculture made possible an improved management of resources long- and well known, with the controlled continuation of a productive and energetically efficient subsistence strategy.

The success of this optimisation strategy is based to a large extent on a long preparatory phase, in which knowledge of cultivated plants that were to be domesticated later was collated, available and alive over many generations and where the use of new resources could be tested. The new source of food was available because annual grasses, i.e. cereal grain species, had a selective advantage over perennial plants under the conditions of a climatic change towards pronounced seasonality and summer dry seasons at the beginning of the Holocene. This rapid adjustment of the wild cereal species to the changed environmental condition is commonly not explained by a gradual adaptation process, but rather in the sense of a 'punctuated equilibrium' (McCorriston and Hole 1991), where a species emerges, selected towards better adjustment, quickly and completely adapted in its characteristics.

The processual dynamics of gradual subsistence change is based on the increasing predictability of the production system and results in a change from r- to k-selection for the cultivated resource (Ellen 1994). In the case of cereals, r-selection denotes that plants which produce a maximum of seeds are favoured, while under the conditions of k-selection plants which produce a limited number of seeds with a high chance of survival are preferred. Domestication finally produces a strong selective pressure on those plants which are more sought after in response to changed human nutritional behaviour and preferences, which are met with higher appreciation and which therefore create increasing demand for higher supplies. The changed characteristics of the productive plants provide a more efficiently usable food source and the population benefits from higher yields per unit area. The new subsistence basis is becoming increasingly more controllable or, in ecological terms, their matter and energy flows can be more efficiently steered.

The domestication of animals as the second corner-stone of an agricultural way of life emerges clearly later than the planned use of plants in the Near East. It is true that sporadic evidence dates prior to 11,000 BP, yet the majority of the early animal bone finds of reliably domesticated provenance only occur from the ninth or late eighth millennium onwards (Legge 1996). The beginnings of livestock domestication are exclusively limited to wild goat (*Capra aegagrus*) and wild sheep (*Ovis orientalis*), whose natural area of distribution completely overlaps with the Fertile Crescent. The foothills of the Zagros Mountains are regarded as their centre of domestication (Hole 1996). Surprisingly little is known about the exact course of these early attempts at domestication. For a while it was proposed (e.g. Uerpmann 1990) that it was triggered by a shortage of animal protein resources at the end of the pre-ceramic Neolithic as a result of excessive over-use of wild stocks of ungulates, incidentally connected with the collapse of this cultural stage (Rollefson and Köhler-Rollefson 1989), which would have prompted humans to look for new sources of protein. Yet it is still a contentious issue whether this was an intended or coincidental process. At present, explanations are in favour of a plausible hypothesis suggesting that human settlements provided an artificial niche for young animals of the wild forms, where they were kept until they reached sexual maturity and then subsequently reproduced. The fact that such a pattern of animal behaviour is actually only compatible with the biology of goats and sheep is in accordance with the findings from faunal bone assemblages (Uerpmann 1996). This would also mean that, for the domestication of animals, the sedentariness of human populations was necessarily pre-adaptive which, in turn, was only made possible through the sustaining availability of vegetable food as part of the necessary basic conditions.

3.1.2.2 Southern Scandinavia

During the seventh and sixth millennia BP, when agriculture had already been
established as a mode of production for a long time in central Europe and the
north German plain as part of the Linearbandkeramik culture (Linear Pottery
Complex; LBK), the area of southern Scandinavia was inhabited by Mesolithic
forager societies of the Ertebølle culture (ca. 7,400–5,900 BP). Archaeological
evidence of contact between both regions through trade and barter indicates
however that there was knowledge of a different subsistence strategy in the
south (Fischer 1982; Zvelebil and Rowley-Conwy 1984; Price 2000). Nevertheless, the adoption of agricultural practices was resisted for some time. This
was likely to do with special local modes of food acquisition of the Mesolithic
populations and stable ecological basic conditions that made the persistence
of a foraging subsistence possible, even though the complex set of causes that
finally led to subsistence change is intensely debated at present.

   Extensive archaeological research suggests (e.g. Rowley-Conwy 1983, 2004;
Price and Brinch Petersen 1989; Price and Gebauer 1995; Price 1996, 2000) that
the people of the Ertebølle culture were sedentary foraging groups, whose
population size and density clearly exceeded those of other European Mesolithic groups and whose settlement pattern was characterised by year-round
occupation of mainly coastal sites. Apart from hunting a wide variety of terrestrial and marine animals and collecting plant food, it has been argued that
the use of seasonally occurring coastal resources played a large role in their
subsistence, a strategy in keeping with evidence of large predominantly
coastal settlements. In contrast to smaller, nomadic groups which follow food
supplies in a seasonal cycle, larger, established foraging groups are in danger
of quickly depleting stationary food resources through excessive use. Particularly from this point of view, the use of coastal food items becomes important,
since the spectrum of usable resources is expanded and thus subsistence
made more secure. That this system, despite the relative spatial proximity to
fully fledged agriculture, enjoyed long-term sustainable and continuous use
has at least in part been explained with the harvesting of seasonal coastal
resources by the Mesolithic populations of southern Scandinavia (Rowley-
Conwy 1984, 1985). The spectrum of food sources exploited by these
Mesolithic people, as identified from faunal and floral remains, shows that
most of the items were available only for a limited period of the year. The
range of foodstuffs encompassed deer and hoofed game, fish, sea mammals,
water birds, hazelnuts and acorns, shellfish and oysters. Recent isotope data
(Richards et al. 2003a) confirmed that, while there was a strong emphasis on
marine resources, some Mesolithic foragers enjoyed terrestrial diets as well.
Supplies were abundant and also balanced in their essential food constituents
for most of the year (based on modern nutritional value tables, e.g. Souci et al.
1981; Bender 1993) – with the notable exception of the lean times during late
winter and early spring.

It is known that in temperate latitudes foraging populations face short supplies of resources during this season, since winter supplies are increasingly being used up and the new vegetation period has hardly begun (cp. Schutkowski 1993a). Under the particular circumstances of southern Scandinavian foragers, it was suggested that oysters would provide an energetically cf. productive resource with an almost seasonally independent stable fat content *Northeastern* and relatively favourable amounts of carbohydrates. Oysters would thus fill *USA* exactly this seasonal resource gap, in which other food components are not available at all, or only in smaller quantities and/or of inferior quality (Rowley-Conwy 1984), and thus conveniently complement the otherwise well established broad-spectrum economy. Archaeologically there is ample evidence of large kitchen middens (køkkenmøddinger) composed of oyster shells, which are characteristic of at least parts of the shore-line settlement sites of the Ertebølle period. Thus, the argument goes that, as long as oyster banks could be exploited as a bridging resource, there was no necessity to change the subsistence strategy in terms of the prevailing ecological basic conditions. This economic pragmatism appears to have been successful, for as long as it was made possible by the conditions of the natural unit.

At the beginning of the sixth millennium BP, however, as part of cyclical changes in sea level, there is a regression of the Litorina Sea (cp. Westman and Sohlenius 1999). The sea level gradually sinks and the Kattegat, i.e. the extension of the Baltic Sea that separates Denmark and Sweden, becomes narrower. As a result, the exchange of water with the North through the Skagerrak north of Jutland is substantially reduced and consequently water salinity within the coastal range drops below 23‰, the critical value for oysters. Concomitantly, the growth and reproduction cycles of the oysters shift as a result of the decreasing water temperature. Thus, the exploitation of oyster banks increasingly lapses. Other shellfish cannot be used to make up for and energetically balance this resource loss, since e.g. cockles and mussels display a more pronounced seasonality pattern of their nutrient content than oysters. The loss of such a key resource therefore further intensifies the already tense nutritional situation in the late winter.

While the consequence of this energetic bottleneck would be a reduction in group size and/or increased mobility, as a rule, the Mesolithic people of the Ertebølle culture have another option at their disposal: to adopt a proven and in the course of more than 1,000 years fully developed agricultural technology from their southern neighbours of Central Europe, which they are principally acquainted with through regular contacts and the exchange of both ideas and materials (Rowley-Conwy 1984).

Possibly, the Mesolithic populations try to compensate the loss of the oyster as a bridging resource by emphasising the exploitation of other food sources. In the long run, however, this is neither energetically worthwhile nor quantitatively compatible with the local faunal stock. Resource compensation, for example by increased hunting of hoofed game, which displays a clear weight and fat loss during the winter months, requires large quantities of lean meat per day and person (Speth and Spielmann 1983) to cover the nutritional requirements. Keeping to the traditional subsistence, termed appropriately by Rowley-Conwy (1984) as 'the laziness of the short-distance hunter', for as long as the basic conditions would permit is, on the one hand, an expression of an efficient optimisation strategy for resource exploitation. But it is, on the other hand, only possible because the long-term stability of the ecological/natural unit or the sustainability, particularly of marine resources, favours such an optimisation strategy in the first place.

Without necessity, a successful mode of subsistence optimally adapted to prevailing external conditions will not be abolished. External factors, like in this case climatic deterioration, change of landscape and a restriction of the resource spectrum through the decline in marine productivity (Larsson 1990), can be regarded as a sufficient set of triggers for larger-scale adjustments of subsistence to the changed conditions. The shift of settlements away from the coast has been associated with resource stress, as well (Madsen 1985; Nielsen 1987), a view that has recently gained momentum for an explanation of the spread of farming through Northwest Europe (Bonsall et al. 2002). Yet, it has been argued that the ecological disaster, which would have forced an abrupt subsistence change, did not take place (Price 1996). Recently, instead, the significance of intrinsic motivations or social determinants of subsistence change are increasingly being discussed as an alternative (and complementary) explanation model, especially for the Neolithic transformation in south Scandinavia (e.g. Jennbert 1985; Price 1996).

The Ertebølle period is characterised by a relatively high settlement density of the ecologically rich littoral zone. The high abundance of available terrestrial and marine resources and their spatial–temporal reliability is the basis of a lasting broad-spectrum economy and a quasi-sedentary population. Social structures develop that are similar to those 'affluent foragers' known from the American northwest coast (see Yesner 1987) as affluence becomes the pre-requisite for incipient social differentiation within the Mesolithic communities. Evidence of *antemortem* trauma in skeletons from Mesolithic cemeteries indicates interpersonal violence, likely as a result of raids and resource competition. By 6,700 BP, items of material culture acquired from LBK communities to the south not only encompass ceramic vessels of different types, but above all artefacts, which belong to the realm of status attributes (p. 358 in Price 1996). Information and trade contacts, which over many generations developed with the established agricultural groups of Central Europe, now facilitate and allow the satisfaction of new needs and social demands.

The readiness to test a new mode of subsistence seems to grow with the increase of prestige gained in return and fosters an initially slow process. However, $\delta^{13}C$-signatures of human skeletal remains (see Sect. 4.2) show a clear and sharp nutritional change from a diet strongly reliant on marine food towards a more terrestrially shaped nutritional basis in the Neolithic populations (Tauber 1981; Noe-Nygaard 1988; Richards et al. 2003a; Fig. 3.1), which has recently been even pushed back in time to around 5,200 BP; and all of which is in accordance with an initial shift of early Neolithic coastal settlement places towards the inland. Yet hunting, fishing and collecting probably remain indispensable for food acquisition, at least in the beginning, when Neolithic settlement activity is characterised by both residential sites and hunting stations. Whatever the scale by which the change in the mode of subsistence is accomplished, it is apparently not connected with a deterioration in living conditions to the extent that it is reflected by skeletal alterations in early Neolithic people compared with those of the late Mesolithic (Bennike 1993).

The Mesolithic–Neolithic transition in southern Scandinavia is complex and likely includes a blend of ecological, economic and social causes. It is not necessarily about actually achieving an improvement in the nutritional basis or gaining more food. It is true; the Litorina regression must not be interpreted in the sense of an ecological crisis, but rather as an event with a constant long-term effect, i.e. the gradual loss of a bridging resource. The eventual subsistence change fulfils a second function, as now food is produced that gains importance in the course of a change of belief systems and social differentiation. The early Neolithic does not, at least according to the number and

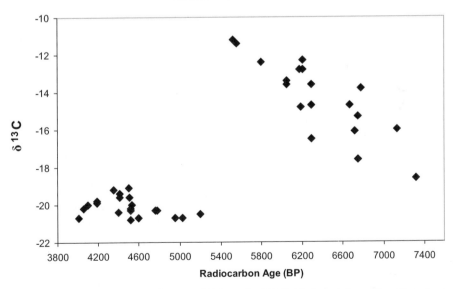

Fig. 3.1. The transition from the Mesolithic to the Neolithic in southern Scandinavia at around 5,400 years BP is characterised by a marked and rapid replacement of marine resources by a terrestrial diet (data from Persson 1999; Richards et al. 2003a)

size of known habitation sites, experience population growth that is connected with the new mode of subsistence. But even though farming and breeding of livestock only become the dominating mode of production relatively late (the oldest reliable radiocarbon date for domesticated cattle is at 5,850 BP; Koch 1998; Persson 1999; cited in Richards et al. 2003a), more than 1,000 years after the establishment of the LBK culture on the southern mainland and, after all, still more than 600 years after the occurrence of Neolithic artefacts in southern Scandinavia, the shift to a terrestrial diet is dramatic, rapid and wholesale, including evidence of floral and faunal remains from domesticated species on a considerable scale. The security of a sustainable and controllable resource situation allows compensation for the high additional costs incurred in terms of work and energy for the introduction of a new mode of subsistence. To do this, it only, as it were, requires the import of an idea, as some argue there is no indication for indigenous development of agriculture (Rowley-Conwy 2004), but regardless, whether the actual adoption of the Neolithic in southern Scandinavia is an internal affair, or whether is was supported by colonising or demic diffusion, the prolonged and close acquaintance with the new subsistence mode appears crucial in facilitating rapid subsistence change.

3.1.3 Maize in North America

The cultivation of maize (*Zea mays*), its domestication and, eventually, its wide distribution across the American continent is comparable with the significance that the spread of wheat and barley had in the old world. For the agricultural centres in the northeast of North America, a differentiated picture is available of the temporal sequence and increasing importance of maize as a staple crop in the subsistence of Native Americans.

Recent radiocarbon dates (Long et al. 1989) indicate that the domestication of maize began during the first half of the fifth millennium BP in the highlands of Mexico. The existence of a second, independent centre of domestication in South America is currently being discussed, but no conclusive evidence has been put forward yet. There is definitive evidence, however, that domesticated maize spread from Mexico both southward and towards the north. It first appeared in the southwest of what is the United States today (Wilkes 1989), with confirmed dates from direct dating and archaeological findings of the time around 3,200–2,800 BP (Smith 1995a). The earliest unequivocal evidence for the east of North America dates from the end of the second or the beginning of the third century AD (Chapman and Crites 1987). But it takes a further 500–1,000 years before maize becomes recognisable as an established subsistence component; and the regional patterns of duration and timing of subsistence change vary to a substantial degree as a function of natural ecotope conditions and socio-political developments.

In the floodplains of the middle Mississippi (Ozarks area, today encompassing the states of Missouri and Arkansas) maize does not play an important role as a basic food for the general population until approximately 1,000 AD. Carbon isotopic ratios of human skeletons from the Woodland and Mississippian cultures (approx. 300 BC to 1,000 AD) show isotope signatures to clearly fall within the range typical of the consumption of (endemic) C3 plants[5] (Lynott et al. 1986). Even after maize is available in the region as a cultivated plant, people keep to the nutritional pattern they are acquainted with and that has been established over several thousand years. It is not before approximately 1,200 AD that increased maize consumption shows up in the skeletal record by significantly elevated $\delta^{13}C$ values caused by the higher intake of C4 plants, here maize. The nutritional basis is being substantially converted with a calculated relative amount of more than 35% maize in the diet; and it takes place relatively quickly, over a period of approximately two centuries only. The beginning of this intensive maize cultivation coincides with the development of larger settlements and the emergence of ceremonial centres.

Such a connection between population concentration and subsistence change also applies to urban centres such as Cahokia (Illinois) that develop during the ninth to tenth century (Bender et al. 1981; Milner 1990; Pauketat 2002). Substantial cultivation and consumption of maize, however, is not limited to these regional centres or exclusively induced by their emergence. Maize is generally the most important staple also in rural agricultural settlements from the turn of the millennium onwards, for example in what is Wisconsin and Ohio today, and takes the place of traditional cultivars with regard to its significance for subsistence (Buikstra and Milner 1991; for Florida, see Hutchinson and Norr 1994). In some areas, e.g. the Nashville Basin (Tennessee, Kentucky), prevailing environmental conditions permit a quicker and more pronounced agricultural intensification, so that the change in subsistence becomes visible in the skeletal isotopic record already around 700 AD. Around 900–1,000 AD, maize becomes the dominating main food component, which is reflected in the most extreme C4 signatures known for North America (Buikstra et al. 1988; see Fig. 3.2).

In the northern areas around the Great Lakes and the east coast, maize is on the whole integrated more slowly into the subsistence system and its adoption as a significant staple takes place later. The temporal delay in relation to mid-continental regions occurred because, first of all, a variety of maize had to be bred that was better adapted to the shorter vegetation period and the

---

5   Depending on whether plants follow the Calvin or Hatch–Slack cycle of the photosynthetic pathway, the characteristic reaction product is a fragment containing either three or four carbon atoms, hence the differentiation into C3 or C4 plants (e.g. Schwarcz and Schoeninger 1991; Ambrose 1993; Katzenberg 2000). Crassulacean plants have yet another metabolism, but these plants are of no importance here.

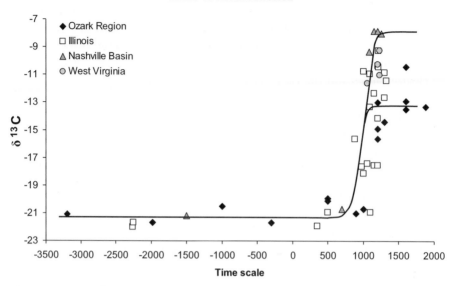

Fig. 3.2. Increase of maize as main dietary staple in the Ohio and Mississippi valleys over time. Low stable carbon isotope signatures indicate a diet based on C3 plants, less negative values represent the shift to substantial consumption of C4 plant resources, i.e. maize. The subsistence change is rapid and accomplished within only a few generations (data from sources mentioned in text). The graph illustrates the steep increase in C4 dietary components

colder climate before maize farming could be attempted on a larger scale (Schwarcz and Schoeninger 1991). This coincides with the observation that in southern Ontario, the northernmost area of maize cultivation, the plant is introduced only around 700 AD (Schwarcz et al. 1985) and that the population keeps a diverse subsistence economy, originally based on the consumption of domestic C3 plants. It is not until the twelfth and thirteenth centuries that maize is well established as a cultivated plant (Katzenberg et al. 1995) and amounts to more than 30% of the diet (see also Vogel and van der Merwe 1977). Recent analyses suggest that a gradual increase in maize consumption can be detected prior to 1,000 AD (Harrison and Katzenberg 2003) whereas, in some coastal areas of New England, maize apparently did not play a major role at all until the seventeenth century (Little and Schoeninger 1995). The increase in the C4 (maize) proportion is accompanied by a gradual rise in population density (Katzenberg 1992).

As in the more southern ranges of eastern North America, population concentration and subsistence change are mutually dependent and develop in parallel. The long lead-time of 20–40 generations that has to pass before maize becomes the dominating plant cultivar is caused by: (1) the make-up of the traditional economies these populations have established, (2) the fact that maize is only one among several domesticated crops and not the first plant to be cultivated either, and (3) to what extent maize is eventually integrated into the existing agricultural subsistence system. For the entire mid-continental range in the east of North America, the cultivation and domestication of native plants is already a firm constituent of a horticultural agricultural system, a long time before domesticated maize arrives in the region.⏋

Already by the second half of the fifth or at the beginning of the fourth millennium BP, people had domesticated four seed plants: the endemic species *Chenopodium berlandieri* (Goosefoot), *Iva annua* (Marsh elder) and *Cucurbita pepo* (Ozark wild gourd), as well as *Helianthus annuus* (Sunflower), which probably entered the region as a culture follower (Smith 1989a, 1995a: see p. 184ff). The former three species are weeds of floodplains and riverbanks as wild forms and, being pioneer plants, are also able to occupy naturally degraded open landscapes.

After the gradual retreat of the Laurentinian ice sheet in eastern North America 6,000–7,000 years ago (COHMAP 1988), the abundance of plant food and game increases due to associated changes in climate and landscape and provides forager groups with the opportunity to seasonally re-inhabit certain areas that were previously covered with ice. This leads to a gradual anthropogenic alteration of the soils, which in turn favours colonisation by pioneer plants. In the end, their occurrence in the settlement areas of prehistoric foraging populations is a consequence of climatic improvement in the east of North America.

The wild forms of endemic species are first collected and later domesticated step by step. In the beginning, however, they do not supply a substantial contribution to the nutritional basis, but rather serve as buffers against seasonal scarceness of resources (p. 197 in Smith 1995a). Only more than 1,000 years after their successful domestication, between 500 BC and 200 BC, the agricultural use of these plants becomes more important for food production, as evidenced by the increased occurrence of macro-remains in archaeobotanical assemblages. For some further 1,000 years, a stable mixed subsistence is established from hunting, gathering, fishing and horticulture based on domestic useful plants.

Even though the experienced farmers most likely know its potential usefulness, maize tends to remain an insignificant cultivated plant after its arrival in eastern North America. The stability of the existing mixed mode of production does not evoke the necessity for a quick subsistence change. Maize is cultivated, but it remains a negligible dietary constituent for a long time and rather derives significance from its function as a status or ceremonial plant. Regardless, it is crucial that the properties of the plant and the knowledge to cultivate it are an active part of the agricultural tool kit over a long period of time. Only this, it seems, can explain why between 800 AD and 1100 AD maize can be widely incorporated into existing multi-crop economies, finally leading to agricultural systems essentially geared towards this one crop – notwithstanding the fact that animal protein and plant food other than maize remain part of the diet (e.g. Morton and Schwarcz 2004). The relatively quick subsistence change is only facilitated by a long preparatory phase.

The spread of maize as new basic resource is accompanied by the occurrence of complex socio-political structures and increased population density. The development of larger settlements with complex social structures and the expansion of maize cultivation are mutually dependent and stabilised by positive feedback. The speed of this change is clearly a function of a long lead-time. Cultural change, the emergence of new cultural institutions in the named North American centres after the turn of the millennium is not possible without the introduction and establishment of maize and the concomitant subsistence change.

3.1.4 Dependency, Continuity and Change

The preceding examples have demonstrated the important role of techno-cultural and social pre-adaptations play as responses to changing circumstances for a successful change of subsistence conditions. In all cases, these adjustments occurred over a relatively long period of time, in which the adoption of the new subsistence strategy could be explored or prepared. Yet, they were regionally limited in their emergence and effects. On a larger scale, the impact may be dramatically different. Supra-regional relations and developments are no recent phenomena: they have existed, flourished and declined in the past as well. It is not uncommon that local populations find themselves tied into dependencies of economic and political power relations – to their benefit but also their harm, if they have little or no influence to exert on shaping these relations. In such a case, they become susceptible to disturbances of equilibrium states and may even be faced with temporary or long-lasting collapse of existing structures. As the following examples show, the possibilities of recovering from these disturbances strongly depend on the extent to which the external constraints that led to collapse are maintained. It will furthermore be attempted to demonstrate that the effects of external factors, which do not emerge from climatic shifts or other changes in natural basic conditions, but which are a consequence of human interference into existing forms of social and political organisation, may lead to mal-adaptive repercussions for the populations concerned, since they did not have sufficient time to prepare for a change in their subsistence conditions by developing new subsistence strategies.

3.1.4.1 Persian Gulf

[handwritten annotation: What were the population parameters of these two transformations 1st to pastoral nomadism then to fish & shellfish dependence?]

During the fifth millennium BP, the Oman peninsula substantially profited from being a stopover along the long-distance commercial route between the cities of the Indus Valley cultures and southern Mesopotamia and from their buoyant economic relations (Sherratt 1980; Cleuziou 1984). In the entire south eastern range of the Gulf, the Umm an-Nar culture had fully developed and flourished in its fortified settlements and characteristic monumental burial architecture (e.g. Frifelt 1975). It is the last known established and sedentary farming culture (Constantini 1978/79 for evidence of barley and millet) before pastoral nomadism spreads in eastern Arabia (Cleuziou 1981). This cultural break during the transition from the fifth to the fourth millennium BP sets off with a decline in urban life and ends with the practical absence of securely dateable settlements from the following Wadi-Suq culture (Edens 1986). Also in the Oman Mountains, well-off oasis towns were plentiful in the fifth millennium (Orchard 1994, 1995), but there is a lack of archaeological evidence for settlement structures during the following millennium. Moreover, this process of disintegration is caused and exacerbated by the contemporaneous gradual decline of the Indus Valley cultures and the urban centres in southern Mesopotamia. As a consequence, traditional trade relations dissolve and the Oman peninsula is turned into a phase of cultural isolation. One promising alternative subsistence option is available, pastoral nomadism, but only because the camel is already domesticated and its resource value is known and available (p. 314ff in Benecke 1994).

The further development of subsistence change as part of culture change can be exemplified for a population of the Gulf coast in what is today the emirate of Ras al-Khaimah (United Arab Emirates). Bone element concentrations of human and animal skeletal remains from three communal graves of the early, middle and late phases of the fourth millennium BP allowed the main food components and thus a probable mode of subsistence to be reconstructed (Grupe and Schutkowski 1989). During the early and middle phases, people most likely relied on a mixed subsistence and their basic diet consisted of animal-derived and vegetable foodstuffs. Considering the close correspondence of climatic conditions along the Gulf coast 4,000 years ago with those found today (Piepenbrink and Schutkowski 1987), dietary reconstruction suggests a combination of subsistence strategies comprising animal husbandry with pastoral farming and seasonal agriculture of small fields on the plateaus of the coastal region. Such a complementary subsistence pattern is common in pastoral societies; and still in recent times farming combined with animal husbandry is frequent and widespread in the northern regions of the Oman peninsula (e.g. Dostal 1985). Towards the end of the fourth millennium, there is a marked shift in subsistence activities towards an emphasis on the exploitation of marine resources – indicated not only by bone chemistry data but also, in particular, by ample archaeological evidence of shell middens along the shore line, bearing evidence for different kinds of fish and large quantities of molluscs of the genera *Terebralia*, *Ostrea* and *Murex* (Glover 1991). These findings strongly suggest a subsistence pattern that required the return to a fully sedentary way of life and, indeed, by the end of the millennium there is again evidence for settlement structures in the archaeological record (Vogt and Franke-Vogt 1987).

The case to be made from this example is that the negative effects of forced, or at least not self-inflicted, cultural and subsistence changes through extrinsic impact could in all probability be substantially buffered and moderated by converting the subsistence basis to the use of long- and well known resources. It can be reasonably assumed that the camel had already been domesticated during the sixth millennium BP, to be used as a pack animal for the transport of copper ore from the Oman Mountains into the coastal regions (Weisgerber 1983; Benecke 1994). If this is the case, then throughout the entire heyday of long-distance trade between the Indus Valley and Mesopotamia, the camel was known and familiar to communities in the eastern part of the Arabian Peninsula for many generations. This long lead-time, i.e. a period of acquaintance, then allowed a relatively quick transition to the new, more peripatetic way of life, which is marked in the archaeological record by an abrupt suspension of settlement activity. Also for the second phase, a long preparation time with the exploration of offshore and marine resources prompted another subsistence change at the end of the fourth millennium BP. This facilitated both a new food acquisition strategy – fishing– to be tested and the existing mode of subsistence – pastoral nomadism – to be supported. As part of this development, food chains became important that would include fodder for camels. Until recently, the coastal areas of the Oman peninsula were lined with dense mangrove forests. In areas where such forests are still present today, mangrove leaves are regularly fed to camels and it is very likely that this was the case in pre-historic times as well. Moreover all mollusc species, which later were so intensively exploited, are endemic to the mangrove forests of the Gulf coast and thus facilitated a gradual habituation to the sea as new dietary provider. Again, long-term acquaintance with the resources and the habitats favoured a long preparatory phase, before the shift towards a new subsistence basis took place.

3.1.4.2 Once Again North America

At other places and other times, regeneration from external pressure on an existing mode of subsistence was not possible, since continuous political and economic exertion of influence became a threat and an endangerment to prevailing living conditions. The cultivation of domesticated maize and its use as a dominant staple crop in North America were connected with a hitherto unknown cultural upswing, for example during the Mississippian in the twelfth to fourteenth centuries (cp. Sect. 3.1.3). But besides maize as the subsistence basis, C3 plants, animal-derived and marine foodstuffs always played an important role as dietary supplements for a balanced nutritional composition.

From the sixteenth century onwards, however, the scene changed. Studies on Native American skeletal populations, dating to the time of the Spanish colonisation of southern North America, reveal that the indigenous population could no longer sustain the basis of their traditional subsistence and were increasingly restricted in their activities by colonial rule. This led either to a much increased or almost monotonous maize diet along with diminished consumption of marine food (for the Atlantic coast of Georgia, see Larsen et al. 1992; for Florida, see Larsen et al. 2001) or to a substantial loss of maize as the main food component (for the Pecos region, New Mexico, see Spielmann et al. 1990), while at other places there is more regional variation in the increasing cultivation of maize before and after the Spanish colonisation (Hutchinson and Norr 1994; Hutchinson et al. 1998). In the New Mexican case, the requirement of Spanish occupying forces for food and work force considerably reduced the quantity of maize that remained available to the indigenous population. It was unimportant in the end whether this was due to the (coerced) delivery of maize to the colonial rulers or the lack of time left to Native Americans to till their own fields– in any case it resulted in a marginal nutritional status and often led to impaired health conditions with skeletal alterations indicative of degenerative joint disease and manifesting stress indicators due to increased work demands from the missions (Hutchinson et al. 1998). In the case of communities along the Georgia Bight coast, it can even be assumed that mounting pressure on an already diminishing subsistence basis in conjunction with the propagation of epidemic diseases finally led to the extinction of the population during the eighteenth century (see e.g. Crosby 1986; Larsen and Milner 1994).

But the loss of traditional subsistence options is only one aspect of the complex social reality. Subsistence systems are embedded in a network of social and ecological relations which form the constituent elements of adaptation to the biocultural environment. A forceful change of subsistence due to external constraints that leaves neither space nor time for testing alternative strategies destroys more than only the nutritional basis. It indeed threatens the balanced totality of political institutions, cultural symbols and natural habitat conditions from which subsistence systems emerge and develop, i.e. the socio-cultural resilience.

And yet the initial economic contact between white colonialists and Native American Nations seemed to be amicable and without threat. On the basis of limited, reciprocal business, there were recognizable advantages for both sides; and furs, hides and food were exchanged for tools. As long as reciprocity could be protected, the system did not interfere with traditional, indigenous ethical concepts, according to which safeguarding of subsistence formed the foundation of life, but not maximisation of resource provision[Only consistent application of the laws of market-orientated economy by the British and later the Americans, which aimed to satisfy the swiftly rising requirements for various commodities, dragged native populations into dependency from which, despite steady resistance, they eventually could not free themselves (for Pawnee, Choktaw and Navajo, see White 1983). Accelerated by a combination of alcohol and loans, all resistance and reluctance, which initially was still able to protect game and food resources from decimation, was finally broken. The primacy of economics led to the shattering of traditional ways of life and to the abolition of a social and ecological balance. The introduction of market relations into societies, where produced food was not monopolised but divided and where generosity and not maximisation was the ethical standard, destroyed social consent from inside and made it vulnerable to influences from outside.] DISEASE !!

3.1.4.3 Tall Seh Hamad, Syria

In historical times, strategies of resource use can also be held in adaptive stability, even if the natural and habitation area are shaped by an eventful political history. The region of the lower Habur River, a tributary of the Euphrates in northeast Syria, exhibits numerous remnants of settlements ranging continuously from the close of the sixth millennium BP into our time. The village of Tall Seh Hamad (the Assyrian Dur Katlimmu) at the lower reaches of the Habur is situated south of the agronomical dry-belt that stretches along the northeast of Syria and is defined by 300 mm isohyets. The climatic conditions do not sustain dry-farming and thus an efficient agricultural use of the location is only possible by extensive irrigation.

Geomorphological studies indicate that contemporary climatic conditions also prevailed in similar fashion during the past thousands of years. Accordingly, there is evidence of irrigation systems since mid-Assyrian times and its likely continuation until approximately 700 years ago (Ergenzinger and Kühne 1991). This continuity in technological and settlement history suggests that subsistence modes and hence nutritional patterns would also have continued to remain very similar.

Possibilities of subsistence and food acquisition are restricted in the area of Tall Seh Hamad; and studies into the geomorphology and climate of the region suggest that environmental conditions allowed for basically two options of land use (see Kühne 1991): agriculture employing extensive irrigation measures along the Habur River and livestock farming in the adjacent steppe areas. While there is archival, faunal and archaeological evidence for the city's flourishing Assyrian period, from which at least qualitative information can be derived about the food consumed; there is no comparable material available for the succeeding historical time-periods. The way in which post-Assyrian historical populations mastered the constraints of their habitat, can largely be reconstructed by chemical analyses only.

Bone element analyses from late Assyrian, Roman and early Islamic times allow investigating this question of continuity or change of a subsistence system (Schutkowski 2005). The data reveal clear indications of a time-related change in dietary components that are most likely the result of subtle adjustments of subsistence strategies to changing environmental conditions. Element analyses suggest that, during Roman times, there is an increasing amount of calcium-enriched foodstuffs, indicating the consumption of milk products, which is in keeping with archaeozoological evidence of pasture animals (sheep, goat, camel, probably also cattle; see Becker 1991a) kept at Tall Seh Hamad. During the early Islamic period, consumption patterns change towards a greater reliance on a diet less rich in calcium, i.e. a broad range of vegetable food, with horticultural products being more important than agricultural crops.

Whilst the main basic food component derived from vegetable items and accompanied by varying amounts of protein supplement, yet there are clear differences detectable in the nutritional behaviour of populations from Tall Seh Hamad through time. This is particularly conspicuous for the Roman period. Archaeological findings of complex burial architecture suggest that, in addition to subsistence change, there are issues related to socio-cultural differentiation within the population.

Reconstructions of the immediate environment of Tall Seh Hamad based on geomorphological analyses indicate that habitat conditions were very similar for the time periods subsequent to the Assyrian heyday as they are today (Ergenzinger and Kühne 1991). Only the use of continuously existing canal and irrigation systems (for which there is ample archaeological evidence), a technology that was available and utilised over a long period of time, allowed maintaining and managing the high level of production of the horticultural and agricultural crops which are so characteristic of the temporal and dietary groups of the population of Tall Seh Hamad. At the same time, the adjacent steppe areas served as pasture for livestock and allowed providing varying amounts of animal protein. Thus the bone chemistry data principally imply continuity of both land use and nutritional basis in post-Assyrian times, although the composition of main basic food components was subject to variation (Fig. 3.3). Subsistence change can be detected both in diachronic comparison and in uncovering apparently slight internal differentiations. At the same time, this implies that an adaptive techno-cultural system can be modified to a considerable degree by gradual change of the subsistence conditions and can be adapted to the natural conditions of the habitat and the changing social conditions, a key feature of social resilience (Adger 2000).

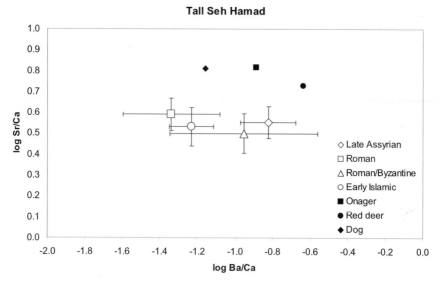

Fig. 3.3. Diachronic variation in nutritional patterns at Dur Katlimmu/Tall Seh Hamad, Syria (see text)

3.2 A Transition Model

The preceding sections introduced case studies of subsistence change from different ecological and socio-cultural conditions and backgrounds. Particular consideration was given to the extent in which pre-adaptive developments and the temporal duration prior to the transition to the new mode of subsistence affected the change and what adaptive or mal-adaptive effects resulted from the length and quality of the preparatory phase. Independent of the respective nature of the causative effects responsible, it should be possible to derive general statements on the course and processual dynamics of change, which can then be connected with the factors actually triggering subsistence change. In the following, a generic model is proposed and elaborated on to analyse the course of subsistence change, which is then refined and modified by identifying and comparing basic patterns of subsistence transitions from the case studies presented.

Zvelebil and Rowley-Conwy (1984; see also Zvelebil 1996) suggested a model to describe the transition from an acquiring to a producing mode of food procurement for those areas where agriculture was not introduced by colonists but originally adopted by the resident population. It assumes spatial and geographical boundaries (agricultural frontiers; Denell 1985) between farming and foraging communities, which represent interactive zones of contact between these two different ways of life and subsistence and which allow forager groups to observe and evaluate the new form of technology and socio-cultural organisation.

The adoption of agricultural technology then proceeds in three sequential phases. In the course of the availability phase, agriculture and animal husbandry are known to forager groups in principle, but there is no or only rudimentary adoption of the other mode of subsistence. Mutual contacts lead to an exchange of goods and information, whilst both sides to a large extent retain their cultural and economical independence. The relative portion of agricultural production stays below 5 %. The following substitution phase is characterised by the increasing take-over of agricultural modes of food acquisition, while a foraging subsistence is still maintained. The amount of agriculture goes up to 50 %. During the consolidation phase, subsistence becomes more substantial and eventually completely reliant on agriculture and animal husbandry. The cumulative nature of this process is illustrated in Fig. 3.4.

Subsistence change, as the preceding examples demonstrate, can be regarded as both an event and a process, irrespective of whether the level of nutrition or the mode of subsistence is addressed. Yet, the time-frame is important and an interpretation of the process has to take into account some more or less gradual change and seek to place the transformation of existing subsistence patterns into a temporal operational sequence. This kind of change of subsistence represents the separation of certain modes of resource

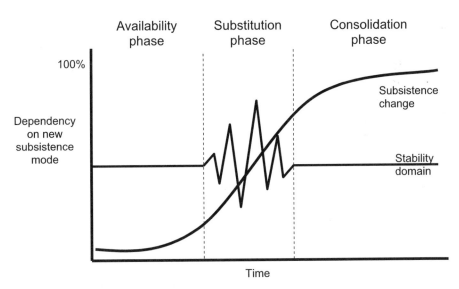

Fig. 3.4. General model of subsistence change. The process is subdivided into the consecutive phases of availability, substitution and consolidation. During substitution, the resilience capacity is exceeded and the system goes through a phase of disequilibrium. The consolidation phase represents a new stability domain. Modified after Zvelebil and Rowley-Conwy (1984)

use and is equivalent to the transition from an existing successful – or at
least durable – to a new food acquisition strategy, whose ecological and economic success still has to be proven, or in other words, the transition from
one stability domain to another. At first, it would still have to be supported
and complemented by the previously practised strategy, since the substitution phase is a time of uncertainty when there is no longer a balance
between economy and ecology. Through the transition towards a new mode
of subsistence, individual components of existing basic conditions of the
habitat and the population are changed and the equilibrium within the system is disturbed (Fig. 3.4). In order for the extent of this inevitable disturbance to remain as small as possible, one can theoretically expect that the
preparatory phase or lead-time (availability phase *sensu* Zvelebil), in which
a population becomes acquainted with the new strategy in the first place,
should be the one with a slower speed and a long duration, whilst the phase
of change, i.e. the adoption of the new subsistence basis, should take place as
fast as possible. The steeper the curve of the substitution phase in Fig. 3.5,
the faster the transition towards the new mode of subsistence takes place
and the sooner the consolidation phase and a new equilibrium state are
reached. This, however, is dependent on the prerequisite that the availability
phase, in which the new subsistence strategy is tested and prepared, lasts for
a long time. Only then is a relatively quick transition towards a consolidated
new strategy of resource use possible.

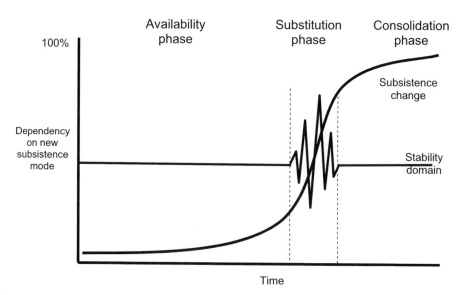

Fig. 3.5. Subsistence change under ideal conditions. A successful and rapid transition is
energetically desirable as it reduces the disruption during substitution, and benefits
from a longest possible lead-time

By contrast, each form of slow and gradual change would have to prove the practicability of operating two different modes of subsistence at the same time, as both would be in a phase of transition; one developing towards establishment, the other towards separation and abolition[6]. A short and steep transitional phase must therefore be assumed as the ideal case scenario. Its sequence, however, will vary as a function of local variations in the natural condition, the innovative potential and the readiness and preparedness of the human population for a change in subsistence. The cross-linking of individual components, which trigger subsistence change and which finally determine its course and its consolidation, is complex and a model can unavoidably only address this in a reductionist way. Especially, the critical phase of transition can probably never be resolved with the desirable amount of detail and differentiation. Succinctness and prognostic possibilities of the model thus depend both on the size of the time-window in which subsistence change can be observed and also on the amount and quality of evidence for a change in the subsistence basis. The varying quality of available data necessarily informs the interpretative options. However, by acknowledging that this is largely beyond our control, a general heuristic model of subsistence transition can still be devised and applied. The relation between the duration of the availability and the transitional phases appears to be the crucial part of the model, since here the decisive conditions can be modelled that affect speed and success of subsistence change.

Analogous models describing the course of growth processes, for example in technical systems (Trömel and Loose 1995), find comparable forms of sequences, as innovation and progress in the development of technical solutions can be represented as a saturation curve in the course of time. The beginning of the growth process starts with an invention, whose improvement and further success are determined, however, by the slowest partial development of the entire process. It can be assumed that also the speed of the temporal operational sequence of subsistence change is limited by inhibiting factors, e.g. energetic efficiency, technological potential or psychological acceptance. Any one factor mentioned, in turn, depends on several other conditions. In view of the complexity of the factors that interlink in the process of subsistence change, it seems to be hardly possible to identify them in detail and characterise them as being responsible for the speed of the total process. This applies in particular to the reconstruction of subsistence change in prehistoric or historic times, which is particularly related to the nature and quality of the data. The increase in size of parabolic telescopes, as an example of the growth of technical systems, appears to be more straightforward to model

---

[6]   There are notable exceptions, for example in pastoral communities, which necessarily have to accomplish two different subsistence strategies to cope with seasonal resource constraints. This rather specific case will not be further pursued here.

than the complex structure of effects of climatic change, food resources, settlement structure and belief systems, which all have to be considered when modelling subsistence change.

What the model allows assessing are common grounds or differences that arise in the process of changes in food acquisition strategies, by viewing different phases of the change separately. As will be shown, the process and course of subsistence change is crucially determined by the length of the lead-time, during which biocultural conditions for the adoption of a new subsistence technology develop. The processual dynamics involved in successful subsistence change bear a strong resemblance to the concepts of cultural selection (p. 334 in Durham 1992) or biased cultural transmission (Heinrich 2001). It will therefore be attempted, aided by the theoretical assumptions on the lengths of availability and substitution phase, to present the application of a general qualitative transition model that is transferable to different forms of subsistence change. The model will thus be characterised by a descriptive element on the one hand, which allows depicting the event of subsistence change, and on the other hand it will be extended by the possibility to assess differences in the processual dynamics of subsistence change.

3.3 Conclusion: Causes and Courses of Subsistence Change

In their variability and multiplicity, the subsistence strategies of human communities are the product of adaptive mechanisms that help use and steer material and energy flows in a given habitat to serve individual and collective interests. On the condition that this makes possible and maintain long-term survival, subsistence strategies represent situations of equilibrium between humans and all other biotic and abiotic components of a given system. The dynamics of these strategies allow constant adjustment to slight fluctuations in resource supply. Serious alterations of ecological conditions, however, can cause and entail change to a subsistence system, unless the response to change is abandonment of the habitat. Possible causes for subsistence change include, e.g. climatic changes, loss of central resources, introduction of new sources of food, changes in settlement patterns or changes in social structures. The question is under which conditions, with what consequences and how successfully do humans choose subsistence change as an option for a change in their living conditions; and which patterns and sequences can be recognised. A number of different scenarios were considered.

During the Upper Pleistocene at the end of the last ice age, the change from a dry–cold to a warm–moist climate altered the natural conditions in large parts of Europe from an open, scarcely forested steppe with characteristic herds of large game to a more strongly closed, wooded landscape, in which the fauna was smaller in form and the flora provided a larger number of vegetable foodstuffs, such as nuts, berries and seeds from different trees and bushes. The different quality of subsistence conditions led to new hunting strategies adjusted to smaller prey and smaller, composite tools serving altogether smaller communities and less durable settlement structures. The archaeological evidence suggests that – probably due to the increasing forestation and a shortage in raw material – the trend towards a reduction of forms in the lithic assemblages clearly takes place before the climatic change. This gradual conversion of technology, together with an equally gradual change in settlement patterns and food acquisition strategies translates into pre-adaptive elements which finally prepare the subsistence change at the end of the late Pleistocene and in the Mesolithic. The shift from a pronounced hunting subsistence to a broad-spectrum economy takes place after a long lead-time.

The transition to agriculture reveals different scenarios. The Neolithic transition in the Fertile Crescent develops from the interplay of climatic changes and the development of optimised resource management. The hunter-gatherers of the Epi-Palaeolithic were already acquainted with wild cereals and this exploitation was intensified when, at the Holocene transition, a moister and warmer climate favoured the propagation of wild forms of cereals and legumes. The evidence of grinding-stones, mortars or granaries is testament to the increasing significance of these resources and clearly emerges before the occurrence of domesticated plants. The enhanced local availability of food leads to permanent settlements which, in turn, support increased cultivation of the more and more esteemed and locally stable resources. Domestication occurs subsequent to this development. These changes in subsistence strategies are accompanied by a gradual social differentiation within the population. Habitual agriculture is the result of an improved and controlled utilisation strategy for a food resource, whose significance and characteristics have been known for a long time. Sedentariness is the condition rather than the consequence of this, which in a comparable way also applies to the domestication of wild animals.

In southern Scandinavia, the richness and availability of natural resources allowed Mesolithic populations to live in permanent settlements, despite their foraging mode of subsistence. Crucially for this affluent lifestyle, among others, was the option of exploiting marine resources as bridging resources in the lean season of the late winter. The transition to a producing mode of subsistence was dependent on three factors: a gradual climatic change that led to a decline in and eventually loss of productivity from coastal resources, an increasing aspiration towards social differentiation within the communities and the long-term knowledge of fully fledged agricultural technology practiced for more than 1,000 years further south in the north German plain. The currently favoured view identifies an intrinsic factor, i.e. the social determinant, combined with changing ecological conditions as being of crucial importance for subsistence change. There does not seem to be autochthonous development of agriculture yet, when finally adopted, the transition is wholesale and proceeds at a dramatic pace. The pre-requisite for this, the technological aspect, has been known for a long time and makes possible the slow adaptation to and rapid adoption of an already established subsistence strategy.

The introduction of maize as a staple crop into the subsistence of indigenous populations in the northeast of North America resembles the preceding example in terms of its temporal operational sequence. Maize was known and cultivated, initially deriving importance largely from being used as a status or ceremonial plant. Yet it became only established as a main cultivar with a delay of 500–1,000 years, depending upon local conditions. Its large-scale adoption, though, went relatively quickly within two to three centuries. Only in the area of the Great Lakes was the process slower, because varieties had to be bred that could withstand cultivation in a colder climate. The wide-ranging use of maize was favoured by the accumulated knowledge and experience of agriculture and horticulture already used to domesticate endemic species for many generations. The integration of new resources into an existing subsistence system thus profited from a long lead-time. The development of more complex social structures and an increase in population density are parallel developments then, which are made possible by the successful introduction of maize as a staple crop.

+ beans + squash

For many Indian Nations, cultural upswing was connected with the cultivation of maize. Despite the new staple crop, however, it was necessary to supplement the diet by other vegetable- and animal-derived products for a balanced nutrition. Comparisons of pre- and post-colonial times suggest that the traditional diet could no longer be sustained, due to the increased forced involvement of Native Americans in trade and employer–client relationships, with consequences that finally led to physical destruction. The external constraints of political and economic structures permitted no adaptation of subsistence to the changed new conditions within a short time. An appropriate advance period, which could have been used as preparation time, was not available.

The collapse of long-distance trade relations between the Indus Valley and Mesopotamia led to the decline and fall of the Umm an-Nar culture at the Gulf coast of the Oman peninsula towards the end of the fifth millennium BP. Large fortified agricultural settlements were abandoned and, during the following period of relative economic insignificance, a new subsistence mode developed based on semi-nomadic animal husbandry and agricultural farming. Most likely the camel was both the condition and basis for this, since it was known to be domesticated and widespread during the fifth millennium. The fourth millennium saw an increased exploitation of marine resources and eventually the return to a permanently sedentary way of life, as indicated by the archaeological record. The familiarity with resources, which later became the basis of the new subsistence, moderated the subsistence change induced by economic dependence and political developments.

The ecological setting of the lower Habur area in northeast Syria is characterized by climatic conditions that necessitate irrigated cultivation, today as well as during the last 4,000 years. Not only has this subsistence technology been continuously known since the Assyrian period, it has also been continuously used. Nevertheless, subtle changes in subsistence through time arise from changing patterns in the relative amounts of foodstuffs extracted from the cultivated habitat. However, the generally prevailing pattern of a mixed economy based on agricultural and livestock farming clearly points to a continuation of well proved and existing modes of land use with enough space, however, for dietary adjustments through time.

These examples provide different starting-points from which subsistence-change set off and was mediated and pre-controlled in various ways. The first general observation is that subsistence-change proceeds more successfully with longer and better preparation of the conversion of the subsistence technology, the resources or the social basic conditions. It has to be examined now: (1) to which extent the presented model of subsistence transition can be used in the cases presented here, (2) to which extent empirical data correspond to pre-assumptions about the course of change and, consequently, (3) to which extent the model is of predictive value.

Within an ecosystemic context, subsistence change will occur if one or more of the several components connected with nutritional supply (=subsistence strategy) are so strongly altered that this results in a serious disturbance of an existing equilibrium. The resilience capacity of a system is exceeded and it flips into another stability domain. Alterations can be either extrinsic, in which case human populations have no influence on causative events, or intrinsic, i.e. intended internal changes due to socio-cultural developments. In both cases, this will affect the spectrum and quality of resource use and require a re-definition of material and energy flows.

It seems to be not only a matter of bare change of established modes of subsistence, whose adaptive value results from a certain temporal constancy and energetic continuity and is known to the human population. It probably equally concerns a change in the socio-cultural basic conditions. Exactly because subsistence is merged into a net of ecological and social connections (see Ellen 1994), change results in and is flanked by an adjustment of social structures to the new conditions. Thus, subsistence change necessarily has to contain a component of societal support and acceptance, because it affects central areas of social organisation, e.g. division of labour or ownership structure.

The balance between these two realms, the ecological and social resilience, is decisive for the relationship between the respective amounts of subsistence change as a process or an event type of change. Moreover, it appears to be a crucial determinant for the success of change. By applying a general heuristic transition model (cp. Sect. 3.2) the process can be divided into individual phases. The duration and quality of the phases on one hand permit a reconstruction of historic situations of subsistence change and, on the other hand, appear to be suitable to project the basic conditions of subsistence change. This reconstructive approach will be described taking the adoption of agricultural production as an example.

The length of the consolidation phase is of subordinate importance for the situation of change. It basically indicates whether and how a new equilibrium is stabilised and can be maintained; and it is thus suitable to evaluate subsistence change once it has taken place. The ideal basic conditions of subsistence change are a lead-time of the longest possible duration and a short substitution phase (Schutkowski 2002a). This is immediately apparent from an energetic point of view, as it is connected with varying efficiencies of subsistence techniques, for example with an increase in labour costs in the case of an adoption of new techniques of food production. Whilst the efficiency of food procurement in a foraging subsistence system can be measured as yield per unit time (cp. Sect. 2.2.3), the basis of assessment in agricultural production is yield per unit area. For the maintenance of food production in an agricultural system, however, a constant input of human labour is necessary, which even at levels of increased energetic yields lowers the total efficiency (see Sect. 2.3). The reliability of produced food resources is traded in for increased (labour) input. Thus the advantage of relatively small opportunity costs in an acquiring mode of food acquisition is offset by an additional input of energy and work force, which is no longer at a population's disposal when agriculture and animal husbandry are adopted and practiced on a larger scale. If it applies that the energetic output of a foraging subsistence strategy produces a surplus that can be related to the additional costs that emerge from the increased labour invested in the new technology of agricultural subsistence, then only on the condition that the amount of surplus equals the costs incurred, or any other reciprocal and complementary connection of these two variables, can a steady and flowing transition towards the new mode of subsistence be possible. Historic examples, such as the transition to the Neolithic in the Baltic region or central Scandinavia (Zvelebil 1996) could be explained in this way, but only on the condition that, to the extent in which the new subsistence strategy is increasingly controlled, the energetic surplus from a foraging way of life can be allowed to become smaller and still the system in transition remains relatively stable.

Empirical data, however, reveal that there is a lot to be said against such a transition during the introduction of agriculture, as cross-cultural analyses of societies practicing agricultural farming demonstrate (Hunn and Williams 1982; based on data from the Human Relations Area Files). It can be shown that, wherever agriculture and animal farming are practised as modes of subsistence, they predominantly contribute either only a small portion (approx. 5–10 %) or a major amount (approx. 50 % and more) of the subsistence basis, i.e. there is a bimodal distribution that separates marginal from substantial agricultural production. Intermediate amounts of farming are rare. This is in

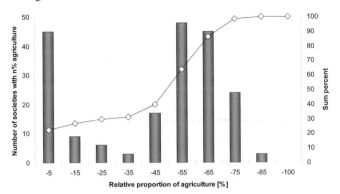

Fig. 3.6. Number of societies practicing agriculture at different rates, expressed as both relative proportions and a cumulative graph. The bimodal distribution indicates that from an energetic point of view either marginal or substantial agricultural farming is sustainable, while intermediate amounts are hardly energetically viable. Data from Hunn and Williams (1982), based on 200 societies from the Human Relations Area Files

clear contrast to all other modes of subsistence, e.g. hunting, collecting, fishing, etc., whose respective contributions to the subsistence may be distributed evenly or simply without any specific pattern (see also Rowley-Conwy 2004). Combining relative frequencies with a curve denoting cumulative percent figures (Fig. 3.6) produces an analogue of the sigmoid curve in the transition model (see Figs. 3.4, 3.5). It illustrates that the range of intermediate amounts of agriculture, which in the model would correspond to the critical transition in the course of the substitution phase, can hardly be detected empirically. In contrast, the contribution of farming to the subsistence in either lower or higher amounts is supported by robust empirical evidence. In the process of substitution, populations go through a phase of dual economy, where on one hand there are additional costs incurred in terms of time, work-force and energy, which result from the new subsistence activities whereas, on the other hand, time, work-force and energy for the traditional subsistence activities are only available to a certain extent, reduced by exactly that amount of additional costs[7]. As far as there are no free valences from the old economy that can be invested into the development of the new mode of subsistence, it can be assumed that for the duration of the transition there will be more unfavourable cost/benefit ratios or energetic yields.

Since agriculture is tying humans and work-force to a certain place, the broad-spectrum economy of a foraging way of life is usually replaced by con-

---

[7]   Such competitive situations for time and energy budgets are known and have been modelled successfully (see Sect. 2.2.1.3, Optimal Foraging Theory). Modelling of subsistence transitions is not available to date.

centrating on a few cultivatable or domesticated species, on which long-term survival is dependent. The previously existing diversity of food options is diminishing, which can consequently lead to an increased susceptibility to fluctuations of the system, because a smaller number of food options are available.

There are important consequences resulting for the organisation of the transition from an acquiring to a producing economy: Only if the yield from agricultural production is large enough to offset fluctuations in the resource supply of the habitat can the new subsistence strategy begin to stabilise itself. This, however, is apparently only possible, if ca. 50 % or more of the subsistence basis are produced from agriculture. Consequently, there is an energetic necessity to operate the substitution quickly, which requires high input and planning aimed at sustainability, in order to stabilise the disturbed equilibrium. The best pre-requisite for this is to utilise a preparatory phase for as long as possible. This was the case in the transition from the Mesolithic to the Neolithic in Southern Scandinavia, where there was long-term resistance to the new subsistence strategy, but when it was adopted, it was rapid and affected settlement patterns, social organisation and dietary patterns. This seems to be part of a general pattern that is characteristic of this transition along the Atlantic coast of Europe (Arias 1999), which includes the tendency of sedentary Mesolithic communities to maintain fundamental aspects of their organisation.

Theoretically, slow substitution may be an alternative, which however only appears achievable if the traditional subsistence is not only still available but also usable for a correspondingly sufficient time, so that constant supplement from the new mode of subsistence is possible. Even though an extended lead-time is not crucial here, it is nevertheless advantageous in the sense of a long test-phase. This alternative scenario seems to have been taking place during the 500 years of Norse settlement in Greenland. Continuous climatic deterioration forced people to increase the inclusion of marine resource into their diet, yet it happened at a steady pace (Arneborg et al. 1999), from some 20 % to around 80 % eventually. The important difference seems to be that, even though the change in diet was considerable, there was no concomitant socio-political change, i.e. whilst the subsistence pattern changed, the overall societal framework, the subsistence strategy in a wider sense, did not (cp. McGovern et al. 1996). This allowed a gradual substitution of food procurement modes and did not necessitate a rapid shift.

In comparison, the Neolithic transition in the Levant rather followed the expectations of relatively fast substitution. An extended advance phase facilitated the emergence of cultural pre-adaptations in material culture, settlement structure and social differentiation. The natural habitat factors changed as a result of climatic and ecotope changes and both made possible and induced the technical management and planning for cereal grains as new resources. Via the stage of logistic mobility, permanently inhabited settle-

ments and finally changed social structures were developing through affluence. Subsistence change is coupled here with a temporally preceding change
in lifestyle. The substitution phase is short in comparison with the preceding
long-term development.

The situation is quite similar for the adoption of maize as new staple crop
in North America. The advance period and the development of domestication
and agricultural practices facilitate a fast substitution, compared with the
duration of the availability phase. And even the conversion of subsistence to a
pattern of seasonal mobility and pastoral nomadism at the beginning of the
fourth millennium BP at the Arabian Gulf coast is favoured by the general
familiarity with the camel as a new resource.

It is the length of the lead-time that becomes the limiting factor of subsistence change to take place as smoothly as possible; and in the long run it is the
decisive factor as to whether changes to the social and political structure that
accompany subsistence change can be absorbed without recognisable distortions. In most of the examples presented here, the speed of substitution is relatively fast in comparison to the advance phase, at least if one is willing to
accept that the time-span of a few generations is brief.

Inevitably, at this point the question of possible causes of subsistence
change arises, as introduced at the beginning of this chapter under the general rubrics of climate change, interruption of a subsistence system, optimisation strategies, spatial expansion and a change in settlement patterns, as well
as socio-political changes. Whilst human interference with the environment,
often with lasting and degrading effects, may contribute to the causes of
change, substantial alterations of mode or basis of subsistence are frequently
only facilitated by large-scale climatic changes uninfluenced and uncontrol-
lable by humans. Facilitate, however, does not necessarily mean realise. The
transition to agriculture in southern Africa prior to the Bantu invasion failed
to materialise not because there were no appropriate plants to domesticate,
but because obviously nobody ever cultivated and domesticated them (Blumler 1996).

What then induces humans to give up an established and thus implicitly
adaptive subsistence strategy and replace it, if the advantage or the benefit
that can be expected does not at least counterbalance the new and additional
costs? In the case of the south Scandinavian Mesolithic, this has been
explained by an increase in social prestige resulting from the assimilation of a
new subsistence technology (Price 1996), even though there is ongoing debate
over this issue (Rowley-Conwy 2004). Similarly, the adoption of maize as a
new cultigen into an existing system of domesticated endemic plant use in
North America seems to be connected with a function as a status or ceremonial plant (Smith 1989a, 1995a). It may thus be assumed that the decision with
regard to the time and process of change into the new mode of subsistence
was governed by those personal goals or collective interests which would
profit from a change in the social organisation by a change in the subsistence

⌊ system⌉ The development of complex social structures parallel to and in connection with subsistence change indicates a top-down manipulative initial control also for the length of the advance phase.

In the case of the Levant, the re-organization of resource management is accompanied by a likewise fundamental change in the social structure, which can eventually be maintained by the use of cultivated plants (Harris 1996b). The development of complex ownership structures and an integration into social networks of barter and trade relations, however, seems to be the consequence rather than the cause of a change of strategies that facilitate managing the use and availability of new resources. The fact that agriculture as a strategy of food acquisition is only stable either on a very small scale or above a substantial level of contribution to the subsistence, however, principally shows to which extent subsistence change requires to be culturally supported. An affluent food situation only favours the adoption of a new strategy.

If there is a motivation for subsistence change, it should preferably be derived from developments in those areas in which change took place due to internal differentiation. Once agriculture, and subsequently animal husbandry, is introduced, it may be expected that the new system will usually stabilise itself with every increase of its contribution to the subsistence. Subsistence change would then be worthwhile, if an energetically more productive or at least continuous alternative to the previous subsistence mode is offered, i.e. if increased resource security or improved resource allocation is facilitated. But, as became clear from the cases of (largely) intrinsic causes for subsistence change (southern Scandinavia, North America), other than just energetic needs can also be covered at the same time, which could not be or at least not better satisfied by the old strategy. In the sphere of social organisation, this is related to issues of ownership structure or status differentiation (see Chap. 4 for connections between social affiliation and resource allocation). Subsistence change then, it seems, apart from the establishment of a new food acquisition strategy, at the same time represents a vehicle for the realisation of cultural goals, as resource use becomes instrumental for the increase in and consolidation of social differentiation and resource allocation.

A fundamental alternative to the emergence of agriculture from prevailing social conditions is the propagation of a new subsistence strategy in terms of colonisation and demic diffusion, i.e. in terms of cultural superimposition by immigrating populations and their biological blending with the indigenous population⌉ This mode of propagation was introduced as 'Wave of advance model' (Ammerman and Cavalli-Sforza 1984) and denotes a displacement strategy with a wavelike expansion of the new subsistence mode. It is known that distinct vegetational signatures and crop packages mark the migration routes of farming practises (Colledge et al. 2004) and that the spread of agriculture into Central Europe followed geomorphological and plant–sociological characteristics of the landscape and first took place along rivers and in the fertile loess plains. Nevertheless, this was more than just an export of ideas,

but rather the knowledge and control of a new subsistence mode and the associated social and political structures that are dependent on humans to be
passed on. This pathway characterised the propagation of agriculture from
the Near East to Europe – and to a yet unknown extent along the European
Atlantic coast. In the direct competition of subsistence systems, this led to a
very successful displacement and abolition of the foraging way of life as a
mode of subsistence in the process of the expansion of the Neolithic way of
life. There is evidence of contact between the last hunter/gatherers and the
first farmers, for barter or exchange of commodities (e.g. Grote 1993), but this
co-existence was only temporary and could not moderate the dominance of
the new subsistence strategy in the long run.

In one way, the spread of the Neolithic in Europe was a diffusion of cultural
characteristics and cultural goals. It was, however, likewise a demic diffusion,
i.e. the physical and genetic propagation of those humans who were the carriers of these cultural traits and who pursued their cultural goals. Their genetic
traces are still present in contemporary populations and contributions to a
change of the respective indigenous gene pools can be demonstrated through
polymorphic serum proteins, mitochondrial or Y-haplotype lineages (Sokal et
al. 1991; Cavalli-Sforza et al. 1994; Richards et al. 1996, 2000; Rosser et al. 2000;
Torroni et al. 2001). Even though only amounting to quarter or less of today's
gene pool (Richards 2003), they represent the genetic echo of those groups of
people who in historic times continuously pushed the agricultural frontier
and thus allow a reconstruction of this propagation. In the course of only
approximately 5,000 years, this development was completed and the onset of a
peasant lifestyle and subsistence was common throughout Europe. Only the
gate to Scandinavia remained locked for a further 1,000 years.

In these cases of demic diffusion, probably after a relatively short transition period, a rapid acculturation of the local forager groups can be expected.
More so, the process can probably be best imagined as a superimposition on
the indigenous population by an élite dominance, where as a consequence not
only was a mature technology implemented and the pertinent domesticated
plants and animals cultivated, but also new and altered social and political
structures were introduced.

Thus there is at least a temporary co-existence of two subsistence modes,
which are performed by different populations in the same habitat. Translated
into the transition model, this is equivalent to a complete export of a subsistence mode on the level of the consolidation phase, which may have been the
case with the arrival of the Neolithic in Britain (Richards et al. 2003b). Actually, the advance phase does not apply here, since the experience needed for
the conversion and application of the new strategy is already available
through the colonisation and has been tested on its long way through Europe.
Besides, the propagation mode of the agricultural technology along certain
soil and vegetation characteristics indicates a strategy where established techniques and knowledge are directly converted. Only at a later stage do the

exploration and development of other, maybe less suitable, environmental settings become necessary. The substitution phase can and should necessarily be short, since only in this way does the new subsistence technology provide the necessary energetic yield to the colonists. The argument that the known and established is being used applies in a similar way here. The subsistence change that takes place during the spread of the Neolithic in Europe is extrinsic, due to the dominance and successful competition of the new subsistence mode.

The biocultural goals, which are pursued along with a change of the subsistence basis, are identical irrespective of intrinsic or extrinsic change. In both cases, the proximate goal is an improved or newly arranged co-ordination of energy flows in the habitat, a new stability domain in order to achieve at least a balanced energetic situation as quickly as possible. Populations whose existing mode of production provides the benefit of an affluent nutritional situation can contribute a high amount of risk reduction during the substitution and consolidation phase. Also, in the case of subsistence changes by gradual colonisation, the indigenous population at least temporarily offers the possibility of improved safeguarding of resources by compensation during the consolidation phase. Resource security and risk minimisation are thus the prerequisites for the – biologically trivial – ultimate goal: the maintenance of the population as the unit that allows the genetic interests and reproductive goals of its individuals to be pursued. Indeed, there are indications that reproductive interests can even support and advance the accomplishment of subsistence change (see Sect. 5.2.4). The ecological range of the variation of this goal is given by the adjustment of established strategies, yet shaped by natural habitat conditions and regional peculiarities. As, however, both the increase in personal genetic fitness and the amplification of social prestige do play and did play a decisive role as motivation for subsistence change, the level at which individual decision becomes crucial for a change of subsistence, which is eventually depicted as an observable phenomenon on the population level (see Chap. 5).

If in principle we accept a generic basic pattern of the transition process, regional solutions are essential for an ecosystemic regularisation of subsistence change for the respective populations. Since subsistence modes are connected by ecological and social factors, genetic and socio-genetic interests cannot be separated: they are mutually dependent. The willingness to sustain the permanence of social agreements and cultural institutions causes the necessity for human populations to adjust subsistence strategies, including their alterations through change. In the end, subsistence change is socially supported. If however along this way, the proximate goal of an improved or re-distributed co-ordination of material and energy flows can no longer be achieved, because the energetic balance is disturbed or the equilibrium destabilised, the ultimate goal becomes unattainable and survival is endangered (see Sect. 3.1.4).

Forced subsistence change, which lacks the possibility of adjusting to the new food acquisition strategy, contains a high hazard potential. Not so much by the new production mode itself, but by the associated changes of the social structure and social safeguarding of the subsistence. [The alteration of basic subsistence and living conditions for Native Americans after their country was colonised by European settlers provides a telling example. The conquest of populations and the systematic withdrawal of their base of life go hand in hand. The time-frame available for subsistence change was quite obviously too short for a successful adaptation to both the ecological and the social changes.] The different rules of world-wide economic expansion were not compatible with a traditional mode of subsistence; and they were too strong and too immediate for most to be offered a genuine chance for developing alternatives.

Today, at the beginning of the twenty-first century, we are confronted with the consequences of a globalisation of markets and the forced abolition of traditional strategies of food acquisition in a similar way. Since this is a development that can hardly be influenced or stopped any more, it likely means a concomitant irretrievable loss of cultural variability, which is exactly the trait that is so characteristic of human communities. By reconstructing historical operational sequences of subsistence change, however, a Human Ecology is able to outline and suggest models that allow extant operational sequences to be described in a similar way. They may then even offer results and possibilities of comparison which are transferable into today's time. Such an archive of pathways and strategic solutions, which served and facilitated human populations to master altered natural and social conditions and associated subsistence change, could thus help develop awareness for a careful examination of biocultural basic conditions of both existing and demonstrably sustainable adaptive subsistence strategies before they are thoughtlessly replaced through political or economical pressure, by publicising allegedly better or more efficient, yet bioculturally displaced strategies.

4 Resources and Social Organisation

As pointed out in the preceding chapter, the adjustment of subsistence to changing basic conditions is, in many instances, connected with a change of the socio-political conditions within a community. In individual cases it may be difficult to make a clear distinction between cause and effect and to either identify subsistence change as being responsible for a change of the social structure or social change as triggering an adaptation of subsistence modes. The circumstances which led to the introduction of agriculture saw both developments so closely linked that they must have mutually caused each other. Conditions in the late Mesolithic of southern Scandinavia even seem to indicate (see Price 1996) that the adoption of agricultural technology, land ownership and resulting new forms of resource availability accommodated the development of social differences within the populations. To the extent that social differentiation is enhanced within a society, the social and political control of resources also increase (see below, Sect. 4.1). In other words, there is a growing influence on control over material and energy flows; and the possibilities of exerting influence are becoming increasingly unequally distributed.

Steering mechanisms, which were in place to organise the procurement, processing and distribution of food under the conditions of the old social structure, are obviously no longer suitable for new forms of food acquisition. Flexible and open social relations, such as in forager societies, which allow an even re-distribution of resources to all individuals through inter-individual networks (see Chap. 2), are not transferable to social structures characterised by individual or corporate possession of resources. Here the availability and utilisation options of resources are bound to differences in social status or affiliation to social groups (see Johnson and Earle 2000). Two examples will outline the range of effects connected with the social control of access to resources. The first, more general, example appreciates the regulation of access to land and the concomitant possibilities of producing food. The second example then refers directly to the effects and consequences of social differences that are reflected in the quantity and quality of foodstuffs available.

The first case is about the social mediation of resource scarcity and about the ways, by means of either social flexibility or rigidity, humans are allowed or refused the use of resources and refers to studies carried out among the societies of the Chimbu and Mae Enga of the central New Guinean highland (Brookfield and Brown 1963, Meggitt 1965; cited in Ellen 1982; see also Wohldt 2004). Both groups practice slash-and-burn cultivation, the widespread and common mode of agricultural farming in tropical areas. The limiting factor of this horticultural mode of production is suitable land, since the top soils are poor in nutrients and, despite being fertilised by the wood ashes, are already exhausted after two or three harvests and thus need a fallow period of up to 20 years for complete regeneration. Cultivated areas must therefore be harnessed through extensive consumption of land. Land, however, is not a commodity of unlimited supply and access must be regulated in the context of social relations. The Chimbu respond to increasing pressure on the resource of land by loosening the criteria for affiliation to kin groups and thus widen access to such a group. Since these descent groups also hold the social rights to exert control over resources that belong to their members, people who lack land are thereby allowed to affiliate with or join hamlets whose territory still contains under-used land. Among the Mae Enga, who live under comparable natural and social conditions, the same external pressure leads to a tightening of inclusion criteria and thus an aggravation of granting access to land within a kin group. Both societies found adaptive solutions for the same problem, the incipient shortage of land as a resource, but in opposite ways and with different consequences. While on the one hand social flexibility leads to an increased dispersion and a more evenly distributed use of the existing land for many, a stronger rigidity in the interpretation of social regulations on the other hand entails the restriction of land use options by a concentration of humans and resources and actually excludes certain parts of the population from resource participation.

The second example aims to identify what the immediate consequences of affiliation to higher or lower social strata are on the possibilities as to which extent certain foodstuffs can be acquired at all or are at people's disposal. In his study of income conditions and the cost of living in late medieval German towns, based on hospital rules and account lists, Dirlmeier (1978; see pp. 76ff, 424ff) reports the amounts in which the different main food components were distributed. Whilst in the fifteenth and early sixteenth centuries, 43 % of all expenditures on nutrition were spent on animal food products in families of wealthy citizens, it was some 30 % for a simple craftsman's household and 14 % only for an orphanage. The proportionate costs of subsistence expenditures for cereal grain and bread were lower or higher, accordingly. For the lower social classes[1], this picture can still be further differentiated: people

---

[1]   Class is not used in the sense it has in political economy but rather as a generic term to denote major groups of varying affluence within a society.

who had to dwell under pronounced deprived conditions spent 50 % on bread or grain and 14 % on animal-derived products, while in lower middle class conditions only 26 % were spent on bread or grain and 23 % on meat, eggs, milk and milk products. Also, food apportionments recorded in the benefice lists of the Holy Spirit Spital in the town of Constance are informative: if only the two extreme forms of benefices, those for priests and the infirm are regarded, the ration for the clergy designated meat allocations of 1.5 kg/day on five days of the week and various supplementary foodstuffs, while for those at the lower end of the social scale only three days with 376 g/day of meat and less diversified supplements are specified. Status and social meaning are thus not only reflected in the possibilities of food procurement, but even in the apportionment of food.

The example from New Guinea represents the regulation of the land/people ratio. As differential resource allocation is exercised through the different interpretation of social responsibility, at the same time decisions are made as to whether certain people have access to resources at all or whether use options are being expanded and attainable for more people. Thus the areas of ecological and social reality are intertwined in the social regulation of resource use and availability in a particular way. In quite another way, the example of late mediaeval upper-German towns attempts to clarify how much dietary conditions, i.e. nothing short of the ability to partake in resource use, is directly determined by the affiliation to a certain social stratum or group. Moreover, it is very likely that the meaning behind different cost ratios is not just about quantities but rather also about differences in quality, at least if the amount of animal protein is considered, which is generally regarded as a useful measure of food quality (WHO/FAO 1973).

Recognisable differences in nutritional behaviour as a characteristic of social groups are common from everyday life experience. Sociological studies relate such habits, those of the upper classes in particular, to a refinement of lifestyle and the development of table manners (p. 277ff in Bourdieu 1997; p. 248ff in Elias 1997), to the display of dietary habits as a sign of wealth (Menell 1986) or to the awareness of healthy and nourishing food (Menell et al. 1992; see also de Garine 1994; Fenton 1997). Just being able to afford different foodstuffs, however, does not necessarily lead to the consumption of food of different quality. In the end, lobsters and oysters do not differ from haddock as a food item per se, whilst certainly they do in price. The social control of access to food resources only becomes evident through the subtle ways of refined distinction.

The aim of this chapter, however, is not to further pursue such culinary refinement. Rather connections between the social organisation and the social differentiation of a population as well as the regulation of use and access to (food) resources will be dealt with. First, a concise overview of different forms of social organisation of human communities is given with an emphasis on the aspect of the respective social or socio-political regulation of resource use (see

Sect. 4.1). At the centre of this chapter will be evidence of differential resource use within past societies. Examples will be derived from bone chemical analyses of skeletal populations of varying date and provenance. It is attempted to show (Sect. 4.3.2) that, with the help of such studies in archaeological sciences, an understanding of resource-related socio-typical subdivision in past populations can be facilitated, which in this differentiated form is not available from other historical sources. Dietary reconstruction is one of the long and well established fields within scientific archaeology and has contributed substantially to our understanding of past subsistence strategies and nutritional patterns. It is the area that is arguably the best and most suitable hinge between an actualistic ecological approach and its projection into the immediate and more distant past, even though palaodietary studies have rarely focussed on the ecology of past populations explicitly but rather pursued a more generic approach of dietary reconstruction. This chapter thus also attempts to reconcile the two aspects and demonstrate that and how basic ecological correlations project and translate into our understanding and exploration of past lifestyle and living conditions. A summary of the present conceptions of the early mediaeval social structure will clarify this as an example (see Sect. 4.1). To facilitate the understanding of analytical procedures, a very brief introduction to foundations and pre-conditions for the reconstruction of palaeodiet from the skeleton will also be provided (see Sect. 4.2).

4.1 The Social Framework of Resource Utilisation

The system of a social organisation of modes of production outlined below largely follows concepts developed by Johnson and Earle (2000; for related ideas, see Service 1962; Fried 1967; Earle 1994) and is meant to demonstrate the increasing control of resources in the development of complex sociopolitical structures. The general principles emphasised here will then be compared with societal features of the European early Middle Ages to examine the extent to which the general model is in agreement with a special case, from which a conclusion will later be drawn about the correlations of social structure and resource access from links between nutritional behaviour and social affiliation.

Some aspects of the characteristics specified in the following sections were already addressed in connection with modes of production and food acquisition (see Chap. 2.1). They reappear here in a different connotation though, as the emphasis is on the evolution of social structures and its effects on use and control of resources rather than on the social basic conditions of food procurement.

The starting point of the considerations is the common notion (e.g. Earle 1994) that there is a developmental succession in the social organisation of human communities from small groups to large, urban societies. Johnson and Earle (2000) differentiate three levels of social organisation within this succession: family groups, local groups and regional polities. On each of these levels, further subdivisions can be made, although a referral to general characteristics for each of them will be sufficient here (see Table 4.1).

1. *Families* or *family groups* represent basic and primary economic units which, as individual units, are to a large extent self-sufficient. Depending on whether such family units are mainly foragers or additionally use domesticated plants and animals, settlement forms can be described as temporary camps or more permanent hamlets. Accordingly, the settlement pattern is adjusted to, for example, seasonal differences in resource density and is determined by a flexible change of aggregation and dispersion of family groups in the habitat. The kinship structure is based on cognate descent, i.e. on the bilateral genealogical derivation from the paternal and maternal line. Since this does not entail kin segmentation within the group and genealogically justified claims on resources, this type of descent allows nuclear families (possibly extended by relatives) an opportunistic, resource-oriented fusion and fission of camps and hamlets. The size of the subsistence group can thus be flexibly adapted to the respective abundance of resources in the habitat. The acquisition of food is divided according to age and sex, yet without a strictly formalised division of labour. If necessary food, especially hunting prey, is distributed between families on the basis of reciprocity. This allows social reconciliation between families to be arranged, which through mutual reassurance helps compensate differential success in food procurement. Also, in times of hardship, extended interfamily social networks serve to minimise risk without necessarily formalising relations. Resources or resource-bearing areas are not normally defended, but use is regulated through flexible customary law in the home ranges. Positions of leadership and decision making are assumed *ad hoc* due to authority resulting from ability and experience and remain without formal or durable change of status of the persons concerned. Decisions as to change of locality or other procedures relevant to subsistence are made by consent.

The sum of characteristics reveals that all areas of social life are designed to fit risk avoidance and risk distribution. The flexibility of strategies used to achieve social organisation allows safeguarding existence in a habitat for both the individual family and the family groups.

2. The social economy of *local groups* is still based on the family as the central unit of subsistence procurement, with the distinction however that risk, technology, trade and conflicts are now socially managed at a higher level. By inhabiting areas where resources are more concentrated or more productive and by increased use of domesticated plants and animals, higher population densities can be sustained. Settlement size varies from about 100–200 persons in acephalous local groups and 300–500 persons in so-called Big-Man societies. Ownership of technological investments relating to resource use is per-

Table 4.1. Socio-cultural characteristics of family groups, local groups and regional polities (modified after Johnson and Earle 2000)

Trait	Family group	Local group	Regional polity
Population density	ca. 0.5 per km²	>0.5 per km²	>5.0 per km²
Resources and subsistence	Wild species; resources patchy, not always predictable; individual or group provision	Mostly domesticated species and stored seasonal food supplies; resources can be transformed; private ownership and collective use	Domesticated species; resource control and concentration; high technological input for food production and trade
Settlement pattern and land use	Camps, hamlets: customary rights of land use	Hamlets, villages: defended territories, corporate rights of possession	Towns, cities: institutionalised land ownership, state-guaranteed rights of possession
Socio-political organisation	Familial and bilateral networks; situative ceremonies; ad hoc leadership	Corporate groups; within- and between-group ceremonies; Big Man or equivalent	Social stratification and regional institutions; ceremonies for political legitimisation; institutionalised leadership; heritable élites

sonal, although the product may be used by many, e.g. fish weirs, whale boats or corrals. The lasting amalgamation of several or many families into a bigger economic unit (local group) is connected with a division of the community into uni-lineal descent groups in terms of lineages (small numbers, known genealogy) or clans (larger, classificatory union of individuals, ancestral line usually not traceable). These descent groups hold corporate functions, entailing for example possibilities to control resources and the legitimacy of their entitlement. Thus, in principle, there is competition over resources within the community. Supra-group entities consist of many local groups and are divided into multiple lineage or clan segments. Stability of such a supra-structure is achieved by ceremonial and exchange relations on the level of the family or descent groups. Local groups constitute themselves as territorial, resource-related units. Rituals publicly affirm their political integration and the meaning of their corporate segments. In acephalous local groups, leadership still remains context-related, e.g. in the case of Inuit whaling captains. In so-called Big-Man societies, an individual represents the local group, who appears as mediator, co-ordinator and negotiator in affairs affecting the group. The Big-Man receives authentication of his office by the community exclusively from the acknowledgment of his personal qualities, abilities and self-initiative; and there is no warranty on the durability attached to the position. The extent of economic and social integration is much higher on the level of local groups. A family can no longer step out of line from the competitive network of social safety devices without the risk of damage. Part of the familial autonomy is surrendered in favour of more security granted by the local groups. Resource surplus is formally distributed through political channels, which allows for continuous mutual official reinsurance. At the same time such "social memory" works towards dispersing the risk of local failures by the benefits of a large, supra-regional network.

 The level of *regional polities* is characterised by an increased or even complete political integration of local groups. Examples are simple and complex chiefdoms as well as early and agrarian state communities. Impersonal bureaucracies and markets gradually replace the significance of kinship and personal loyalty. Subsistence is still based on the exploitation of rich resources, although an increasingly high technological input is aimed at an ever stronger independence from the original natural conditions, for example by irrigation, long-distance trade or market orientation. The necessity to feed large populations leads to a shortening of food chains through the cultivation of grain and/or tuber crops and thus an associated reduction in the energetic costs of nutrition. The development of central structures is strengthened, both in terms of settlement patterns and social organisation. Although local groups remain important in organising subsistence on the local level, regional bodies increasingly gain influence. Within the society there is further social differentiation and stratification, accompanied by the formation of élites. Territoriality and thus control over resources is increasingly expressed in terms of the possession titles of élites and institutions as well as private property with nationally guaranteed and formalised rights of use. The reality of social differentiation and the existence of supra-personal institutions gains public authentication through ceremonies and thus works towards stabilising the system. An intensification of subsistence economy is the necessary condition for the development of this kind of social complexity, yet it is not sufficient: it is just as necessary that measures of intensification can be controlled. Accordingly, social stratification of the society also implies differential control of resources. As a consequence, then, there is the long-term possibility for a concentration of political power on the level of institutions as a result of controlling the central management of resources.

It is evident that an increasing development of social complexity is regularly accompanied by likewise increasing control of resources. Modelling the social organisation of modes of production thus connects the functionality of social institutions of human communities with the functionality of the habitat-typical strategies of food acquisition or the use of resources suggested by the conditions of natural units. This implies that certain subsistence strategies are bound to corresponding patterns of social structure. It seems reasonable to assume that there is some sort of human ecological principle that can be generalised. Indeed, extensive cross-cultural studies seem to principally support this. Spiro (1965) found that the development of political economy in human societies is represented along an almost universal chain of causality, with societies that produce food by agriculture and/or livestock farming displaying social differentiation, which in turn is a condition for the emergence of institutionalised political units; a connection that has been confirmed numerous times since.

Nevertheless, it must not be overlooked that even a "developmental model" of the social organisation of resource use is only a heuristic approximation in which the material world, like in other models, is summarised in categories. Inevitably this is connected with a decrease in the significance of traits that may be characteristic of an individual case, a certain region or time. The question arises as to what extent a model developed on the basis of most complex and multi-layered data of extant or sub-recent populations is transferable to conditions prevailing in historic populations; and thus whether the underlying ecological principle, according to which the function of social institutions is connected with the function of typical strategies of food acquisition, also applies universally and to a particular historic time-period. Can the social structure and economy of past societies, which for obvious reasons can only be deduced from archival, historic and archaeometric sources, be brought in line with this model? In other words: is there a reflex of the social on the natural basic conditions? This question will be explored in more detail, taking the European early Middle Ages as an example in the first place. Other findings from a variety of different socio-cultural and temporal settings will be introduced, however, to underpin and broaden the argument.

4.1.1 Social Organisation of Subsistence in the European Early Middle Ages

What is known? Current historical and archaeological evidence about the social organisation of the early medieval society suggests the following (cp. Steuer 1982, 1984, 1987; Boockmann 1988): the structure can probably be best described as an open, ranked society. Families constitute the basic elements, which in late Roman tradition are to be understood as *familiae* in the sense of household communities. Despite having equal rights, rank differences do exist, both between families and within large groups of families. In contrast to the tightly bound social arrangement of late antiquity and the strict stratification of society during the later Middle Ages and the following epochs, however, no such corporate structure exists in the early Middle Ages (c. 5th–8th century AD). It is true, though, that at the top of the society there is a ruling family (for example, the dynasty of the Merovingians) with unrestricted highest rank position and succession. All "subordinate" families, be they officials or free peasants, however, form a multi-layered make-up of different societal or social ranks that can be denoted as vertical structural differences. There are only faint hints as to a parallel structuring within the society, e.g. in terms of vocational groups.

There is no aristocracy by birth that would denote the affiliation to a legally and strictly defined corporate group, yet there are a few noble families of different, but variable rank, because the social position of individual families or family members, be they noble or not, can change quickly and is in constant re-arrangement due to personal and non-hereditary privileges awarded by the ruling family. These are an expression of rank and reputation resulting from success in times of war and from economic possessions. The differentiation is thus based on acquired status. Proximity to a centre of power can be conducive and help in gaining rank and reputation. Independent of a person's starting position, however, the opportunities for advancement are evenly distributed and social mobility is high. Accordingly, the loss of proximity to the centre of power can also lead to social decline, as is known from written sources of contemporary witnesses (e.g. Gregory of Tours; Weidemann 1982). Gain, as well as loss, in terms of status acquisition can be high. Rank conditions are not fixed, but unstable due to temporal shifts in the status structure. This graded rank system of groups of families is correlated with possession and use options of land of varying size which, depending on the size of a *familia*, also includes "rule" over people. The union of families to form a village can be interpreted as a territorial unit. Differential conditions for availability and use of resources result from the property structures within a village.

The principle of "allegiance" is considered a characteristic trait of the early medieval social order and is referred to in the term "Personenverbandsstaat" (Steuer 1989), denoting a small number of leaders and their loyal families. The assignment of personal privileges elevates the beneficiaries to a certain rank, binds them to the emitters of these privileges, i.e. usually the king or highest ranking noble, and in this way secures loyal fellowship during times of war. At the same time, it grants the right and privilege of free landholders to carry weapons and to use these weapons in allegiance for the enhancement of rank and reputation. Individual wealth, bestowal of office and property of land become the determinants of rank position in Merovingian times. At death, these personal privileges expire and the material attributes of acquired rank are made public in terms of grave goods: they are visibly demonstrated to the community in the context of funerary rites. It is the relative instability of the social order that necessitates an overt display of the acquired status position, as it were as proof of authentication for future generations. This characteristic trait was abandoned when, with the introduction of Frankish Carolingian rule, a stratified social order was established in which former aristocracy by possession ("Besitzadel") is now confirmed through hereditary titles as a new social stratum, therefore making the demonstration of status through grave inclusions redundant.

At the beginning of the early medieval period, food production is based on mixed subsistence derived from the so-called field-grass economy (Abel 1967), with agricultural and livestock farming alternating. Later, approximately by the middle of the seventh century, the intensification of agricultural production increases, notably by crop rotations that allowed improved utilisation of soils and arable land. The variety of archaeobotanical evidence of crop plants suggests that agriculture was primarily subsistent and orientated towards risk-spreading and that horticulture, in particular, was of importance (Willerding 1980; Rösch et al. 1992; Rösch 1997). Later, while animal husbandry was still maintained, a shift towards the intensified cultivation of grain could be observed, which basically reflected increased settlement activity, as the predominant cultivation of cereals shortened the food chain and thus energetically alleviated the supply situation.

With that in mind the question is: in which way can the characteristics of the early medieval social structure and mode of production be brought into agreement with the general pattern of the development of increasingly complex social structures; or in more general terms, how can social organisation and resource use be linked? Early medieval conditions seem to show a mosaic structure of characteristics, the organisation of which originates from both local group and regional polity levels of organisation. It becomes clear in the first place that genealogical units – extended by persons of affinal kin, i.e. the *familia* including inhabitants of the farmstead who do not have blood relationships with the owner's family – are in control of self responsible authorisation of availability and the use of field and pasture. The size of the property obviously correlates with social position, i.e. the acquired social rank, implying a connection between social differentiation and resource use. Nevertheless, the social position is not fixed by heritage and is not even guaranteed during lifetimes. The social order is, as it were, defined by its instability. The agglomeration of several *familiae* into a village, however, can be regarded as corresponding to a resource-related, territorial unit of the local group. The authentication of possession titles results from the distribution of privileges. Towards the outside world, this is presented by the possession of attributes indicative of status, such as sets of jewellery or weaponry, which are an expression of individual differences in wealth and with rank, which is publicly affirmed in the context of social rituals, here the funerary ritual.

A preliminary evaluation of this comparison shows that, in the same way as contemporary populations do not follow an ideal typology, also historical realities contain subsets of different aspects and characteristics, as is revealed by this mosaic pattern. The uncertainties of the categorical contours which result from comparison with the general model are possibly to be seen as a sign of societal change. The early Middle Ages can be regarded as a distinct transition from the firm structure of late antique tradition to the new and still-forming structure of the Carolingian Frankish empire that succeeds the Merovingian rule (Boockmann 1988).

It does not come as a surprise, then, that the early medieval social order equally displays characteristics of, also in a modern sense, rank societies (e.g. Dumond 1972; Peregrine 1996) and stratified societies (Plotnicov 1996). It represents a social form in which, as in ranked societies, there are institution-alised differences in the achievement of status positions and in which people are principally self-responsible for making their own living. At the same time, rank positions define social status, by which differential access to political power and strategic resources are legitimised for high-status individuals (Fried 1967). On the other hand it also applies to what Plotnicov (1966) describes as characteristic of stratified societies, i.e. that children begin their life with the social advantages and disadvantages of their parents' respective social position, because they are born into an existing rank position of the family. However, there is the difference that such status is not fixed in the early medieval society, but is changeable through the personal skills of the individual or the combined efforts of the *familiae*.

Moreover, stratification is also seen as the cause of incipient surplus pro-duction in the subsistence mode. And, indeed, intensification of agriculture (see Rösch 1997) develops from subsistent production by the end of the early Middle Ages, along with stratified social structures and incipient but clear corporate professional differentiation. In this sense, the early medieval society with its richness of facets even bears evidence for transition as an expression of social change. The initially postulated functional relationship between social structure and resource conditions can thus be confirmed for the early Middle Ages and the general model applied. Available information about the internal and external order of the early medieval society also shows that, for historic time periods, there is a fundamental connection between forms of social organisation and resource use. Whilst these findings first and foremost apply to early medieval societies in Continental Europe, there is growing evidence that the temporal equivalent in Britain, the Anglo-Saxon period, bears enough resemblance and analogy (Härke 1997) to extend the general model.

Status as a function of wealth and possession implies differently distributed possibilities of resource control and use within the community. Yet this alone does not specify to what extent persons or groups make use of their rank differences in order to utilise acquired privileges for the differential access to resources, i.e. to what extent a change in their ways of life becomes visible. It can neither be ascertained in what way the graded social rank system bears possible consequences for nutritional behaviour, nor likewise whether consequences result at all, in view of social instability and mobility. Is status equally reflected in the funerary evidence and in food habits and do rank-related behavioural patterns become recognisable? In other words: what are the possibilities for examining the hypothesis of a correlation between social rank and food quality?

In the same way in which the amount of ecological information contained in historic sources was explored, the same question can be applied to the most direct source available for the analysis of human living conditions and lifestyles from the past: the human skeletal remains. The following section will, therefore, deal with foundations and potentials of analytic methods suitable to detect main food components from archaeological skeletal remains to the extant deemed necessary to prepare the following discussion of correlations between nutritional behaviour and social affiliation.

4.2 Reconstruction of Nutritional Patterns

Like any other organisms, humans are part of their habitat's food webs, yet their nutritional behaviour is not just based on their partaking in natural flows of matter and energy but also on their manipulating these very flows, which they steer through planned interference in order to secure their long-term use. These interferences, even though at significantly different levels of depth, happen independently of the mode of production in human societies, since they occur both in extracting and producing subsistence modes. Typically, the outcome would be nutritional patterns adjusted to conditions of the natural environment (see Chap. 2).

In living populations, the possibilities for an assessment of such patterns are naturally direct and immediate. Not least because not only are the foodstuffs themselves, as it were the material part of nutrition, accessible but also the biocultural framework of food, i.e. its production, availability and meaning. For all historic periods, however, this framework is only tangible by inference or reconstruction which should, if at all possible, include the respective natural environment as well as the societal basic conditions – because whilst the proximate goal is to reconstruct diet and nutrition, the ultimate epistemological aim is not, but rather the conditions of food procurement in a population, i.e. its subsistence.

For those historic time periods where written sources are available, archival information allows a reconstruction of such conditions in principle and can succeed in causally linking socio-political, natural and subsistence factors. For obvious reasons, this is much more difficult when written documents are scarce or altogether lacking. But methods and techniques of archaeological sciences have advanced substantially (for a recent overview, see Brothwell and Pollard 2001; see also Pernicka and Wagner 1990; Herrmann 1994), so that the instrumental analysis of a vast range of source materials provides detailed knowledge and understanding of past dietary behaviours and subsistence techniques on an unprecedented scale. Investigations of plant macro-remains (e.g. Van Zeist et al. 1991; Jacomet and Kreutz 1999) or faunal remains (e.g. Davis 1987; Benecke 1994; O'Connor 2000) are commonly employed and offer qualitative means for a reconstruction of habitat and production conditions. Largely indirect indications of nutritional status are obtained through palaeopathological analyses (e.g. Roberts and Manchester 1993; Aufderheide and Rodriguez-Martín 1998; Schultz 2001; Ortner 2003) by assessing possible negative effects of living conditions and activity patterns from skeletal alterations at the macro- and microlevel.

Whilst all these techniques offer valuable approaches, those methods are more preferable and particularly suitable which facilitate the direct analysis of nutritional patterns and thus achieve the integration of past human populations into the food chains or food webs of their respective habitat by providing a more immediate ecological context. Chemical analyses of archaeological remains, human, faunal (and ideally floral) meet both criteria, because there is a direct relationship between the elemental composition of skeletal remains and the diet consumed.

Bone chemical studies involve analyses of both the organic matrix and the mineral component. In the former, a reconstruction of nutritional patterns is commonly derived from the measurement of stable carbon and nitrogen isotopic ratios, usually from bone collagen; and novel isotope systems, such as sulphur, have recently produced promising results as well (Richards et al. 2003 c). More recently, though, bone apatite and tooth enamel have gained importance in carbon analyses (Sullivan and Krueger 1981; Ambrose and Norr 1993; Harrison and Katzenberg 2003; Lee-Thorp and Sponheimer 2003). The named elements are subject to a systematic fractionation of their isotope ratios between trophic levels and tissues (for example $\delta^{13}C$, $\delta^{15}N$ or ^{43}S; expressed in parts per thousand), dependent on the respective specific geochemical and climatic habitat conditions as well as different metabolic mechanisms (cp. Schoeller 1999). It is always the heavier isotope, for example ^{13}C or ^{15}N, which is discriminated against in metabolic pathways, as it is energetically less costly for an organism to process the lighter isotope. For carbon, two different models have been proposed to explain the diet–tissue relationship: (1) the linear mixing model, according to which all ingested carbon, irrespective of the type of food it was derived from, is incorporated equally into all tissues and (2) the macronutrient routing model, which in contrast assumes that dietary protein is preferentially built into tissue protein, while carbon derived from carbohydrates and fats is built into the carbonate of bone apatite (Schwarcz 2000). Whilst this has an effect on the reconstruction of which parts of the diet can be detected by the analysis of carbon isotope ratios from various tissues, the routing of nitrogen is virtually not affected by such processes (Ambrose et al. 2003).

In general, there is a 2–3‰ shift towards more positive values in carbon ratios between trophic levels, whilst collagen $\delta^{13}C$ is enriched by 5‰ relative to the diet and apatite (or rather carbonate carbon) by 12‰. Isotopic ratios of humans feeding largely on a diet based on C3 plants, i.e. all temperate cultivars, would show collagen carbon values of around −21‰, whereas a diet based on C4 plants, such as maize, sorghum or millet, would produce a signature around −7‰ $\delta^{13}C$. Nitrogen isotopic ratios increase by 3.0–3.4‰ from one trophic level to the next. These shifts are largely caused by the amount of animal protein consumed as plants only contain comparatively little protein, so that their contribution to the $\delta^{15}N$ ratio may be easily masked by already small proportions of meat. Despite this non-linear relationship, higher nitrogen isotope ratios in the consumer's tissues do reflect enhanced consumption of meat. In this way, isotope ratios allow for example the identification of the relative amount of foodstuffs from marine or terrestrial biotopes, freshwater resources and the consumption of C3 or C4 plants (see e.g. Schwarcz and Schoeninger 1991; Ambrose 1993; Katzenberg 2000; Sealy 2001).

Trace element concentrations are measured from the mineral part of skeletal tissue, the hydroxyapatite, or – to denote the diet-related incorporation of foreign elements – the bioapatite (for overviews, see Sandford 1993; Lambert and Grupe 1993; Sandford and Weaver 2000). Analyses focus on elements such as strontium and barium, mainly because they are said to be the least affected by the diagenetic processes of the burial environment (for further details, see Sandford 1993; Sandford and Weaver 2000; Fabig and Herrmann 2002). While in the past other elements suggested to be indicative of animal protein consumption, e.g. zinc or copper, were proposed for palaedietary studies, there is mounting evidence that their use should be strongly discouraged, as too little is known about the translation of dietary into skeletal concentrations, aggravated by the fact that both are essential elements and as such subject to homoeostatic control (Pfannhauser 1988; Ezzo 1994a, b). Assuming that people in the past were consuming local produce, trace element concentrations reflect local geological signatures in the first place; yet in principle the relative distribution of trace elements in the main food components is retained in the skeleton. But, due to effective discrimination mechanisms both between trophic levels and tissues, the element calcium is always preferentially incorporated into the storage tissue of the skeleton at the expense of other (trace) elements in the course of metabolic processing and absorption of nutrients, a process referred to as the biopurification of calcium (Elias et al. 1982). The consequences for the reconstruction of palaeodiets from trace element contents are that foodstuffs poor in Ca, e.g. meat, even if consumed in larger quantities, are masked in the element/Ca ratio of the skeleton by the simultaneous presence of small quantities of Ca-rich food, e.g. milk products or legumes (see Burton and Wright 1995). The element/Ca ratio in hard tissues thus reflects the relative content of dietary calcium in the first place, for example the content of strontium or barium relative to calcium, but it is not suitable to deduce the relative amounts of plants versus meat in the basic food. Therefore, for a reconstruction of subsistence strategies on the basis of multi-component diets, i.e. those typical of omnivorous humans, it is useful to know the average Ca content of the various food components which were accessible to a population or could be utilised.

Also, and this holds for stable isotope analyses as well, comparative data from faunal remains ought to be obtained and information be available as to the general range of resources used for food procurement in the habitat and by the population under study. Only then can strategies of food acquisition, mode of production and indications of differential resource access be derived from trace element distributions in archaeological bone. Combining both kinds of element analyses, isotope ratios and trace element concentrations, is most promising due to the complementary nature of reflecting dietary signals in body tissues.

4.3 Biocultural Correlates of Differential Access to Resources

4.3.1 Ecological Context

As described earlier, general correlations between forms of socio-political organisation and mode of production can be derived from cross-cultural comparison. Plasticity or rigidity of organisation allows large and flexible or only minimal reactions of the entire community to changing resource situations and thus possibilities of adaptation to a sufficient or scarce resource supply. Due to these general connections, it can be assumed that food production and social differentiation of a society is mutually dependent, so that social status among other things, is defined and stabilised by control of natural resources and means of production. Control, however, also denotes the pre-requisite for differential access and use of (food) resources. But the important question that can help clarify the mode of resource availability is: (1) whether just a variable refinement of consumed food is concerned – sufficient food supply of the population provided – or (2) whether differential access to resources results in measurably different consumption of certain main food components of different nutritional quality by groups within the population, i.e. whether socially determined differences within a community leave measurable biological traces. Does the skeleton reflect such consequences of granted privileges or lower status; and how well is bone chemistry able to depict the natural and social reality of historic communities?

It is worth noting, though, that employing stable isotope ratios and trace element concentrations for dietary reconstruction is not about the identification of trophic level effects in the mode of nutrition of human populations. This is not to be expected in the habitually omnivorous species of *Homo sapiens*, apart from a few exceptions such as people of the circumpolar region or vegans. Rather, this is about evidence for intra-population differences in nutritional behaviour relative to biological and cultural factors. In applying the fundamentals of general ecology to human circumstances, there are two aspects to an ecosystemic classification of this correlation.

First, the utilisation of energy flows within a food chain: due to the pyramidal organisation of food chains, the transition from one to the next higher trophic level results in a ca. 90 % loss of energy (see Chap. 1.2). In other words, each trophic level additionally employed in food production increases the total input of necessary energy by about ten-fold. Crop farming in comparison with animal husbandry is thus energetically more favourable, because the latter can achieve more yields per assigned energy unit and therefore has a more efficient cost/benefit ratio. Accordingly, the production of a comparable quantity of animal-derived food products is connected with a substantially higher input of energy. In addition, working animals, for example horses, may even become food-competitors, since they require high-energy fodder, such as grains in order to perform properly. In a scenario of socially different resource allocation, it can therefore be reasonably assumed that lower social status is connected with an energetically more favourable food production that is based on a shorter food chain, or in general with the consumption of food resources whose procurement requires less energy input. Higher status and thus the availability of more and better economic means, would accordingly allow food production along a longer food chain and/or with higher energetic input, so that animal husbandry and principally also an enhanced consumption of animal-derived foodstuffs can be expected.

Second, and this aspect is closely connected with the former, socio-cultural characteristics of populations have to be viewed as components of the respective human ecosystems. If it can be shown that nutritional behaviour varies as a function of social status, then this would allow the establishment of socially dependent resource control and thus allow denoting the actual determinant for at least part of the material and energy flows in the habitat of past societies.

The question of group-typical patterns of food consumption has occasionally been taken up in the context of bone chemical analyses, although its coverage is altogether remarkably thin, and in the majority without ecological reference[2]. It is the merit of early studies, such as those on pre-Columbian Native Mexicans (Schoeninger 1979) and Native Americans of the Dallas Culture (Hatch and Geidel 1985), to introduce the idea of status-related consumption of high-quality food items and claime evidence in support for past societies. Even though probably right in their general conclusions, the interpretations put forward have to be regarded as over-simplified in hindsight, due to a lack of understanding of the complexity of biopurification and the incorporation of elements into tissue stores (see Burton and Price 2000). Other results are contradictory, as no clear connections between social status and access to high-quality food could be established (Blakely and Beck 1981; Brown and Blakely 1985). The first indication of interrelations, even though inconclusive, between diet and status in early medieval populations was published by Grupe (1990). Even though trace element signatures are able to reveal status-related dietary behaviour, as will be demonstrated below, a breakthrough seems to have been achieved with the more widely applicable and accepted analysis of stable isotope ratios. Some of these examples will be incorporated in the following section to broaden the argument in favour of the detection of such patterns. Despite the fact that such studies are still relatively scarce the issue of socially different nutritional behaviour and, in particular, its ecological context and consequences can be regarded as one of the keys to understanding resource allocation and use in past societies and is thus of ongoing interest and deserves further attention.

4.3.2 Linking Dietary Patterns to Social Ranks

The following observations are based on results derived from trace element and stable isotope analyses of early medieval populations from southern Germany (for methodical issues see, e.g. Schutkowski et al. 1999) and will be complemented by directly comparable studies covering various other time-frames and historical settings. A brief description of the archaeological context will set the scene.

For the best part of the late fifth to the eighth century A.D., the mode of subsistence in rural communities can be considered primarily subsistent and aligned to the direct supply of the local population. The general pattern of land use resulted in mixed farming with agriculture, animal husbandry and horticulture (e.g. Jäger 1980). Archaeobotanical finds provide evidence for various grains, roots, tubers and leafy vegetables, as well as legumes (e.g. Willerding 1980; Rösch 1997). The former settlement area of the Alamanni in

---

[2]   Even though disputed for interpretational and methodological reasons, results of early trace element analyses are mentioned here, because they still allow a case in point for the detection of socially varying food consumption.

the German southwest is known for the widespread cultivation of spelt wheat (*Triticum spelta*; Gradmann 1902) and other cereals. Among the domestic animals were cattle, pigs, sheep, goats and poultry (Abel 1967; Fingerlin 1974; Knaut 1993; Benecke 1994; Kokabi 1997). The consumption of freshwater fish is likely to have occurred, but on the whole not as a main dietary component. Likewise the consumption of larger quantities of sea food or anadromous fish can be excluded, not least because of the geographical location of the examined locations, but also because the long-distance trade of for example salted herring does not gain notable importance before the establishment of the Hansa.

Based on these foodstuffs known to have been available to early medieval populations, a breakdown of main dietary components and their typical element contents is possible (Table 4.2). Ideally, such data would be based on archaeological samples from the respective sites or areas, which however is hardly feasible due to the almost invariable decay of this kind of organic matter. Therefore, the table presents pooled literature data taken from modern food analyses, predicated on the assumption that relative differences between food items and trophic levels justify such an actualistic approach. Thus, for the time being, the values indicated in the table are the most direct approximation to the skeletal trace element supply by certain food items. Analyses of entire food web data of the examined locations are not available, mainly because the cemeteries and settlements for the same site are rarely discovered and excavated, which results in a general lack of the faunal remains immediately utilised by the population under analysis. As far as faunal remains from grave inclusions are available, though, these are included in the analyses. Supra-regional food webs that could serve as useful points of orientation have only recently become a matter of serious attention and are thus largely outstanding, yet the principle is depicted in Fig. 4.1 for trace element distributions. The pattern of isotopic abundances in the different habitats is shown in Fig. 4.2.

Table 4.2. Element concentrations in different foodstuffs (modified after Schutkowski et al. 1999)

	Ca (mg/g)	Sr (μg/g)	Sr/Ca
Meat	200	1.1	0.0055
Roots and tubers	420	2.6	0.0062
Cereals	700	2.0	0.0028
Pulses	2,120	4.7	0.0022
Leafy vegetables	7,560	2.1	0.0003
Milk products	8,070	5.0	0.0006
Milk	9,600	0.5	0.00005

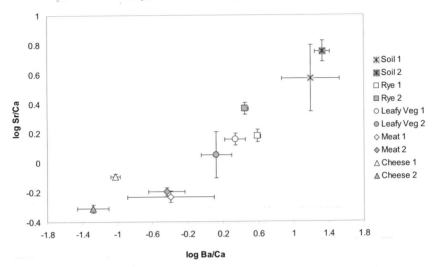

Fig. 4.1. Trophic level effects as a result of biological purification for various foodstuffs (data from Fabig 1998). Even though the overall discrimination effect applies, the data reflect discernable differences in element supplies between two localities (locations *1, 2*)

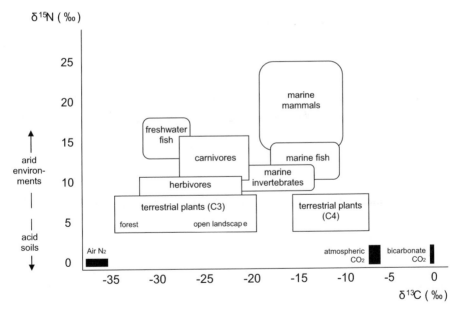

Fig. 4.2. Isotopic abundances of terrestrial and marine food resources

The following case studies will present examples of various forms of dietary differentiation within past populations. They can be observed at different levels which relate to the principles of social organisation outlined above and encompass the extended family and the reflection of individual status on diet.

4.3.2.1 Early Medieval Alamanni

The site of Weingarten is situated at 500–550 m above sea level on a plain constituting the foothills of Lake Constance, a natural setting providing favourable conditions for agricultural production (Dongus 1991; see also Sect. 2.2.7). The early medieval Alamannic cemetery revealed some 800 burials, displaying fine examples of social differentiation as reflected in sophisticated grave inclusions (Roth and Theune 1995). Overall, and this holds for early medieval cemeteries in general, men of higher status would be buried with sets of weaponry, the most prestigious being a combination of a spatha (double-edged sword), a seax (single-edged shorter sword), spear and shield, occasionally also a 'franziska'. Accordingly, such assemblages are often referred to as warrior graves. Graves of women of equivalent status would be equipped with precious metal, e.g. gold or gilded brooches and other jewellery (for a detailed discussion, see Steuer 1989; Siegmund 1998; Härke 1990). These differences appear in all age groups, including children, indicating the importance and meaning of status within the different family groups and their display of acquired rank.

The overall picture of dietary supply points towards the consumption of a diet based rather homogeneously on typical C3 components, such as domestic cereals ($\delta^{13}C$ –19.8±0.7). The relatively wide spread of carbon isotopic values suggest that the diet may have included some additional C4 components, such as millet, a dietary supplement which would have been known from Roman times and which would have been available through trade relations. On the whole, the population displays $\delta^{15}N$ values that inconspicuously place it within an omnivorous range (8.6±0.6), whilst some of the higher nitrogen values may indicate a supply of freshwater fish and the particularly low values point to low animal-derived protein intake (Fig. 4.3). Trace element data appear to confirm the general picture, the spread suggesting a variety of foodstuffs ranging from low (e.g. cereals) to higher (e.g. milk and milk products) amounts of calcium-enriched items. Comparative values from animals of known trophic position can be taken in support, as they display values in the expected ranges of herbivorous (e.g. cattle) or omnivorous (e.g. pig) animals (cp. Schutkowski et al. 1999).

Early medieval cemeteries are characteristic in their display of individual wealth, expressed through a wide variety of grave inclusions. Not only are they believed to denote the social standing of *familiae* but also to demonstrate

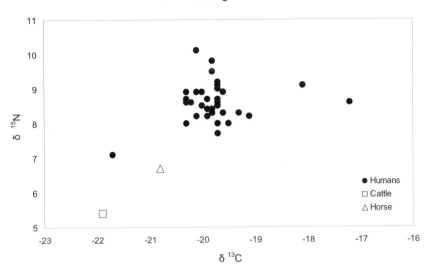

Fig. 4.3. Stable isotope signatures indicate a typical omnivorous diet at Weingarten (data from Schutkowski et al. 1999)

gender-typical and personal rank conditions. Overall, adult individuals of higher or lower attributable status do not seem to vary distinctly. However, individuals of elevated social standing display a much closer range of isotopic signatures, indicating a more homogeneous diet, perhaps as the result of choice and taste preferences. In contrast, individuals of lower ranks display a much wider spread of ratios, with values suggesting supplements of C4 and freshwater fish components for males, and one individual in the female group with an almost herbivorous signature (Fig. 4.4). It is not unreasonable to assume that, while the latter is likely to be the result of having to exploit a wide variety of available and accessible resources, the former is an indication of being able to afford cuisine and dietary quality not being dependent on protein quantity. Men of different status do not exhibit significantly different diets, apart from the general spread of their isotopic ratios. High status women, however, display a tendency towards a slightly elevated consumption of animal protein.

The group of sub-adult individuals shows a comparable spread of trace element values, indicating some overlap between social groups. However, there is a general tendency towards the consumption of mineral-enriched foodstuffs in children of higher social standing, i.e. those displaying status-indicative items in their grave furnishing. On the assumption that this is best achieved through milk or milk products (cp. Burton and Wright 1995), the consumption of more animal-derived items translates into a congruence of nutrition and status, i.e. the socially varying access to resources, nutritional behaviour and social placing, in these individuals as well.

Weingarten

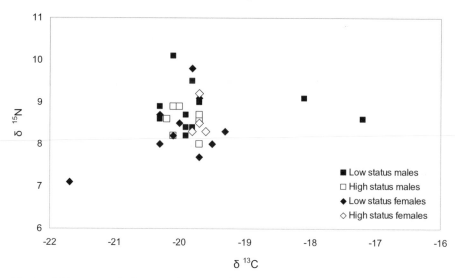

Fig. 4.4. Weingarten: the isotope signatures for males and females of higher social status indicate a much more homogenous diet than those for their lower-status counterparts

statistics? looks confused

Kirchheim unter Teck, at around 300 m above sea level, is located southeast of Stuttgart at the northern rim of the Swabian Alb. It is placed within a valley 5–8 km in width and the habitat benefits from favourable climatic and soil conditions (Dongus 1961). Its early medieval cemetery contains 167 individual burials (Däcke 2005), the majority of which have been subject to some kind of bone chemical analysis.

Irrespective of element signatures typical of the locality, Kirchheim in comparison with other locations in southwest Germany (cp. Sect. 2.3.7) is distinct in its isotopic and trace element values, which both point towards higher amounts of animal protein in the average diet. While the carbon values are typical of a C3-derived diet ($\delta^{13}C$ –19.4±0.3), the nitrogen signatures of $\delta^{15}N$ 9.5±0.6 are not only higher compared with Weingarten but are also slightly less varied. Trace element ratios overall point towards a diet relatively low in mineral, which with all due caution, could corroborate the enhanced animal protein intake indicated by the nitrogen values, yet not in the form of milk or milk products (cp. Schutkowski 1987; Burton et al. 1999).

In terms of social differentiation of dietary patterns, there is a general tendency for individuals of higher social standing to consume more animal protein yet, unlike at Weingarten, the spread of values is more pronounced in the high-status group. It is interesting to note, though, that this inhomogeneity is largely caused by a wider range of carbon values in males. Males, in general, do not display a clear distinction in their dietary patterns irrespective of their

Kirchheim women

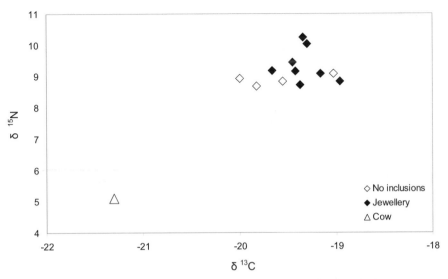

Fig. 4.5. Kirchheim unter Teck: women of higher social standing generally consumed more animal protein

locks statistically insignificant

burial furnishings, yet the high-status group encompasses the individuals with both the highest and the lowest nitrogen values. Women, in contrast, whose grave goods contain items indicative of elevated rank positions, exhibit higher nitrogen values and thus indicate a better and prolonged access to animal protein sources (Fig. 4.5). Among the sub-adult individuals, it is again the high-status girls rather than boys[3] whose trace element values suggest less consumption of calcium-rich foodstuffs, which in this context is likely to be consistent with more meat consumption.

The location of Wenigumstadt belongs to the natural unit of the Reinheimer Hügelland, a mountainous area of moderate altitudes which extends from the northern edge of the Odenwald to the banks of the lower Main River (Klausing 1967). Even though climatically less favoured than the Bergstrasse area west of the Odenwald, its extended loess top soils provide good conditions for agricultural use. Out of a total of 268 burials, a random sample of 96 adult individuals was analysed for general and status-related dietary patterns. Only trace element data are available.

Taking into account locality-typical expressions of element values, but also considering available comparative data on southwest German early mediaeval subsistence strategies (see Chap. 2.3.7), Wenigumstadt appears to represent a

---

[3] For methodical issues of sexing sub-adult skeletal remains, see Schutkowski (1993b), Sutter (2003).

dietary pattern characterised by larger amounts of foodstuffs relatively low in mineral, i.e. mainly containing lesser amounts of calcium. Contrasting the human data with comparative values available from horse bone, it becomes likely that, apart from the usual cereal grain component, the consumption of animal protein (yet not in terms of animal-derived products) has formed the basis of the general dietary regime, in particular as the horse, being a work animal, was fed high-quality fodder most likely in terms of cereals.

Irrespective of dietary differentiations within the population, other biological data reveal an interesting pattern. On the whole, individuals whose grave inclusions place them at the lower end of the social scale achieve a significantly higher mean age-at-death than those of elevated social status (45.6 years vs 40.1 years). This observation translates into gender-related patterns as well, in that low-ranking males live longer on average than their high-ranking counterparts (46.8 years vs 36.0 years). For women such differences cannot be confirmed.

Apparently, neither the privileges which led to elevation into a higher rank position nor the concomitant change in dietary conditions leads, at least for men, to biological success measurable in terms of longevity. A possible explanation can be assumed against the background of early medieval social structures: The higher the rank of the *familiae* a man belonged to in the community, the closer his family would be to the king and thus the more important his allegiance in times of war (p. 519 in Steuer 1982), a potentially risky return for awarded privileges. This may explain why higher-ranking men, but not women, suffer from reduced ages-at-death compared with lower social ranks. Empirical evidence fro this would be conceivable through increased trauma-related lesions in the skeletons, yet such data are not available at present. Beyond that such an explanation would have to remain limited to the whole of higher-ranking *familiae*.

On the level of individual groups of families, there is evidence for more far-reaching differentiations in nutritional behaviour (Schutkowski 2002b). Archaeological findings suggest that three distinct areas of burials can be identified in the eastern part of the cemetery which are not only spatially separated but which are distinguishable by the construction of graves. Element concentrations of the interred individuals reveal clusters of similar bone chemical composition[4]. Whilst without genetic underpinning there is no conclusive proof, the element signatures lend themselves to interpreting these groups of inhumations as groups of people with the same dietary history, or rather with the same possibilities of access to dietary resources, i.e. as family units (Fig. 4.6). Notably, Families 1 and 3 display characteristics indicative of more amounts of calcium-rich foodstuffs than family 2, which

---

[4] Inhumations with similar element concentrations are found scattered throughout the remaining cemetery area, indicating that the detected clusters are most likely not the result of adjustment of element contents due to small-scale diagenetic effects.

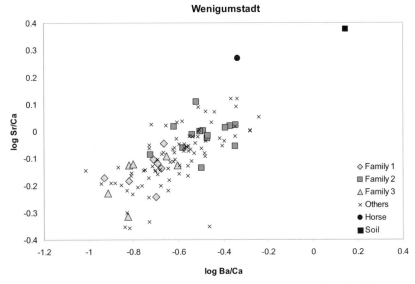

Fig. 4.6. Wenigumstadt: groupings within the cemetery interpreted as families show distinctly different dietary signatures

points towards increased consumption of animal-derived food products. It is true the groupings represent only relatively small numbers, yet if the interpretation of family units is upheld it becomes obvious that such groups cannot be enlarged in individual numbers at will. Rather, the evidence suggests that they are the result of non-coincidental conditions which are tied to nutritional supply.

   This result is important. For the first time, the combined evidence of bone chemical and archaeological findings allow the demonstration of family-bound nutritional habits in historic populations. However, this goes beyond the mere illustration of culinary preferences, which are commonly known from the everyday life experience of family-typical nutritional habits. Rather, the groups of families differ in their element contents in such a way that they suggest distinctions of food quality. Based on the established correlations between social standing and possibilities of food preference, such differences can be reasonably interpreted as control and differential access to resources. Since the hierarchical arrangement of the early medieval society is based on a family-bound rank structure, it at the same time demonstrates that the family unit reveals itself as the constituting element of the early medieval society, even through its resource-related and status-dependent nutritional behaviour. *Familiae* can thus be identified as behavioural–ecological entities. More so, because rank differences are also connected with control of resources generally, e.g. land, but also personnel, so that familial differences in food quality depict the socio-historical side of status differentiation within the society.

4.3.2.2 The Wider Picture

More recently, the issue of socially diverse nutritional patterns in the past has been gaining momentum and there is mounting evidence of clear correlations between resource availability and social standing from various time periods and cultural settings. Unlike the previous case studies, the following examples are somewhat different in that they demonstrate the translation of rank and power relations into resource control in societies with a much more rigid stratification than the more permeable and mobile social organisation of continental early medieval times. Yet, the principle and overall ecological context remain the same, despite the fact that the evidence is still scattered and naturally unsystematic owing to the variety of provenance and lack of systematic investigation.

The historically oldest case reported so far dates back to the first millennium BC from the La Tène period of Iron Age Bohemia (Le Huray and Schutkowski 2005). A notable feature of the cemeteries at Kutná Hora and Radovesice (Valentová 1991; Waldhauser 1987) is the lavish display of rank and standing in the burials of males, which led to their description as warrior graves, characterised by sets of weaponry of different quality and composi-

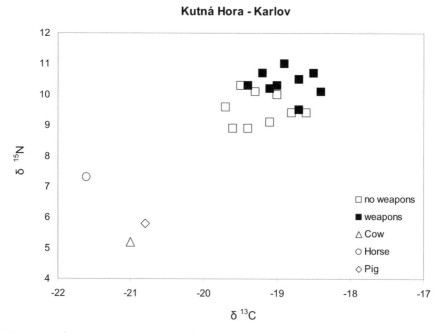

Fig. 4.7. In the Iron Age community of Kutná Hora, the dietary behaviour of individuals from so-called warrior burials can be distinguished from those individuals without weaponry inclusions (Le Huray and Schutkowski 2005)

tion. Those with the richest display of their warrior status during life show nitrogen isotope signatures that indicate more consumption of animal protein that their counterparts of lower ranks (Fig. 4.7).

Cahokia, the only historic Native American city north of Mexico situated in what is Illinois today, sustained a thriving community of up to 20,000 inhabitants during its heyday towards the twelfth century AD (e.g. Pauketat 2002). Ruled as a theocratic chieftainship and featuring a strict socio-political hierarchy, its burial mounds reflect status- and gender-related differences in health and diet. Notably, at mound 72, females were found to display clear differences in the relative amount of maize consumption as a function of status. Whilst high-status women generally show $\delta^{15}N$ values typical of animal-derived protein, low-status women obtained most of their protein from C3 or C4 plant sources, with their bulk diet mainly derived from maize (Ambrose et al. 2003; Fig. 4.8).

The Romano-British site of Poundbury Camp, Dorset, revealed various examples of status display in terms of burial architecture, the most prominent being mausolea built for the most closely acculturated people who had achieved highest social standing within their community. Again, those individuals who could be identified as belonging to the elevated end of the social hierarchy were clearly discernable in their dietary behaviour, reflected in the

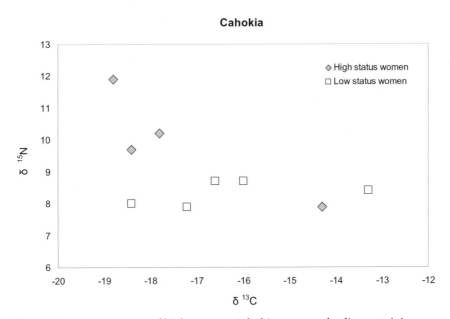

Fig. 4.8. On average, women of high status at Cahokia consumed a diet containing more animal-derived protein. Lower-status women had to rely largely on plant food (data from Ambrose et al. 2003)

increased amount of animal protein over those of lower social ranks (Richards et al. 1998).

Quite similar to the examples from southwest German early medieval sites, these case studies provide evidence to underpin the general pattern of diet as expression and even display of status. While this may be true in particular in those cases where allegiance and support to a prince in times of feud and war was an expected obligation, the expression of social inequality in terms of dietary difference is wide-spread and engrained in the social fabric of both the political and the ecclesiastic realm (e.g. Mays 1997; Polet and Katzenberg 2003). The spectrum, however, ranges from an obvious and overt display of resource control to the subtle use of rank-related differences in the access to qualitatively different food components, which results in the special quality of the status-difference in consumption.

4.4 Conclusion: Status-Dependent Resource Use

This chapter discussed correlations between social structure or the internal social differentiation of populations and the availability of food resources. Analytical results were presented that indicate differences in the access to and use of nutritional resources, with an emphasis on conditions during the early Middle Ages. In an attempt to apply both the socio-historical and the analytical evidence first within the general context of social organisation and resource allocation and afterwards as a particular case in point, its observable manifestation in early medieval society is examined.

From the viewpoint of socio-cultural evolution, three general levels of social organisation in human societies can be distinguished which are characterised by an increase in the complexity of regulating central areas of life, such as food procurement: family groups, local groups and regional polities. Human communities organised in family groups retain the possibilities of flexibly reacting to fluctuations in the resource supply within their habitat by a mobile or semi-sedentary way of life. Additionally, this is facilitated by family relations, from which usually no genealogically justified claims for possession of resources are derived. Between families, food is re-distributed according to the principle of mutuality, which helps compensate for varying individual success in food procurement. Patch-choice of food resources is decided by consent as it were, representative of an absence of political institutions by which subsistence activities could be regulated.

The social economy of local groups likewise builds on the family as the basic subsistence unit, although decisions on the co-ordination of food production lie with political institutions. Individual families form larger units, the local groups, which constitute territorial and resource-related associations, whose kin-based organisation into clans or lineages give rise to rights of use and control over resources. Compared with family groups, local groups are characterised by an overall more-pronounced economic and social integration into supra-familial structures.

This concentration continues to increase at the level of the regional polity and leads to the formation of central political structures, which in the sense of regional bodies oversee subsistence activities on the local level. Within the society, increasing social differentiation and stratification take place which find their expression in the possession and control of land and resources by private individuals or institutions.

In a similar way, characteristic traits of social organisation during the European early medieval era can be connected with the possibilities of access to resources. The *familia* as the central element of early medieval society reserves the self-responsible rights of disposal and use over farm and pasture land. Rank differences between families are an expression of privileges awarded by a ruler and the size of the respective property correlates with exactly this acquired social position. Rights of disposal and control of resources therefore vary as a function of social arrangement within the society. This implies that social organisation and resource use are directly connected. Within the social space of early medieval funerary ritual, the respective rank of the deceased person is demonstrated by certain ensembles of status-indicative attributes, e.g. sets of weaponry or precious metal jewellery. Grave inclusions are thus a reflection of status during lifetime and allow the reconstruction of rank differences *post hoc*. Existing differences in food quality for such recognisable social groups within a population can be equated with differential control over and access to resources and thereby increased consumption of high-quality food in terms of animal protein. The question of social variability of dietary behaviour is thus closely connected with structural differentiations within populations.

Furthermore, there are indications that overall an association of dietary behaviour and status may be more pronounced in women than in men. For men, a more even distribution of social ranks across dietary groups suggests larger independence in the availability of food, i.e. a less-strict social governing of access to food resources, or rather a less-pronounced direct translation of status differentials into measurable quantities of high-quality food. Rather, as some examples suggest, affiliation with higher social ranks allows more selectivity or discernment in the choice of food: status allows choice, which leads away from having to put up with what is readily available. Similar explanations have been suggested for the Anglo-Saxon site at Berinsfield (Privat et al. 2002), where poorer people likely consumed freshwater fish and meat from low-maintenance domesticates, whereas wealthy men could afford to provision themselves with meat from animals involving resource-intensive husbandry, such as cattle and pig. The general finding that women seem to show a stricter adherence to socially different nutritional patterns may have the somewhat straightforward explanation that women as a result of the sexual division of labour are generally "closer to the source" and therefore in direct control of food resources. Yet, even though more difficult to prove, their affiliation to families of varying ranks may find its expression in domestic differentiation as opposed to the political allegiance expected of men. Nevertheless, evidence of gender-typical consumption patterns entails the perspective of behavioural–ecological approaches to past populations, which in this form would otherwise only be ascertainable for historic times with extensive written sources. This holds true as well for evidence of family-related nutritional patterns.

Such findings without contradiction can be put into the context of general connections between social status and resource conditions. As members of families of prominent social position, the majority of high-ranking individuals accordingly reflect a qualitatively better nutritional basis than individuals of lower social status. Since rights of disposal and use of land are regulated through such graded rank systems within a society, differential control and access to resources becomes depicted in the socially different nutritional conditions. The examples discussed here provide this evidence in terms of food resources. Cultural and social constructs thus become biologically comprehensible.

At the same time, the results show clearly that by way of chemical analyses of skeletal remains basic effect-mechanisms can be discovered within past human ecosystems, since in this context differential access to resources also means that material and energy flows in these stratified societies were subject to manipulation and control governed by social affiliation. In this way, the interference of humans with their environment or the controlling of these interferences can be directly connected with the socio-cultural condition in human populations. It appears to be crucial that this can be successfully accomplished for historic samples, which for obvious reasons can only provide less detailed information and possibilities for exploration. Nevertheless, parallels and common grounds in basic patterns of resource acquisition can be observed with recent populations. This result underlines the potential implicit in the transferral of ecological questions to historical contexts.

The correspondence of dietary behaviour and social affiliation in (the examined) populations is non-random; and congruence was found to be very high, yet not complete. The absence of total correspondence comes as no surprise and can be explained plausibly both in terms of physiology and mentalities. For the early medieval examples, given the high social permeability and mobility of the society, in which the individual acquires his place in the social rank order by the assignment of privileges, archaeological findings do not permit a definitive statement about the original social group of the individual. A characteristic socially indicative ensemble of grave inclusions only reflects the current social positioning at the time of death, which an individual may have reached through either social advancement or decline. Since it is known that humans acquire their nutritional habits in the course of socialisation and that family traditions play an important part here and can be readily identified (Murcott 1990; Menell et al. 1992), a change in social status need not necessarily involve a change of habits, i.e. in this case a change in the nutritional basis or dietary habits. An alternative explanatory model would be that social mobility not only entailed a change in the way of life but also in nutritional preferences and behaviour. However, due to the pace of bone metabolism and turnover – bone collagen isotope values reflect the dietary intake of the last 5–10 years – element analyses can only depict longer-term food habits. It is thus well possible that a change in nutrition did take place but had not yet resulted in measurable effects in the skeleton at the time of death. Which of the two explanations applies in a specific case cannot be decided; and from a biological point of view both are equally plausible.

In those cases where larger numbers of non-adult individuals are available for analysis in early medieval assemblages, the congruence between social affiliation and dietary behaviour is clearly pronounced. Assuming this is a valid general observation, this is important for two reasons.

On the one hand, not only adults but likewise also children and adolescents can be assigned certain social rank positions due to the quality and specific composition of their grave inclusions. They thus reflect the social position of the *familia* from which they originate, as sub-adults from high- and low-ranking families were in each case buried with grave goods of comparable quality, respectively. If one considers the social mediation of food habits within the family – which would also mean nutritional possibilities due to status – patterns of congruence between nutritional behaviour and rank are first of all (indirect) evidence for the different circumstances of the families due to their differential rank positions. Children born into a better-off family accordingly profited from more-favourable circumstances, while children from families at the lower end of the social scale were in a less-privileged situation.

On the other hand, the results from childrens' skeletons can generally confirm those of the adult individuals: status-related nutritional differences in sub-adults are an expression of the varying possibilities of resource use by families due to rank differences. Adults were in a position to be able to change their rank position in the course of their life both upward and downward, although social descent was naturally avoided (Weidemann 1982). While social ascent would not necessarily be expressed in a change of food habits, social descent quite probably led to a change in living conditions and thus to a concomitant shift in the nutritional basis. In children, rank-correlated nutrition should thus reflect a very close correlation to the possibilities of use and availability of resources. Childrens' dietary patterns would thus become, as it were, a direct reflection of differential living conditions in the early medieval society. In behavioural ecological terms, this is equivalent to differential resource allocation of parents to their descendants, mediated by their economic possibilities and circumstances.

In a quite different, yet theoretically consistent way this correlation can be corroborated by the findings of family-typical nutritional patterns. Areas in a cemetery interpreted as family burials are in agreement with groupings found in element distributions. The separation of element signatures not only identifies families in their dietary behaviour or food traditions, but the groups differ as well in the quality of the food consumed in a way which strongly suggests differential resource-availability. This represents the analytically and methodically independent possibility of recognising and verifying what social historical findings conclude to be the hallmark characteristic of early medieval society, i.e. an arrangement of the society into *familiae* of different social ranks. The *familia* as the constituent unit of the early medieval society can thus be regarded as an economic, social and nutritional entity.

It remains for future research to demonstrate the immediate biological consequences of the different resource access at the population level. For contemporary societies and historic communities with suitable archival records, there is clear evidence of correlations between status and reproductive success (e.g. Voland 1998; see Chap. 5). Archaeological populations principally hold the same potential, provided the analysis of ancient DNA succeeds in developing into a generally approved tool for a reconstruction of kinship and genealogies in prehistoric communities (e.g. Hummel 2003; but see Cooper and Poinar 2000). There is thus great expectation that, once the possibilities of palaeogenetic screening are unequivocally established, the link between social affiliation, nutritional behaviour and fitness maximisation can be demonstrated.

At present, however, stable isotope and element analytical studies are already able to demonstrate that only a differentiated view of both biological and socio-cultural classification systems successfully allows the linkage of empirical analytical data with behavioural–ecological categories in past societies. This entails the perspectives of fine-tuned analysis of nutritional habits with regard to social and societal conditions, promisingly leading in principle to assessing the ecological basic conditions of historic populations. The detection of differential nutritional behaviour in past populations makes possible the detailed description and interpretation of general and group-typical circumstances, including socio-ecological differences. Bone chemical analyses can thus make substantial contributions to outlining lifestyle and living conditions in historic times and their ecological and socio-historical determinants.

5 Population Development and Regulation

The preceding chapters have taken into consideration a systematic account of food acquisition strategies, changes of subsistence bases and differential access to resources in human communities. Different modes of resource use were interpreted as the result of biological and cultural adjustments to the ecological default conditions of the respective habitat. The success of such adjustments can be measured to a large extent as to how and what extent populations, by interacting with all other system components, succeed in steering matter and energy flows for the pursuit of individual goals and collective interests in such a way that – ideally – long-term survival in the habitat is ensured. Effects which arise from this with regard to the size and composition of populations have only been very briefly addressed so far.

The outcome of utilisation strategies is reflected in different patterns of distribution and dispersion of human communities. Populations can therefore be characterised by the criteria of spatial demarcation and density. However, spatial continuity or discontinuity is only one aspect of population structure. A further central characteristic of populations is their variable internal structure. What may look like a simple and uniform composition at first sight, is in reality often a lack of structural homogeneity. Populations can be divided by vertical arrangement into social strata or differentiated horizontally, e.g. as in the case of societies that are organised in clans, lineages or other corporate groups. An ecological view of populations will thus ask for their structures and will try to analyse their emergence from the mutual influence of natural environment and cultural mediation and the respective different contributions of the two.

The concept of population is closely connected with a system of hierarchical classifications. Depending on the level of explanation, the classification includes larger or smaller groups; and the categories used to look at populations overlap. Considering natural, biological and cultural determinants, four central of aspects result that affect the structure of a population.

1. The demographic aspect stands for the perhaps most abstract category according to which population structures are defined by fertility, the number of births and deaths (natality, mortality), disease rates (morbidity) and

population movements (migration). These factors have an effect on the size of a population and its age and sex composition in a given temporal sequence, i.e. population dynamics.

2. The genetic aspect refers to the extent to which a particular population partakes in the overall gene pool. It thus describes the consequences of founder effects or genetic drift, but also genetic kinship relations within and between populations, for example as far as the effect of geographical conditions for gene flow are concerned, i.e. the presence or absence of natural boundaries, such as rivers or mountain ranges.

3. The sociological aspect essentially considers the socio-culturally defined internal differentiation of human communities and their effect on population structures. This includes for example social, status or occupational groups, religious communities, but also kinship groups (whose composition, depending on classification, may not necessarily be based on genetic affiliation). Biologically defined groupings, such as those derived from age and sex, also have sociological significance.

4. The ecological aspect finally refers to the degree to which populations inhabit the same and/or similar habitats and share a common resource pool. Effects on the population structure result both from comparable and/or different environmental conditions the respective populations are exposed to and from the extent by which strategies are developed in order to adjust the size and composition of the population to the respective habitat.

Summarising these aspects, human populations can be characterised as a group of individuals of the same species forming a reproductive unit and inhabiting a common habitat, so that their members are subjected to the same environmental factors and use a common resource basis. Populations and their individuals react to changes in the environment by genetic, physiological and cultural adjustments of their behaviour. A population thus possesses characteristic traits that result from its composition shaped by age and sex affiliation, kinship and social groups, birth and death rates, as well as patterns of spatial distribution.

In the context of an ecological view, these different aspects of population are regarded as a result of the interaction of a certain species, in this case humans, with the environment and are expressed through the presence, distribution and composition of a population in a given habitat (p. 123 in Hardesty 1977). The change in population characteristics, such as density, size, fertility or mortality, can thus be regarded as being directly dependent on the abiotic, biotic and cultural factors of the environment. These include variables such as climate, soil quality and geomorphology, as well as the supply of animal and plant resources and the distribution or allocation of resources within a community. These factors, in turn, are directly correlated with the possibilities of use and availability of resources in socio-ecological (Zwölfer 1987), i.e. human-shaped and affected, ecosystems. This applies in particular to cultural factors, including for example technological skills that influence the efficiency of utilisation strategies, but also socio-cultural arrangements regulating the division of labour. Since these factors eventually aim at long-term survival in the habitat, adjustments to the ecological conditions of resource availability thus have an effect on cultural solutions for the basic biological tasks of self-preservation, survival and reproduction. In order to achieve this, genetic information alone is not sufficient, but rather cultural responses decide on the selective advantage and success of optimal resource use. Populations can therefore be explored by considering the ecological context of strategies of resource exploitation and the effects this has on structure and composition, i.e. effects of local environmental conditions and the respective reactions to certain stressors occurring in the habitat.

One of the key factors, the availability of food resources, has important consequences for the size of a population. This connection was phrased by Thomas Malthus in his "Essay on the Principle of Population" (1798). He postulated that populations and the natural resources[1] they live on grow at different speeds. Because resources could only increase in a linear fashion while populations had the innate capacity to grow exponentially, it was inevitable that, at a certain point in time, the population would grow at a faster rate than could be supported and sustained by the available resources. The diversion of these two growth curves thus leads to so-called checks and results in a counter-balance of this development. Such checks are significant events and can, for example, occur in the form of an epidemic disease, war or famine which leads to the decimation of a population, with the outcome that resources and commodities become available again at levels sufficient to maintain a population that is now reduced in numbers. Moran (2000: p. 83) summarises such external events influencing populations as extrinsic factors, e.g. supply shortages and climatic changes, and complements them by the population-effective factor of disease vectors (micro-predators *sensu* Groube 1996). Additionally, next to these determinants, intrinsic aspects must be considered which affect the size and distribution of populations. They result from factors as different as the possibilities of resource use through affiliation to a social group, the distribution of food within a family or a household, the preference for children of a certain sex or contraceptive measures and birth control. In principle, organisms possess the propensity and capacity of unlimited procreation. Hardly any species, however, uses this potential to its full extent. What is regularly found instead are so-called natural population sizes as a direct result of selection mechanisms, which in human populations not only encompass biological, but naturally also socio-cultural characteristics. Population density and growth is thus subject to delimitation under the principles

---

[1]   Malthus actually considered food resources only, but the logic of the argument lends itself to using a wider notion of resource. For a reconsideration of Malthus' population theory see Seidl and Tisdell (1999).

of natural selection. It follows that any measure taken by an individual, any behaviour and characteristic that within a population is successful through preferential selection and that has a dampening effect on population growth can be considered regulatory.

None of the intrinsic or extrinsic factors is equally or constantly effective over time in any population. In changing combinations, however, they contribute to maintaining the overall stability within an ecosystem. Particularly, the intrinsic factors lend themselves to causing effective mechanisms that make possible a flexible adjustment of population size or population density to the available resources. If this leads to local equilibrium, it can be regarded as a hint towards population regulation (Wood 1998). It has to be examined, therefore, under which circumstances a regulation of population density can be expected to occur at all. How is regulation correlated with characteristics of populations such as density or size and thus with the available resource supply? And finally, at which level is regulation adaptive, i.e. who benefits from flexible stabilisation of the population? Two short examples may help to set the frame.

The ecological basic conditions of the Greek island of Karpathos (Vernier 1984) only allow limited productivity that cannot be increased even through the employment of machinery or fertilisers. The social structure of the population is traditionally characterised by a strict hierarchy, at the top of which there are big farmer families whose possessions are passed on through the generations via primogeniture in a bilateral mode of inheritance, thus ensuring the concentration of resources within the family. This maximisation strategy results in marriages usually taking place only between firstborn children and under keeping of mutual economic advantage. Later-born young men, whose economic foundation is endangered anyway considering the scarce resources, usually try to meet the limited opportunity of marriage by outmigration, whilst for later-born girls, celibacy marks the way of reproductive self-restraint. They generally stay unmarried and serve as maids free-of-charge on the farms of their eldest sisters. Of all the behavioural options they theoretically have at their disposal, this helper-at-the-nest strategy is the one that both safeguards the reproductive options of the sister line and protects the control of resources in the best possible way under the given circumstances (see also Crognier et al. 2001). This form of kin selection is prepared and encouraged by principles of education, which are clearly geared towards the anticipated traditional distribution of power and social roles. Culture thus becomes a means for the accomplishment of family reproduction strategies. Thus, the successful perpetuation of biogenetic and sociogenetic characteristics, whose evolutionary biological success is stabilised by cultural selection, leads to the development of what has been termed an adaptive behavioural phenotype (Vogel and Voland 1987).

For the Eipo from the highlands of New Guinea, it has been postulated (Schiefenhövel 1986, 1989) that dynamic population regulation could keep the population at a stable equilibrium below the resource capacity of the habitat. Data identifying the total area of land cleared for horticulture indicate that the overall size of this area stayed about the same, at least for a period of 40 years prior to the field study. It led to the assumption that this area of land, accordingly, could sustain an equally constant and stable population. However, the sex ratio of those children surviving until the third year of life is clearly skewed in favour of boys, suggesting that manipulation of the secondary sex ratio can be made responsible. Indeed, this is supported by evidence of the preferential infanticide of girls. The readiness to kill a female newborn child – an action that is not socially penalised – is enhanced with increasing age of the mothers. This eventually leads to demographic zero growth as the number of postmenopausal women approximately equals the number of daughters surviving into reproductive age. In this way, the higher number of deaths occurring in young men, for example as a result of interpersonal violence, is re-balanced and results in a more equilibrated sex ratio in late adulthood. Further cultural rules, which for example affect sexual abstinence or coitus frequency, add an additional stabilising element. Under the given ecological conditions and resource availability, this so-called autochthonous system of self-adjustment functions as an adaptive element for the control of population growth.

At first sight, the examples seem to present two principally different views regarding the adjustment of population density to a given resource situation, as on the one hand population development is explained through individual or collective kinship reasons of fitness maximisation, whilst on the other hand collective self-regulatory mechanisms are made responsible for population adjustment to the ecological basic conditions. Yet, both explanations are not necessarily mutually exclusive, but rather are the expression of different points of views, namely on which level, but not whether, certain behaviour is being selected as adaptive. An ecologically orientated view would thus ask for those adaptive mechanisms which result in the adjustment of equilibria and maintenance of stability domains, whereas an evolutionary biological view would look for the fitness advantages of individuals. In both cases, however, the population aims at a state of relative ecological stability. Even with the Eipo, this may in the end be the result of individual weighing of pros and cons, of costs and benefits, which result from the raising of more or less children. Clearly without having to endeavour a group selectionist explanation (see Sect. 5.2.3) the adjustment of population densities can thus be understood as the cumulative effect of individual or family reproductive decisions, which in sum lead to observable equilibria. Proximate causes are varied and may be a reaction to resource supply or include safeguarding children as retirement insurance or work force; and the ultimate cause is invariably linked to fitness.

Differences and common grounds of strategies of adaptation will be discussed in the following, i.e. selected examples exploring the flexibility and variability of responses used by populations and their individuals in order to utilise resources under given or changing environmental conditions and how this affects population structures. The examples refer to the central areas of life and function that were the subjects of the preceding chapters. They will show connections between modes of subsistence and patterns of fertility, change of subsistence strategy and reproduction, or socially different access to and availability of resources within a population and differential reproductive success of individuals or groups.

5.1 Populations and Carrying Capacity

To understand the conditions under which mechanisms of population regulation become necessary at all, it is helpful to know about density distributions and the growth patterns of populations. Depending on the abundance in which resources occur in a given habitat, different dispersion patterns result for the individuals of a population (Fig. 5.1; cp. p. 174ff in Begon et al. 2003). The simplest case of a random distribution is of only subordinate importance, as it would imply equal probabilities of occupying a given point in space, which is not only highly unlikely in a species as necessarily social as humans, but which would also imply the random distribution of resources in a habitat. In contrast, a distribution approximating uniformity is to be expected if resources are evenly distributed in both frequency and quality. This scenario has been proposed for Neolithic farmers (Zwölfer 1987), whose small villages

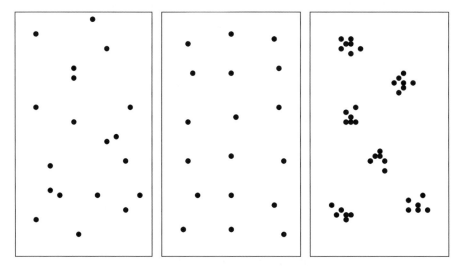

Fig. 5.1. Random, near uniform and aggregated distributions of individuals or populations in a habitat

and hamlets form patches distributed in a statistically almost uniform manner across a relatively large area with comparably good soil quality. Most frequently, human populations are to be found in aggregated distributions. Such distributions occur in direct association either with local concentrations of resources or with other more irregularly distributed materials or energy sources. Human settlements are typical and at the same time very complex examples of aggregated distributions, as they develop through the positive reciprocal effects of individuals within a population. Aggregated distributions reflect the most non-random dispersal pattern and create distribution patterns that are best described in terms of effective density rather than average density, as agglomeration is the characteristic trait.

The connection between resource distribution and spatial distribution patterns of populations has a bearing on the developmental potential, i.e. the growth of the respective populations. Two basic forms can be differentiated (Fig. 5.2). In the ideal condition that scarce resources or interaction with other organisms, predators or diseases do not limit the growth of a population, this results in exponential growth. Age-specific birth and mortality rates are constant. Normally, the exponential growth of populations is temporally limited, since continuous uniform use exhausts the resource capacity. However, as the example of world population development shows, exponential growth can be demonstrated as a general pattern over a long period of time, even though

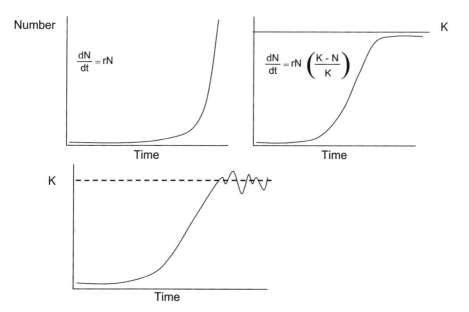

Fig. 5.2. Patterns of population growth. *Top left* Exponential growth. Logistic growth. *Top right* Ideal representation. *Bottom* Realistic representation, with oscillations of population size around the carrying capacity (*K*)

close examination reveals that it is a cumulative rather than a continuous process (see below). Also, world population growth has recently been projected to approach a plateau by 2100 (Lutz et al. 2001). Yet, for shorter periods the pattern of exponential can be sustained, e.g. for the population increase in Central Europe between 700 AD and 1300 AD (Grupe 1986).

If under conditions of exponential growth a limiting factor is set against the theoretically unlimited resource availability, the curve changes and turns into a saturation phase, i.e. growth becomes logistic. This change in the slope of the curve is the result of external or internal factors causing human population density on a certain level to change towards equilibrium. This level is affected by the growth limit of the respective habitat, i.e. the availability of resources and the ability of the individuals in a population to use these resources. In other words, carrying capacity first of all determines population size in equilibrium (Rogers 1992).

Evidence for the logistic growth of human communities is rare and scholars have struggled to provide empirical support for Malthus' mechanistic assertion, largely because the basic assumptions are quite rigid (see above). Forager societies are generally believed to live below the capacity limit of their respective habitat. In view of relatively large home ranges and a flexible group composition for an optimal use of existing resources, this seems very likely (however, see Sect. 5.2.1). Similar conditions have also been described for horticultural societies (see Rappaport 1968; Bayliss-Smith 1974; Brush 1975; Ellen 1982). The Tokugawa period of Japan (1720–1850; Rosenzweig 1971) provides a rare historic example showing that delimitation of population growth can also occur in complex societies, even though it is achieved through the deliberate control of the quantity and availability of resources in a given system. Archival sources reveal that during this time Japan was almost completely isolated from the outside world and was characterised internally by a strictly organised, rigid feudal system. Being dependent on the use of domestic food resources alone, even larger-scale population losses caused by regularly occurring famine were considered acceptable by the ruling classes and resulted in the population being kept at the level of approximately 26–27 million. Hence, in principle, there was a regulation of population size by the delimitation of resources, although the actual limiting effect could only be maintained through the prevailing rigid domestic power relations. Accordingly, political restriction, which approved of a policy of shortage and hunger, made the Tokugawa period a time of substantial unrest and rebellion among the rural population (Vlastos 1986). The carrying capacity of a habitat, as this example shows, is thus a function not only of biological or ecological, but equally so of socio-cultural conditions.

Growth delimitation, however, rarely remains permanently below the capacity limit. In many ecological systems, oscillation around the natural growth limit of the habitat is observed. This has been described as a non-equilibrium pulsing state, where maintenance of the system (respiration, R)

on average does not exceed production (P), so that the carrying capacity is reached when $P/R \sim 1$ (Barrett and Odum 2000). Likewise, the excessive use of resources can lead to a temporary or long-lasting drop in the capacity limit. This would be the case when the carrying capacity has been exceeded to such an extent that the system has lost its resilience and flips from one ecologically stable equilibrium to another (Seidl and Tisdell 1999). The respective capacity of a habitat is thus a density-dependent, growth-limiting factor bearing the consequence that populations in the long run are resource-limited in their growth potential. The competition for limited resources within a habitat can thus involve the necessity for density-dependent population regulation.

In human populations, the dynamic adjustment of population density to the capacity limits in a habitat is much more frequent than a homoeostatic adjustment. Facilitated in particular by the fact that learning processes, the formation of traditions and behavioural changes can lead to acquiring new resources or an increase in the efficiency of resource use, a continuous or precipitous expansion of the carrying capacity limits occurs as a result of the interaction of cultural and ecological capacities. Taking this into account, the population trend from the Palaeolithic to modern times demonstrates that world population growth can be seen in a more differentiated way, i.e. no longer only as simple exponential growth, but as logistic growth with variable capacity limits, connected in a sequence of consecutive stages (Fig. 5.3). It is true that, on an individual basis, each of these stages starts with exponential growth, because any expansion in resources, i.e. a reduction in resource pressure, leads to an initial, rapid population growth. This growth, however, stabilises after a time, leading to a saturation plateau (see Groube 1996). In fact, the momentum of rapid population growth implies downsizing of population density after an initial overshoot (Barrett and Odum 2000). For human societies, therefore, quite a special form of logistic growth is the regular case.

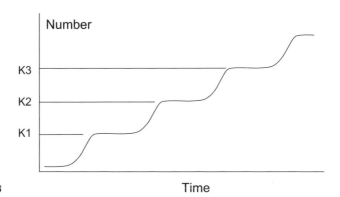

Fig. 5.3. Logistic growth with variable capacity limits *K1–K3*

Butzer (1976) provides an example of population growth which can be interpreted in this sense of a shift of variable capacity limits. In his investigation on population trends in Egypt since the early Neolithic to approximately 1000 BP, projected from data for settlement densities, the utilisation of water from the Nile is taken as the basis for a constantly improving adjustment of irrigation efforts. In the initial phase of agriculture, humans simply made use of the Nile annually flooding its banks and depositing extremely fertile silts in the form of fluvial sediments. People lived on the high terraces and cultivated arable crops within the range of the alluvial fans or riverbanks. During this time of exclusively natural irrigation, it is thought there was only a small increase in the population size. Approximately during the first half of the fifth millennium BP, the natural irrigation system was replaced by an artificial, technologically steered system. By excavating the branches of the river and digging connection channels in order to steer the water-flow and, additionally, distribute it onto the fields by means of levered buckets, the annual inundations could be better controlled and the water better channelled to where it was actually needed. This, in turn facilitated up to three harvests per year. During this transitional phase, agricultural produce could be at least partly increased and, accordingly, Butzer extrapolates a moderate (and, in relation to the previous period, an increased) population growth for this time-period from the rising numbers of settlements. The increase in population size could be continuously sustained in the following time-periods right into the Roman epoch of ancient Egypt. The amount of cultivated land was expanded and in particular the introduction of mechanical lifting devices, which even in times of lesser Nile flooding ensured that sufficient amounts of water could be brought on the fields, facilitated further population increase. The assumed population development is directly connected with technological innovation and the resulting shift of the capacity limits in a habitat. This principle can be illustrated by adding hypothetical capacity limits to the outline of population growth suggested by Butzer (Fig. 5.4).

This example clearly illustrates that the principle of carrying capacity in human ecosystems primarily has a meaningful descriptive or heuristic value. As far as the relationship of population dynamics and cultural or technological developments and the connection of these two phenomena are concerned, carrying capacity is suitable as an illustrative means, in order to qualitatively represent the efficiency of resource yield and the stability within a system. However, it is very difficult to exactly calculate the carrying capacity of a habitat in any particular instance. The difficulties encountered are of a principal nature, because the dynamic character of carrying capacity entails that, if at all, only a very short-term and partial view of subsistence developments and population trends can be accomplished. The comparison of carrying capacities of different populations in equally different habitats is even more difficult (see Ellen 1982). The diversity of conditions under which at any given time an optimal – or mal-adapted – population size may be observed with respect to

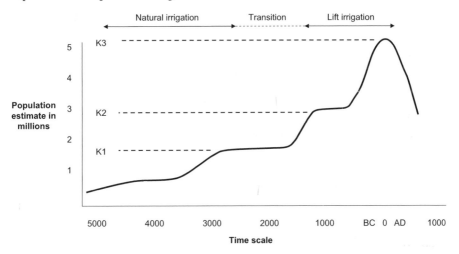

Fig. 5.4. Schematic representation of estimated population growth in ancient Egypt (modified after Butzer 1976). Increase in population size is facilitated by technological progress, which leads to expansions in carrying capacity (*K1–K3*)

the carrying capacity of a habitat leads to basic problems of comparability, because such an attempt would also have to consider annual or seasonal fluctuations in resource supply or work intensity, i.e. a temporal depth for the availability of resources. Equally, carrying capacity also depends, among other things, on the kind of existing political or administrative controls in a society, on how land use is regulated, or whether differences in property and the distribution of land may lead to substantial differences in the respective carrying capacities. In other words, carrying capacity needs to take into account individual impact, as for example affluence would reduce the number of people that can be supported by a given resource base. Carrying capacity has thus been considered to be two-dimensional (Barrett and Odum 2000), because it represents density and per capita demands in a habitat.

From a human ecological viewpoint, differentiating two kinds of carrying capacity appears to be more important than calculations, instead. First of all, ecological carrying capacity is given by the resource situation in the habitat itself, i.e. population density is assessed at a given time in relation to resource capacity. But human exertion of control, e.g. in terms of technological or social changes, introduces a cultural carrying capacity into the system (Moore 1983; Seidl and Tisdell 1999). If human populations at a certain level of population density are adjusted by the carrying capacity of a habitat, then there is a connection both with the availability of resources and with the ability to actually use these resources. Ecological and cultural carrying capacities are thus inter-related. The cultural capacity limit, in terms of technological innovations, however is in itself a density-dependent phenomenon: as population growth can gradually lead to resource depletion and shortage, efforts and considerations are promoted to again raise the ecological carrying capacity and the capacity limit in a given habitat by a technology momentum (for rural populations, see Boserup 1965; for industrialised countries, see Simon 1977). That is, technological innovation and the shift of the capacity limit by a cultural exertion of control would be an effect of rather than a reason for population growth (Rogers 1992). Population growth is then, in turn, a possible response to technological change, showing that these components are mutually dependent. It appears difficult, however, to specify in individual cases which of the two factors – population growth or technological innovation – is to be regarded as the independent variable in this mutually reinforcing system with variable capacity limits (Cohen 1975).

By considering both the time factor and the interdependence of cultural and ecological capacities in a given habitat, it can be summarised that carrying capacity of a habitat corresponds to the population size which at a given point in time can be regarded as the optimal adaptive state or density of a population. Because carrying capacity is determined both by the resource situation and by behaviour options, it is dynamic and uncertain by nature (Cohen 1995). It relates to different more or less stable states of the environment which are brought about by a complex set of factors, including institutional arrangements, technology, consumption patterns and human goals (Seidl and Tisdell 1999). The carrying capacity of a habitat is thus crucially determined by the subsistence strategy of the human population; and its effects on population development, in turn, affect any adjustment of the capacity. Each considerable change in human population size leads to a dissolution of the equilibrium state that once existed and requires adjustments, in order to re-establish a new steady state. Therefore two different options are available in principle: a flexible shift in the capacity limit with the possibility of renewed logistic growth, or a regulation of the population to or below the carrying capacity of the habitat.

5.2 Mechanisms of Population Regulation

In the following sections, regulation of population is understood as the effect of any form of change in the composition or size of a population in connection with the availability of resources and their utilisation options, without asking in the first place at which level the changes are caused and bear an influence on demographic parameters. Thus, intrinsic and extrinsic factors can be equally decisive for an adjustment of changed numerical relations within a population.

First, connections between the size and possibilities of development of populations and diseases, as external regulation factors, are briefly addressed. For pre-historic times, in particular, the outline of a historical epidemiology has to be considered a desideratum; and it is outside the scope of this book to even provide a draft of it. The following remarks are therefore limited to those resource-related topics dealt with in the preceding chapters and they aim at establishing general connections with e.g. modes of subsistence and subsistence change. Moreover, the emphasis of the chapter in general is on internal mechanisms of population regulation and their effects on the emergence of population structures, so that the aspects of disease, environment and population are discussed first, in order to allow reference to those aspects at later stages.

The chapter will further examine such cases of adjustments of population structure that, under conditions of natural fertility, can allegedly be attributed to self-regulating mechanisms, in which the population as a whole is said to possess adaptive feedback control options, so that exceeding the capacity limit is avoided. The sequence of examples is arranged in such a way that, besides the actual or alleged regulation on the population level, the amount and importance of individual decisions for regulation mechanisms becomes increasingly evident and can be demonstrated.

5.2.1 Environment, Disease and Population

In ecological terms, the emergence and propagation of diseases is regarded as an evolutionary interaction between pathogenic organisms, the human host and given biotic and abiotic environmental conditions that allow this interaction to develop. In a human ecological context, the environment is extended by socio-cultural components which are used to change natural habitat conditions and to transform them into systems shaped by human activity that form the ever-changing basic conditions for the co-evolution of pathogens and humans. As a reaction to specific external conditions, organisms aim at a state of internal homoeostasis, as adaptive mechanisms are selected which sustain survival – with or despite the pathogen – and which ensure proper functioning of physiological and biochemical pathways. If situations develop in which an organism can no longer react adequately to the stressor, i.e. if there are no means of adaptive reaction available or they cannot be developed quickly enough, this leads to the outbreak of disease. Diseases thus develop from a lack or failure of adaptation to existing environmental conditions under the effect of pathogen stress (Goodman et al. 1988). In the context of co-evolutionary interrelations of environment, host and pathogenic organism, therefore, balancing selection towards steady states between the components involved can be expected.

The diversity of the environments which human communities were exposed to and which they created in the process of their biocultural development led to regionally and temporally different disease patterns that, in turn, were expressions of this human/environment relationship. Depending on natural habitat conditions, contact with other populations, deliberate alteration of natural units and collateral impact, resource supply and the resulting subsistence techniques, these environments make possible varying population densities. Growth and density distributions of populations should thus show correlations with the emergence and occurrence of certain diseases and allow a conclusion to be drawn on ecological basic conditions, because significant transitions in human ecological history are connected with particular disease patterns (McMichael 2004).

Estimations of human population densities at the end of the Palaeolithic come to approximately ten million, but it may have even been fifteen million under the conditions of a fully utilised carrying capacity in those parts of the earth occupied at the time. This suggests a small growth rate of below 0.02 % per year on a global scale for the longest time of human development (Hassan 1983); and other authors suggest growth rates as low as 0.008 % (Pennington 2001). With the transition from an acquiring to a producing mode of subsistence during the spread of the Neolithic, there is evidence for higher population densities. Growth rates are likely to have been at least 0.05 % and, under particularly favourable resource conditions, even rates of 1 % can be assumed (e.g. for southwestern Europe; Jackes et al. 1997). Accordingly, during this transition period there is a rise in world population size that is closely related with altered resource situations and subsistence patterns. At present, though, it cannot be exactly clarified which process caused this development (Cohen 1975), i.e. to what extent population growth has to be regarded as the independent variable in this system, which through population pressure caused a technological change and a shift of the capacity limit (Boserup 1965), or rather whether a new subsistence mode and technological progress accommodated larger populations. There is no question, however, that they were mutually dependent (see Chap. 3).

The interruption of relative demographic equilibrium at the end of the Palaeolithic is said to have been caused either by increased fertility, by reduced mortality or by a combination of both (Landers 1994), with the latter explanation being the most probable. It is remarkable, however, that in the context of the transition to agriculture generally, in contrast to the forager societies that previously occupied the habitat, at first an increase in mortality and reduced life expectancy can be observed (see Cohen and Armelagos 1984). Cross-cultural data of living and sub-recent pre-industrial populations support this result in as much as mortality figures between foragers, horticulturalists and pastoralists do not significantly differ (Hewlett 1991). As long as there are no conditions of intensive agricultural subsistence present, this seems to be a generally confirmed trend.

The development and adoption of agriculture as a subsistence technology were accompanied by changes in lifestyle that crucially affected the occurrence and persistence of diseases. Humans began to permanently live in larger settlements that became more and more densely populated over time. At the same time, this brought about an increasing proximity and contact between humans and between humans and animals. To the extent in which critical concentrations of population numbers were reached, diseases could therefore on the one hand be more effectively transmitted and on the other hand become effective as density-dependent external modulators of population densities. Here, in particular, population-effective viral and bacterial infections and parasitic infections have to be mentioned. Thus, as a function of density distributions of human populations, certain predictions can be made as to the disease profiles connected with the development and change in modes of production (for the general context, see Anderson and May 1982; Harrison and Waterlow 1990; Swedlund and Armelagos 1990). Impaired hygiene and poor nutritional quality, often concomitant with crowded living conditions, likely exacerbated this situation.

Since the emergence of certain modes of production has to be viewed within a framework of socio-cultural evolution (see Sect. 2.1), predictions of correlations between patterns of population dispersal and disease should be possible for earlier historic times, as well. In principle, evidence of disease can be ascertained in human skeletal remains. While palaeopathology traditionally employs morphognostic and microscopic methods to diagnose skeletal change caused by disease (Ortner 2003; Schultz 2001), DNA analysis of archaeological skeletons targets the detection of pathogen DNA and thus provides an approach to even identify skeleton-mute or sub-pathological infections that left no traces of skeletal alteration. Such molecular-biological assessment has, with varying success, been attempted for tuberculosis, the plague and leprosy/treponemal disease (e.g. Baron et al; 1996, Burstein et al. 1991; Rafi et al. 1994; Greenblatt 1998; Wiechmann and Grupe 2005; Zink et al. 2005) and, while still sometimes riddled with technical problems (e.g. replicability), it offers unprecedented potential for the elucidation of past disease patterns. But since it is far from being widely applied or used as a screening device, the majority of palaeopathological information is still derived from macro- and micromorphological inspection.

Under the conditions of a foraging mode of subsistence, or in general in pre-agricultural communities, which feature small group sizes and relatively high spatial dispersion, the prevalent diseases are those that have a slow speed of infection, low virulence or those that use intermediate hosts (Ewald 1994). Furthermore, the pathogens of importance and able to stay endemic among human populations scattered across vast areas were those which can survive for a very long or even an unlimited period of time inside a host organism, for example chickenpox or hepatitis. In general, enzootic agents and vectors are thought to have prevailed. Due to this endemic persistence, however, there is high resistance against the pathogen within the host population, so that such a kind of host/pathogen relationship is commonly regarded as phylogenetically relatively old and widespread (Meindel 1994). Also, parasitic infections such as bilharziasis may have played a part in tropical areas.

[The transition from foraging to farming provided new opportunities for random contact of mutant microbes from domesticated animals and pests (e.g. rodents, flies) and humans.]While the majority of such encounters would have been without consequence, those that were successful had significant impact. With increasing population densities and a permanently sedentary way of life, the propagation of the majority of today's known infectious diseases, e.g. measles or rubella, but also typhoid and dysentery, were promoted. The survival of the pathogens causing these diseases requires host population densities of at least, often well over, 100,000 individuals in order to sustain a sufficient number of infected individuals within the population. Transformations of the environment in connection with agricultural techniques may have favoured the propagation of certain pathogens. Malaria, for example, is more frequently found in areas where slash-and-burn agriculture is practised than in unaltered tropical environments, because forest clearings create a more suitable microhabitat that provides the malarial mosquito (*Anopheles* spp) with more favourable conditions for survival (Mascie-Taylor 1993). In some areas, mechanisms leading to an improved persistence of the pathogen have caused genetic adaptations in the form of balanced polymorphisms, as in sickle cell anaemia (see p. 221ff in Harrison 1988), which through selective adaptive advantage allow human populations not only to cope with the pathogen stress but also to continue their occupation of a certain habitat.

During periods of rapid geographical expansion of anatomically modern humans, it can be assumed that the speed of their propagation was also favoured by the fact that they advanced into areas with only low incidence rates of infections and that pathogens, which could become dangerous to humans, were thus virtually missing (Dobson 1994), even though the reverse cannot be excluded because pioneers may have encountered novel pathogens to which their immune system was not adapted. The migration and mobility of humans into areas already inhabited, however, was frequently accompanied with substantial health risks for the indigenous population. The increasing concentration of humans promoted a growing mobility between local populations related to the production of (stationary) food resources within more densely populated areas, as transport and exchange of goods and commodities attained increasing importance. This, in turn, only created the conditions favourable for infection and new and quick paths of transmission arose for diseases that would not have persisted under conditions of low-density distributions. Increased military and commercial contacts between the historic empires of the Old World facilitated the swapping of dominant diseases, or rather pathogens, leading to a trans-European equilibration of infectious disease agents (McMichael 2004). The spread of the Justinian Plague in the sixth century and the Black Death during the fourteenth century in most of Europe provide a manifest illustration of the devastating effects of novel pathogens, with all known dramatic consequences for the decimation and survival of populations (e.g. McNeil 1976; Keil 1986), particularly in times of major demographic and social change. Likewise, during the age of exploration and imperialism, the export of pathogenic organisms led to a trans-oceanic spread of diseases on an unprecedented scale, affecting South Asia and the Pacific, Australia and the Americas. On every occasion, conquistadors and colonists hit populations which were not immunologically adapted to the new pathogens and which were thus fatally hit by ailments established as common infectious diseases in the Old World, for example measles, smallpox or influenza in the case of the Americas (see Crosby 1986). As a consequence of increased trade and travel, the spread of infectious diseases has today reached a global scale, with enormous implications for biocultural adaptation in times to come.

Overall, pathogen stress and density are closely related with ethnic diversity, i.e. they constitute a biogeographic patterning which follows latitudinal gradients. It affects population size to the extent that chiefdoms and states in areas of high pathogen stress are also numerically small (Cashdan 2001). This may be one of the reasons why European demographic replacement of indigenous populations was only successful in temperate zones but not in the tropics (McNeil 1976; Crosby 2003).

Effects of subsistence change to agriculture on the spectrum of human disease have been a particular focus for some time and can meanwhile be demonstrated for more than 20 regions of the earth, based on palaeopathological findings (Cohen and Armelagos 1984; Cohen 1989; Armelagos 1990; Belfer-Cohen et al. 1999; Cohen 1994; Eshed et al. 2004a). Degenerative joint disease or an increase in robusticity, conditions which can be rated as indicators of increasing physical workload due to agriculture do not reveal uniform results that would suggest skeletal manifestations of increased physical demands and activity. Those infectious processes that lead to bone changes can be more clearly assessed, though. Both non-specific osseous reactions, e.g. those caused by osteomyelitis and disease-specific skeletal alteration (e.g. from tuberculosis), are found at higher rates in agricultural populations. It has been argued that the correlation between increased frequencies of infections and permanent sedentariness was common knowledge from everyday life experience and might thus have prevented human communities from increasingly adopting a sedentary lifestyle. But this was not the case: instead, this mode of living occurred and spread widely, which is regarded as evidence of an increasing desire either for and reliance on a secured food supply by provision of stocks or for an improved possibility (or necessity) for defence, while impaired health conditions had to be accepted (Cohen 1994). Yet, there may have been other reasons, for example related to improved reproductive options (an avenue that will be explored later in this chapter).

An increased prevalence of *cribra orbitalia*, the porotic pitting of the orbital roof of the skull has been advocated as evidence of the general degradation of food quality during the transition to agriculture (Cohen 1989, 1994). Such bone changes are commonly and readily connected with long-lasting dietary deficiencies or parasitic load that lead to anaemia – even though other aetiologies are known and well founded (see Ortner 2003). In particular, an increased consumption of food rich in fibre, such as cereal grains, is noted among the causes for a decreased absorption of iron from the diet. However, specifically in the case of iron, a high plasticity in the intake of this trace element from the diet has to be taken into account, so that at least faint bone manifestations of this lesion do not permit one to conclude a severe state of deficiency (Schutkowski and Grupe 1997). In addition, higher frequencies of other lesions indicative of stress situations that may result from synergies of deficient nutrition and infections, e.g. disruptions in the mineralisation of the tooth crown (enamel hypoplasia), point to the fact that agriculture and sedentariness did not (or not right away) lead to an improved capacity of buffering ecological or generally resource-linked crises, but on the contrary caused an intensified susceptibility to such events.

Despite these trends, which indicate an initial degradation of the nutritional situation and living conditions during the transition to agriculture, substantial geographical variability has to be assumed. Particularly, for the transition phase from a foraging broad-spectrum economy during the Mesolithic to Neolithic agricultural production, there is a gradient stretching from the northeast to the southwest of Europe, in the course of which subsistence change took place without recognisable damaging consequences for the health of the populations, as far as this can be derived from the skeletal record (e.g. Meiklejohn and Zvelebil 1991; Jackes et al. 1997), a trend that is beginning to emerge for the Levant as well (Eshed et al. 2004b). This corresponds to an only slight reduction in body height, or even no change at all, which has been proposed as a useful indicator of developmental and growth conditions of human populations and the correlation of economic prosperity and biological well being (see e.g. Steckel 1995; Bogin 2001). It also seems remarkable that, in the course of increasing differentiation within societies, social differences in infestation frequencies and disease rates begin to emerge, with a better health status of socially high-ranking individuals being related to nutrition of better quality (e.g. Cohen and Armelagos 1984; Walker and Hewlett 1990; see also Chap. 4).

While this suggests variances in health that show up through differential access to resources within the population, it has been attempted to explain general characteristics of social organisation of human populations likewise as a reflection of habitat-typical disease patterns, which are incidentally a corollary to the prevalence of overall environmental determinants, which govern the dispersal of human communities (Cashdan 2001; Collard and Foley 2002). Stress caused by pathogens is part of those environmental conditions that are characterised by a lack of predictability. It is therefore regarded as one of the drivers for the evolution of sexual reproduction, since thereby genetic variability is ensured among offspring with the prospect of more possibilities of plastic reaction to uncertain habitat conditions (Hamilton et al. 1981). It has been postulated that disease stress and marriage systems are causally connected in human populations with the effect of spreading the risk of genetic burden (Low 1988, 1990). Based on data collated from 186 societies world-wide, this hypothesis was examined and the influence of different pathogenic organisms (*Leishmania, Mycobacterium leprae, Plasmodium, Schistosoma, Trypanosoma*, filariae, spirochaetes) was examined. It was demonstrated that the number of highly polygynous marriages increases with increasing pathogen stress, whereas monogamy, polyandry and low-level polygyny decrease (Fig. 5.5). Since the three latter forms are missing completely under conditions of high pathogen stress, a threshold effect dependent on the stress rate has be assumed to be in place here, triggering a higher occurrence of polygynous marriages only when the combined exposition to pathogens exceeds a certain level of potential detrimental health effects. It was furthermore observed that sororal polygyny, i.e. the marriage of several sisters to the same husband, decreases under the same conditions and, instead, the exogamous choice of women is more widely spread. Therefore in conditions where the habitat provides a high potential of pathogens that may lead to a substantial impairment of reproductive possibilities and options, socio-cultural strategies of marriage patterns may be in place to secure that the genetic variability of the offspring is as high as possible, so that reproductive success can be optimised even if there are casualties due to prevalent infections. This can be regarded as further evidence that, even with increased selection pressure from pathogens, cultural responses to ecological constraints can be achieved and established, which in the interplay of pathogen

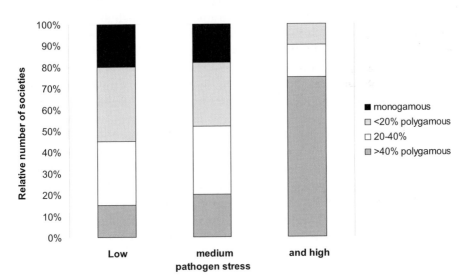

Fig. 5.5. Variation of polygyny in 186 societies, depending on the level of pathogen stress (modified from Low 1990)

and host in sum lead to balanced co-evolutionary conditions, without individuals losing out on their reproductive interests.

It follows, however, that even under the limiting conditions of the substantial external stress situations that may affect populations, there is a reaction norm of potential fertility, which appears to be always at or above the one needed in order to maintain population size, i.e. which allows populations to approach the capacity limits of their habitats. Which mechanisms become effective then, or how big is the influence of intrinsic control options of population regulation, if a constant adjustment of population density to the resource options of the habitat is necessary, in order to prevent overexploitation of existing resources?

5.2.2 Culture and Ecological Equilibrium

The view of regulation as a systemic characteristic of populations corresponds to classic systems–ecological concepts of equilibria between a population and other components of the habitat adjusted by negative feedback (e.g. p. 576ff in Begon et al. 2003). Accordingly, in human populations, behaviour patterns and socio-cultural institutions in their entirety – seemingly – serve to maintain stability and the propensities of self-regulation. Such connections have been considered as explanatory models for neo-functionalistic or culture–materialistic interpretations (see Sect. 1.1), by which the observable homoeostasis between population and environment is described as advantageous for the population. Empirical observations have repeatedly been explained in this way and taken as evidence for the existence of homoeostatic characteristics. Three particularly illustrative cases, which at the same time show different modes of subsistence and socio-political organisation, are presented in the following and will be examined critically as to how they fit with such a functionalistic interpretation. By their order of arrangement, the examples will clarify that, apart from the characteristic of an alleged regulation attributed to the population as a demographic unit, increasingly individual decisions can be identified as causal factors for regulatory measures. Besides the obvious systemic interpretations, it will thus have to be shown that models, which are compatible with theoretical basic assumptions of evolutionary biology and thus less idealistic and supportive of group selectionist behaviour, at least allow a plausible alternative explanation of the empirical data, if not even the only conclusive reason. Among the classic examples of this kind is Rappaport's (1968) study of the Tsembaga Maring in New Guinea (see Sect. 2.1).

5.2.2.1 Of Pigs and Men

The population size of the Tsembaga, one of approximately 20 Maring-speaking local populations in the highlands of New Guinea, amounted to approximately 200 persons at the time of Rappaport's field study. The group was divided into five clans which shared common rights to use natural resources; and they maintained close marriage relationships with one another and were engaged in and allied through violent conflicts against neighbouring groups, but they were equally connected by the arrangement of feasts and rituals. One of these rituals, the *kaiko* celebration, gave rise to the assumption of autochthonous mechanisms of population regulation among the Tsembaga (p. 153ff in Rappaport 1968). The *kaiko* is a constituent of a ritual cycle which is characterised by repeated flashes of aggressive conflict in inter-group warfare and intermittent periods of peace of 5–20 years' length. Superficially, there are various causes for these arguments; but the ultimate reason lies in competition for land as a resource.

Tsembaga subsistence is essentially based on horticultural produce. Since the fertility of tropical soils is exhausted quickly, new gardens must be laid out regularly and the previously cultivated areas taken into fallow for many years, thus creating a pattern of shifting cultivation. In addition, domesticated pigs are kept, a task predominantly carried out by women. In order to habituate the pigs to the settlement, piglets are hand-raised and, even though allowed to feed in the forests on their own, they are regularly supplied with food.

The size of the herds is subject to cyclic fluctuations and mainly adjusted through the *kaiko*, since during the ritual, with the exception of a few juvenile animals, all pigs are slaughtered. With increasing herd size, however, the increasing numbers of pigs, which are fed with tubers from the garden, develop into serious food competitors for the human population. Immediately prior to the *kaiko* observed by Rappaport in the 1960s pigs consumed about half of all cultivated sweet potatoes and over 80 % of the manioc. Just to accommodate this supply in fodder, the average garden area therefore had to be more than one-third larger before the *kaiko* than thereafter. The meat of the animals culled during the ritual only serves as food for the Tsembaga to a small extent: the predominant part is sacrificed to supernatural powers or distributed to allies, who thereby can be compensated for their support and assistance in past inter-group conflicts. Thus, on the whole, the nutrition of the Tsembaga can be regarded as poor in animal protein (see Sect. 2.2.2). Where, then, is the adaptive advantage to accumulate large herds of pigs, which moreover eat a large part of the arable crops and are thus energetically more costly than beneficial to the human population?

After all allies have been compensated, the excavation of a certain plant (*rumbim*) marks the end of the peace period and a set of taboos are waived, which during the past years prevented violent conflicts taking place or resources being over-exploited. Hostilities break out and are usually not terminated until the disliked parties agree that the conflict has taken enough casualties and a higher death toll would be unacceptable. The *rumbim* plant is set and a new period of peace begins. At the same time, this is the beginning of a new round of pig breeding, with the known consequences of resource shortage. Rappaport (1968: p. 224ff) explained the function of the conflicts by the fact that this is the only way by which the scarcest resource, i.e. land, can be re-divided and re-distributed among the local populations as part of a cyclic operational sequence. Clans are being reassured of their support by alliances, which are renewed by reciprocal distribution of pork meat on the occasion of the celebrations. The number of war casualties, estimated to amount to 2–8 % of the total population, affects the size of the work-force available and the defensive capability of the community. However, whether these losses at the same time represent a control of population increase is questionable, particularly since it is usually men that get killed who, in contrast to women, are biologically speaking not a limiting factor for reproduction.

Rappaport thus interprets the linkage of population density, resource situation and ritual as an example of density-dependent regulation. But what is the trigger to embark on a new phase of conflicts, i.e. what initiates a new *kaiko* and thus a new ritual cycle? Which signal is suitable to give information on the ecological quality of the habitat? It is the size of the pig herds which is said to be the indicator, although in quite a subtle way. It is not the number of animals which is decisive, but the increasing complaints of the women over the amount of additional work they do in the gardens, as ever more pigs must be fed. The workload, as it were, thus represents the sensor for a threatening critical constellation, in which too many pigs take too many resources and the humans and pigs compete for land, the increased use of which is subject to a taboo. In the logic of an ecosystemic interpretation, it follows that by way of the *kaiko* ritual both the human population and the pig herd size are adjusted in a feedback loop and an excessive use of the habitat is avoided. The ritual thus becomes an unconscious mechanism that allows the quality of the Tsembagas' habitat to be maintained, without pushing the population to or beyond the borders of the carrying capacity and leading to habitat degradation.

Independent of how likely at all it is that such a thought-out mechanism, as a collectively unconscious action, can stabilise the human population below capacity limit, the question is how lasting such an adjustment might be. Can it be reckoned that, through the *kaiko* ritual, the ecosystem is balanced on a long-term basis? Simulation experiments showed (Foin and Davis 1984; see also Shantzis and Behrens 1973) that fluctuations in both human population size and pig population will lead to equilibrium only on the condition that it can be assumed that the regular war losses exactly equalise the increase in human population size achieved in the course of the peace period. Every other constellation would either entail the loss of pigs, a decrease in the human population or both. The only stability domain which is still rendered possible by the simulation cannot be conceived as durable under real conditions, as it would presuppose that the allegedly crucial determinant variable, the pig population, can be assessed so accurately that the timing of a _kaiko_ can be specified precisely in each case and would not be fixed too early or too late. The ritual is thus highly unlikely to function as a device of population regulation.

In more far-reaching simulations (Foin and Davis 1987), the level of the ritual was left behind and an attempt was made to include fundamental ecological variables which are of importance altogether for the subsistence of highland populations of New Guinea and which allow general connections between resource use and homoeostatic adjustments to be established. Among other factors, the average duration of garden use, the natural succession on the fallow areas, the different growth rates of the human populations and the food energy consumed on average were considered. On the assumption of altogether small fluctuations within the system it showed that Maring populations aim at lasting equilibria, which are usually achieved within two generations. However, in the case of disturbances, it takes a considerable length of time to enter a new stability domain. A loss of 5 % of the population, for example as a consequence of inter-group conflict, would entail a recovery period of 80 years, even for a system under otherwise stable and continuous basic conditions, i.e. a considerable decrease in resilience capacity (cp. Holling 1973).

Even if equilibrium can be assumed to develop afterwards in the model, this scenario normally evades scrutiny under real empirical conditions. In fact, the populations at any point of the observation are in disequilibrium, which is maintained by external factors such as the fragility of the habitat and the resulting resource shortage. At the same time, however, the population constantly and in the long-term aims at a state of equilibrium within the system, whose achievement is possibly even facilitated by the _kaiko_ ritual, although hardly by its functioning as an integrative determinant. Dissipation or homoeostasis depends on the respective point of view. An ecosystemic interpretation is confronted here with an unsolvable dilemma since, depending on the viewpoint, an adaptive stabilising function may either be plausibly accepted or rejected. Is there nevertheless a possibility to explain the existence and function of the ritual in a behavioural ecological context?

The social organisation of the highland populations of New Guinea into clans and lineages is connected with the control of resources. These corporate groups are based on kinship relations; and kinship relations also exist with other groups through exogamous marriage beyond the own local group. The ritual, whose central constituent is the distribution of pork meat for the paying-off and renewal of social obligations, strengthens the co-operation between certain groups regarding the subsequent redistribution of resources in the course of conflicts. It can be assumed that in the end there is a strategy hidden behind the complex ritual, which by means of kin selection provides a reproductive advantage to certain sets of the gene pool within the population so that it provides members of certain kin groups better access to resources and thus better chances for the raising of their offspring. The ritual would then serve the confirmation of co-operation and alliances of individual kin groups and the establishment of a stable strategy, since under conditions of sharp and hostile competition over resources between the local groups, co-operative strategies could indeed have a selective advantage (see Peoples 1982). The fact that the ritual is not held at regular intervals but rather if and when required lends support to this argument. Likewise the herd size, which should actually serve as a sensor for the coming arrangement of the celebration, is artificially manipulated by the additional purchase of pigs, so that a new cycle can be initiated more quickly. This would however only make sense if the resource/land ratio for certain individuals or a certain kin group became marginal, so that the reproductive possibilities could become impaired. Such an interpretation could then, in agreement with evolutionary biological concepts, also explain regulation visible on the level of the total population, which aims at equilibrium in the sense of ecosystemic feedback.

5.2.2.2 Camels and Age Groups

A second precedence of population-stabilising mechanisms has been observed and described for the Rendille, a community of pastoralists living in the arid lowland of the Kaisut and Chalbi Desert southwest of Lake Turkana in Kenya (Spencer 1973; Fratkin 1986, see also p. 229 in Moran 2000; p. 45 in Harris and Ross 1987). Their social organisation has a patrilineal structure and is arranged into clan or lineage groups. The smallest segmental unit is the extended family, which is made up of several nuclear families (Sato 1980; also for the following data). The Rendille live in a dual-residence system of settlements and herding camps, whose composition and number alters with the seasonal changes in environmental conditions. Their herds predominantly consist of camels, but they also keep sheep and goats (Fratkin 1986). A peculiar fertility depression has been observed among the Rendille that was interpreted as a response to a low carrying capacity and limited resources in the habitat that would allow a lowering of the reproduction rate to be stabilised by cultural implementations (Roth 1993, 1999).

Apart from numerous other characteristics they share with neighbouring pastoral societies, the Rendille possess a complex categorical system of the age organisation, which consists of age sets as well as age lines across generations. This system differs between the sexes. Male individuals belong to both different age sets divided according to age roles such as boys, warriors and elders (= married men) and age lines. Females, however, usually only belong to age sets, which are likewise differentiated according to social roles.

Every 14 years, a new age set is constituted by a collective circumcision ceremony. The formation of such age sets is subject to a complex set of rules. The initiation of boys takes place according to the birth sequence of their brothers and is scheduled in such a way that boys become members of the third age set following that of their fathers. In this way an age set line develops, in which every third age set is linked according to a father–son–grandchild pattern. This cyclic structure of age sets thus leads to equally cyclic age set lines, which tie-in all the Rendille. One of these lines is named *teeria* and represents a group of high-status individuals, as it consists of firstborn children only.

Daughters, whose fathers are members of this group, are subsumed under the term *sepaade*. The majority of Rendille girls marry when young warriors are at an age corresponding to the age span, which results from the addition of two age groups to that of the daughters' fathers. *Sepaade*, however, have to wait for the time span of a further age group and in addition are only allowed to marry after their brothers do. Thus, the date of marriage of *sepaade* women can be postponed by up to 14 years.

As a result of these regulations, average marriage ages for men are 31.8 years and for women 25.1 years (at the time of the field study). Population growth of the Rendille was calculated by Sato (1980) to be 1.7–2.6 % per year. With a total population of 423 individuals at the time of the study, this would correspond to an increase of approximately 7–11 persons per year. Such small growth rates of Rendille however seem to be peculiar. If marriage ages of neighbouring pastoral societies are taken as a comparison (e.g. Somali, Samburu, Gabra), it shows that men marry in their late twenties, while women get married between 15 and 20 years of age. Compared with these figures, Rendille women above all clearly enter reproduction later. At the same time, unmarried women are subjected to strict pregnancy prohibition. If a woman nevertheless becomes pregnant, then the illegitimate child is either aborted or killed after birth.

Also regarding inheritance regulations, the Rendille differ from their neighbours. Among the Samburu, the owner of a herd divides livestock among his wives during his lifetime, so that after his death it can be passed on to his sons. Among Rendille, however, women are not entitled to possess herds, but instead in the case of her husband's death the herd is transferred to the firstborn son. Later-born sons are only given some camels on the occasion of their wedding, which leads to the fact that they usually migrate to the neighbouring Ariaal or Samburu. Eventually, this also forces women to marry outside their group.

Thus, a whole bundle of cultural measures is present among the Rendille that leads to a reduction in fertility, from the age set system right up to direct interference in population control. The tuning of the different mechanisms is essentially oriented towards the reproductive cycle of the camels. Although Rendille also keep sheep and goats, camel herds form the most important part of their pastoral subsistence and represent the most vital resource; and they are of outstanding significance for the Rendille not only with regard to food, but also in terms of transport and, particularly, as a sign of individual wealth. However, camels are very demanding in requiring space for their pasture area; and their populations, in comparison with other herd animals, have quite a low growth rate. It has been suggested, therefore, that the ecological conditions of camel husbandry are reflected in the social structures of the Rendille.

With a maximum of 2.6 %, the average growth rate of the Rendille population is still below the 3.4 % of the camels. In addition, the mean camel generation length of 13.6 years corresponds well with the 14-year cycle of the Rendille age sets. This striking correspondence leads to the conclusion that, by applying a cyclical age set regulation, a homoeostatic feedback mechanism would have been implemented which allowed maintaining both the human and the camel population in a balanced growth relationship. Can this behavioural pattern of the Rendille thus be explained in the sense of population regulation as an interrelation between humans and resources aiming at equilibrium (Roth 1993, 1999), or is it an example of reproductive renouncement of individuals for the advantage of the group?

If at the level of the total population parameters indicative of reproduction are compared between *sepaade* and all remaining women, it shows first of all that in both groups the birth intervals on average come to 2.8 years (Roth 1993). A comparison of the reproductive life histories of *sepaade* as opposed to non-*sepaade* women, however, results in a clearly reduced total fertility of only 61 %[2]. These differences in fertility are caused by earlier or later marriage ages, respectively. On average, *sepaade* women marry eight years later than other women and therefore cannot compensate for their reduced total fertility. Marriage age, hence, appears to be the actual trigger for steered biological reproduction and leads to a difference of 2.4 parities on average between the two groups (6.1 compared with 3.7; Roth 1999). This is also reflected in differing reproduction rates (Fig. 5.6). It is true that, compared with other women, *sepaade* women have reduced reproductive possibilities, but even their growth rate is just as clearly well below that of the camels. The size of the human population thus remains in an order of magnitude that obviously even at longer terms allows the vital resource of the camel to be spared and, at the same time, keeps both components of the system at a stable or balanced level.

How does this behaviour of the Rendille fit at the individual level? How does it accommodate the necessity to have a male heir in order to keep the herd possession within the family? Rendille are peculiar in this respect as well, as they are said to operate male infanticide, which is apparently used whenever the birth of a boy brings an endangerment of primogeniture

---

[2]   Total fertility is regarded here as the average number of living offspring born to women who have completed their reproductive lifespan (p. 103 in Ulijaszek 1995).

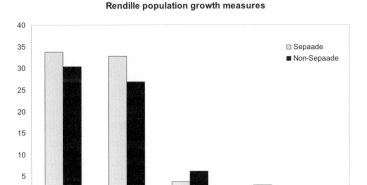

Fig. 5.6. Outcome of socio-culturally governed female fertility among Rendille pastoralists (data from Roth 1999). Higher mean age at childbearing leads to reduced fertility of sepaade women compared with their non-sepaade peers

inheritance. This can however be quite a risky strategy. Although a father without sons leaves his herd to the sons of his brothers, which ensures that the stock remains within the family, the actual occurrence of such a case is feared (Roth 1993). Particularly in marriages with *sepaade* women this fear is not unfounded, because these relationships significantly more frequently remain without a male descendant due to their altogether lower fecundity. At the individual level, the tradition of *sepaade* thus seems to cause a relatively uncertain inheritance constellation with the risk of endangering the passing on of resources within the immediate family association. The advantage for an individual is not so easily recognisable, in particular not for *sepaade* women.

Obviously, the Rendille have established a mechanism of population regulation by means of which cyclical age classifications facilitate a decrease in the reproductive potential at the group level, something that specifically holds for the *sepaade* women. This may indeed be maladaptive in terms of individual fitness expectations as well as with regard to rights of inheritance. Nevertheless, despite individual disadvantages for reproduction this behaviour is retained by social agreement and cultural standardisation. The question, however, seems unresolved as to why *sepaade* women would tolerate the clear loss in fitness or where the benefit is to limit one's own reproduction, even if this only contributes to a reinforcement of the altogether low growth rate that can be observed.

An alternative explanation for the emergence of the *sepaade* tradition, which is provided by the Rendille themselves, first of all has less to do with regulation but more with those times when aggressive territorial behaviour

and raids were constituent parts of safeguarding resources. During such conflicts, men (warriors) were away from the natal group for longer periods of
time, so that the tasks of herding were transferred to the *sepaade* women, who
became renowned for their herding skills but who, of course, in this way were
removed from the marriage pool[3]. Considering that the Rendille are aware of
the fertility-reducing effects of their cultural strategies, which in sum lead to
population regulation (Sato 1980; Roth 1999), it would not make any difference whether the origin of the tradition has to be searched for in territorial or
social behaviour. How can the consonance in human and camel population
dynamics be explained without permitting the development of maladaptive
behavioural patterns? For the men married to *sepaade* women, the way out of
the fitness dilemma is to frequently embark on polygynous marriages (Roth
1993) in order, on the one hand, to reduce the danger of passing away without
leaving a male heir and not being able to retain the concentration of resources
within one's own family and, on the other hand, to likely increase their reproductive success. Here, the individual advantage seems to achieve its ends. For
sepaade women, no such possibility for compensation is reported. However,
Rendille women today, after raids and territorial conflicts with the neighbours no longer play a major part, have pleaded for an abolition of the *sepaade* regulation, which eventually was agreed by Rendille elders (Roth 2001).
Up to this point, the problem of reproductive self-restriction could obviously
not be resolved.

In sum, it has to be noted that Rendille at several levels of behavioural
options discovered possibilities of how to achieve a balance between herd size
and population size in view of scarce resources in a limited habitat. Strict
marriage procedures, direct population-controlling interference and a close
tuning of social behaviour with the reproductive cycle of their most important resource led to stability and balance. Direct reproductive advantages for
individuals are not to be constituted so easily. Due to the inheritance regulations, only firstborn sons occupy a privileged position, which allows them an
improved distribution of resources to their descendants and thus, in the sense
of life history strategies (e.g. Hill 1993; Hill and Kaplan 1999) leads to an
increase in their personal fitness. At the level of the entire population, conscious cultural control connects the concentration of herds and the delimitation of reproduction.

Their habitat does not offer Rendille the same possibilities of flexibility as
are enjoyed for example by the neighbouring Ariaal (Fratkin 1986). The Ariaal live in a more favoured natural area and can much better compensate for
fluctuations in the habitat through a more variable composition of their herds
of livestock. The Rendille however, due to the ecological conditions of their

---

[3]   Incidentally, these skills contributed to the persistence of the habit to postpone mar-
      riage for *sepaade* women (Roth 2001).

arid environment, can only employ a clearly more specialised subsistence strategy with fewer options. This is at least a proximate cause for the accomplishment of adaptive rigid socio-cultural mechanisms of regulation by which a selective restriction of reproduction is established and maintained at the community level.

5.2.2.3 Carrying Load and Carrying Capacity

While for the Rendille the marriage age of certain socio-cultural groups can be seen as a crucial determinant for population growth, for !Kung communities it is the birth interval, which at the population level was proposed to facilitate the adjustment of population density to the resource conditions of their habitat. Howell (2000) and Lee (1972, 1979) reported that !Kung women would give birth every four years on average. Under the conditions of high birth rates generally assumed for pre-industrial populations, such a relatively long period could be regarded as a sign of unutilised reproductive possibilities. Why can this behaviour be observed nevertheless, or rather why would it be adaptive? First the proximate mechanisms will be explored that allow the maintenance of long birth intervals at all and afterwards whether there is an adaptive benefit for the !Kung under the given environmental conditions.

Mechanisms of population control find their expression in behavioural responses to certain habitat conditions and entail demographic effects. Physiological mechanisms include amenorrhoea as a natural consequence of pregnancy and lactation or disruptions to the functioning of the ovarian cycle related to energetic supply shortages. Such factors can then have effects on, for example, seasonal birth frequencies, which are related with certain subsistence activities in the yearly seasonal cycle.

Lowered or suppressed fertility during lactation (lactational amenorrhoea) is based on changes in hormone concentrations determined by the frequency of breast-feeding (Ellison 1991). The process of nursing induces the release of prolactin and the increased titre of this hormone has an inhibiting effect on the production of prostaglandins and is thus controlling the ovarian cycle. In this chain of events, it is not the nursing event itself, but the frequency at which it provides the appropriate stimulus which is crucial. A frequency of at least six lactation phases per day is deemed sufficient in order to maintain the prolactin concentration at the critical level for more than a year (Delvoye et al. 1977). For each month a baby is exclusively breast-fed, the post-partum amenorrhoea is extended by a further 0.8 months (Goldman et al. 1987). A study investigating these interactions with !Kung revealed a frequency of 13 events per day on average (Konner and Worthman 1980). Such frequent and continued nursing with a subsequent slow reduction of its frequency plus an incipient provision of solid food supplements could thus lead to birth intervals of more than three years with !Kung for these physiological reasons alone, even though re-examination of the data suggests that the period has rather to be reduced to some two years (Pennington 2001). Similar evidence for this kind of lactation-induced fertility suppression is also reported for the Gainj from New Guinea (Wood et al. 1985).

There are data of varying detail regarding the question as to the influence the nutritional status of the mother has on variances in fertility. The negative effects of severe dietary deficiencies on fertility are indisputable. For example, nine months after the 1944–1945 famine in the Netherlands, the so-called Amsterdam tulip winter, a reduction of fertility of up to 50 % was observed (Stein and Susser 1978); and, after the flooding disaster in the middle of the 1970s in Bangladesh and the subsequent period of famine, similarly strong declines in the birth-rate occurred (Mosley 1979). Weak chronic malnutrition however does not seem to result in a considerable reduction of fertility, while at the same time a good nutritional status can be responsible for a shortening of the lactation period (Bongaarts 1980, 1982; Lunn et al. 1984; Ellison 1995; Ulijaszek 1995). In marginal habitats with varying resource conditions, like for example that experienced by the !Kung, it is however conceivable that the energetic reserves of the mother can support her and the baby, but are insufficient for a further pregnancy. Therefore, longer birth intervals would be adaptive, because they lead to a better supply for the mother and concomitantly the child, which in the long run means a higher survival rate of the offspring. This fitness-orientated connection is supported by a generally positive correlation between probability of survival of a child and the length of the birth interval (Ellison 1991).

Nutritional energetic supply is also known to influence ovarian cycle functions (e.g. Rosetta 1995). It can be regarded as sufficiently established that the ratio of body fat to total body mass is crucial for regular ovulation (p. 106ff in Ulijaszek 1995). Furthermore, it is known that high energy consumption, e.g. as a result of high activity levels, can lead to the suppression of ovulation. Female gymnasts are a blatant example. For !Kung women, Bentley (1985) was able to show that, even given the relatively low frequency of gathering trips, it is possible to explain their low average total fertility by the connections between activity patterns, hormonal functions and reproductive capacity (see also Spielmann 1989).

For mothers, the energetic costs of lactation result from the energy content of the milk and the energy necessary for its production. Since lactation requires some 500–1,000 kcal/day, the overall energetic situation can prevent the critical body fat/body mass ratio being achieved. This may then have effects, e.g. on the resumption of ovulation. Additionally, a seasonal patterning of birth rates can result from the seasonally varying availability of food resources. Thirty-two percent of the births among !Kung occur approximately nine months after the end of the dry season, i.e. conceptions take place accordingly during a period of relatively favourable nutritional conditions

(Howell 2000). Comparable correlations between resource supply and birth rate were described for other ecological conditions and modes of production as well (e.g. Condon and Scaglion 1982; Bailey et al. 1992; Ellison 1994; Bronson 1995).

Quite early on, Lee (1972) suggested a synthetic view on possible causes for the mentioned physiological mechanisms of fertility depression among the !Kung. He indicated that the double demand on !Kung women, who have to carry both the collected food and their children, was sufficient to explain the long birth intervals. Activity patterns, energy consumption and nutritional status were combined here in one relatively easily measurable variable, the carrying load. It was observed that, if the birth interval was shorter than the assumed optimal length, the risk of death for child or mother could increase, e.g. by a combination of marginal nutritional situation and infection stress. A longer than optimal interval, in turn, would represent a sub-optimal utilisation of reproductive possibilities. Blurton Jones (1986) examined the adaptive value of the empirically observed average birth interval among !Kung and suggested it was optimal rather than too long for maximised reproduction, as one might have presumed. Intervals that are shorter than the proposed optimum would result in reduced total reproductive success, while the mothers who only raised a child approximately every three to four years also achieved the highest fertility (Fig. 5.7). The determinant that led to the evolution of this strategy is said to be the average carrying load of the mother. Since children are only able to keep pace with adults during the food-gathering excursions after about the age of four and thus have to, be carried at a younger age, an

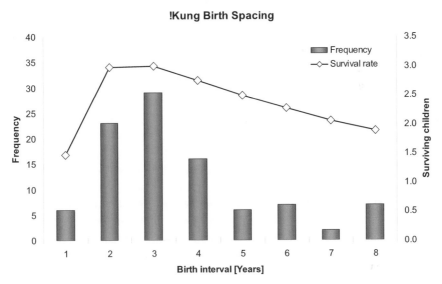

Fig. 5.7. Frequency of different birth intervals among !Kung and its relation to the average number of surviving children (data from Blurton Jones 1986)

additional child would increase the carrying load of the mother enormously as she would have to carry two children. To carry one child plus the collected plant food, however, was observed to be perfectly manageable. !Kung women who changed to a more permanent pattern of residence and moved to cattle-posts were found to show a reduced birth interval, by eight months on average, since their work and carrying load was reduced and, additionally, more suitable vegetable food was available to wean their children at an earlier point in time (Lee 1980). However, whilst the fact seems to be undisputed, it is not clear whether high-fertility women were drawn to cattleposts because of better circumstances to raise more children or whether the conditions of sedentariness allowed for more offspring to be raised (Howell 2000; Pennington 2001).

Under the given conditions of their habitat, !Kung may thus have developed behavioural patterns that work towards adjusting the number of offspring to the resource situation. There seems to be a practical tangible advantage on the individual level for the mother as a result of reduced carrying loads. On the population level, this may lead to keeping the population below or just at the carrying capacity of the habitat and preventing overexploitation of resources.

It is to be examined though, to what extent this pattern is generally applicable, i.e. whether low fertility as found with !Kung is a general characteristic of forager societies. The immediate neighbours of the (Dobe) !Kung, such as the Ghanzi and Ngamiland !Kung show quite different patterns (Pennington and Harpending 1988; see also Pennington 2001). It is true that direct comparisons of birth-interval data should be met with caution due to the different research designs of the studies. It can be shown however that – *contra* Blurton Jones's prediction – women who give birth to more children also have more surviving offspring and an increase in family size, i.e. a higher frequency of pregnancies, does not necessarily mean a reduction in fitness.

A comparison with the East African Hadza is even more striking (Blurton Jones et al. 1992). As opposed to 4.7 surviving children on average with !Kung, the total fertility rate of the Hadza amounts to 6.2 on average and is thus comparable with agricultural societies (see below, Sect. 5.2.3). The Hadza however live in a richer habitat with higher yields from gathering and hunting activities; and the habitat can sustain a higher population density and increased population growth. Hadza women carry their children less and their children are able to contribute to their own supply of food to a much higher extent than for example !Kung children. By the age of five, they already manage to provide themselves with about half the total food they need. !Kung children, however, depend on their mothers more heavily, since resource patches, e.g. Mongongo groves or water, are more scattered and, particularly during the dry season, are less well attainable (Blurton Jones and Sibley 1978). Besides, there is substantial predator pressure, which would additionally increase the risk for children collecting food on their own. However much they are adapted to the ecological constraints of their habitat, !Kung with their low fertility nevertheless seem to represent the special case of a particular local adaptation rather than the pattern for a general interpretation of population dynamics in forager societies. In fact, recent comparisons of demographic data collated from field studies on extant hunter-gatherer societies reveal enormous variation in total fertility, ranging between 2.8 and 8.0 (Pennington 2001). Whilst some of this variation is attributed to the refugial habitats foragers are often confined to and other explanations take into account the adverse effects of infectious disease on fertility, a lot of the observable variation is likely to be a direct response to prevailing ecological conditions.

In all three case studies, be it the Tsembaga in New Guinea, the Rendille of East Africa or !Kung of the Kalahari desert, a close linkage of assumed (Tsembaga, !Kung) or actually existing (Rendille), population-effective behaviour for the specific conditions of the resource distribution in the habitat is apparent. With horticulture and its extensive land-use strategy of slash-and-burn agriculture, land as a resource is the limiting factor – both regarding the human population and the livestock, here pigs. In the arid habitat of the Rendille, where herds of camel practically represent the only insurance for survival, a co-ordination of human and animal reproductive cycles for a balance between these two populations provides an adaptive strategy. With !Kung, finally, the workload of women has a part in limiting the reproductive potential as an effect of resource availability and the resulting physiological consequences. On the population level, there is thus a fundamental connection between access to resources and reproductive conditions, i.e. population dynamics. Any habitat provides a certain capacity limit and, on the population level, stabilising behavioural patterns can be observed by which matter and energy flows are co-ordinated in such a way that the capacity limit is not exceeded. Such functional explanations are sufficient, as long as proximate causes are concerned, to explain a certain behaviour pattern or whether proximate mechanisms are concerned that may lead to certain behaviour. Proximate explanations deal with physiological and psychological mechanisms as well as the cultural knowledge that in each case cause a certain behaviour (Cronk 1991a). Especially the Rendille example, and probably also that of the !Kung, however demonstrate that a demographic pattern which can be observed on the population level in the long run represents the sum of individual strategies by which individuals can maximize or at least optimise their life-time reproductive success (Low 1993).

What looks like homoeostatic regulation on the population level – and in sum even may be so – however arises as the result of individual reproductive strategies whose 'goal' it is to raise as many offspring as possible under the given ecological basic conditions. It does not develop as a self-regulatory characteristic of a population, i.e. the population as an entity does not pursue regulation. Ultimate causes rooted in the the evolution of behaviours, whose consequences will show in demographic patterns, lead to an evaluation of fitness on the level of individuals, not groups. They relate to the adaptive meaning of behaviour. Although selection on different levels is conceivable, it effectively takes place on the level of individuals and genes. Therefore the question remains whether under these circumstances we can still regard regulation as a homoeostatic mechanism.

5.2.3 Population Regulation – Myth or Reality?

Section 5.1 introduced the term carrying capacity and its connection with Malthusian checks, elements of counter-steering which depress growth, if the population density exceeds the potential sustainable by natural or otherwise available resources provided by the habitat. Events such as an exhaustion of food resources, epidemic diseases or competition for resources become effective as extrinsic density-dependent factors, while for example natural hazards constitute density-independent factors, which can likewise lead to population depression. If these processes occur or are observed within a system, then the assumption of feed-back or stochastic factors which affect the adjustment of population parameters is compatible with systems-ecological or population-ecological points of view.

Likewise, all organisms in a given habitat have both the ability and the propensity for unlimited proliferation. Their instinct-driven behavioural repertoire prepares them to reproduce as successfully as possible. The population development of humankind over time is an apt example, even if its exponential growth resulted from a sequence of connected series of logistic growth processes, which became possible by adjustment and/or extension of the respective capacity limits (see Sect. 5.1). This is apparently contradicted by the fact that, where encountered, natural populations show relatively constant sizes and densities. They obviously are in a resource-related equilibrium or their density oscillates around a relatively stable capacity-dependent value.

Can intrinsic, self-regulatory mechanisms thus be made responsible for this which, in line with ecosystem theory, would lead to homoeostatic control of populations? As the preceding examples showed, such an explanation would have to consider the interplay of different behavioural patterns and socio-cultural arrangements and conclusively combine them within a very complex fabric. This is not always possible without contradiction. Application of the parsimony principle however can explain the relative constancy of population sizes with fewer assumptions by referring to the observation that not all organisms actually fully utilise their possible reproductive potential. Either not all individuals reach reproductive age and are thus excluded from reproduction, or the opportunities for reproduction are, for whatever reason, unequally distributed in the population. If within a population these differentials are based on just as different adaptive valences, then the better-adapted phenotypes with better strategies of resource use will be those that are 'pre-

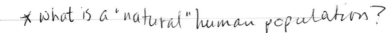

✗ what is a 'natural" human population?

ferred' by natural selection and whose genotypes will be represented in higher frequencies in future generations, since they possess better chances of reproduction. The adjustment of equilibria and establishment of stability domains would then be the result of strategies by which the personal and/or the inclusive fitness are maximised – an explanation completely compatible with the principles of evolutionary biology.

This aspect can be conveniently linked with ecological principles because human populations, in their local and temporary developments, show some form of logistic growth and pursue a k-strategy of reproduction. Compared with other organisms, humans have only a few children and raising their offspring requires and is characterised by a high parental investment. A k-strategist would therefore seek to successfully raise not the maximal, but the optimal number of descendants to ensure the best possible means of securing genetic representation in future generations. It thus may only appear as if a population is stable below the carrying capacity, when in fact the population density has not reached the capacity limit yet due to a low growth velocity of the overall population.

There is no need to invoke old-style group selectionist explanations according to which some individuals within a community would sacrifice their own reproductive options for the benefit and well being of the whole, or limit them to the extent that a collapse of the population due to resource shortage is avoided (see Wynne Edwards 1962). Even if phenomenologically there appears to be such an effect of social behaviour on the stability of a population, it would first have to be shown where there is benefit to the group, in comparison with benefits for the individuals of the group[4]. A level, however, on which group advantage is evaluated as the evolutionary unit of selection is not corroborated by any theoretical default assumption; and even altruism has a selective advantage only if there is recognisable prospect of remuneration for the individual (Hamilton 1964; Trivers 1971). What at first sight looks like group-supportive behaviour is resolved in the personal maximisation of fitness and the pursuit of individual reproductive interests (Alexander 1979).

Nevertheless, the restriction of birth rates can be a very effective means of achieving this optimisation, as the conditions of raising offspring and providing them with improved chances of survival are enhanced. At the same time, parents increase the chances to spread their own genes. In this sense, !Kung are an example for the optimisation of reproductive strategies under the given environmental constraints and potential of the habitat: for the individual, reproductive success is highest if birth intervals are large and thus infant mortality is kept low. However, here, reproductive success is not maximised via

---

[4]   Recently, modified group selection models have been introduced, yet their validity is subject to ongoing discussion (Boehm 1996; Wilson and Sober 1994).

birth rate but by 'weighing' costs and benefits in the conflict to raise a new child and thus to possibly endanger the life of the preceding child. The success of this strategy is measured as the balance of costs incurred in raising a child and the benefit for inclusive fitness arising from it (Lack 1954; Williams 1966; Trivers 1972; see also Curio 1988); and selection will always favour those individuals who tend to maximise their inclusive fitness. Such a mechanism by no means contradicts the idea of density-dependent resource restriction in a habitat, but only demonstrates the evolutionary biological and theoretically consistent way in which the regulation of population density takes place and has to be viewed.

5.2.3.1 Manipulation of Sex Ratio

Among all the proximate mechanisms by which human populations are able and known to establish adjustments of their reproductive behaviour, the withdrawal of parental care, including its most extreme appearance, the killing of a child, appears to be directly counter-selective – quite independently of numerous other social, psychological and cultural implications which are connected with such behaviour. How can a strategy serve fitness goals, if the substantial energetic investment of at least 80,000 kcal, which a pregnancy requires on average, is obviously wasted when a newborn child, a baby or an infant is killed? Is infanticide the touchstone for an evolutionary biological view of reproduction, or is it a usual, although more expensive part of the behavioural repertoire by which humans exert control over their reproduction, dependent on population density and available resources?

Culture-comparative data available from the Human Relations Area Files reveal (Daly and Wilson 1984; Scrimshaw 1984; for medieval evidence, see also Kellum 1974; Coleman 1976) that infanticide is common world-wide and can be observed in about half of the societies examined in this respect. If the data are differentiated according to the circumstances under which infanticide is practiced, it is often linked to children who originate from extra-marital relations and to cases in which congenital defects or severe illness would limit the offspring's probability of survival. The most frequently recognisable reason however has to do in the broadest sense with difficulties resulting from the current resource situation of the parents (Daly and Wilson 1984). The option to kill a newborn child in such instances appears as a possible and also socially sanctioned strategy (e.g. Minturn and Stashak 1982), if the ratio of available resources and the number of offspring or dependants would entail a reduction of either the personal or the inclusive fitness. By taking into consideration past and future reproductive opportunities, the individual decision over further investment in a child would thus weigh the quantity and quality of resources that are at their disposal in the sense of a maximisation of lifetime reproductive success. The sex of the child can be of substantial importance for such a decision. Among the many examples known, such as those from China and India, a much-discussed case of this kind is the preferential female infanticide described for Inuit.

Since Inuit live in a harsh habitat under extreme ecological conditions, the preferential killing of girls is interpreted as a functional adaptation, because men play the central role in food procurement and, since they are exposed to an increased mortality risk on their hunting trips, the manipulation of the sex ratio in favour of boys would be an adjustment to the difficult resource situation. The claim of such connections however is based on very limited direct evidence (Smith and Smith 1994); and possibly there is actually only one case of first-hand information (Schrire and Steiger 1974; for Netsilik, see also Guting 1987). Likewise, non-uniform results come from stochastic simulation experiments, according to which on the one hand the continued killing of girls at rates of more than 8 % would lead within a few generations to the collapse of the local population (Schrire and Steiger 1974), while on the other hand even a 30 % shift in the sex ratio would not threaten the existence of a population in the long run (Chapman 1980). In view of the fact that small population sizes are known to be particularly susceptible to chance fluctuations in their demographic characteristics (see Bocquet-Appel 1985) the question arises to which extent under the given ecological conditions would Inuit populations be able to bear an intentional killing of female individuals, as they are the resource which represents the limiting factor for reproduction.

In the absence of other information, data on girl infanticide is in the majority derived from historical censuses, in which boys appear in significantly higher numbers than girls (e.g. Freeman 1971; Riches 1976; Irwin 1989). Well documented records (for Canadian Inuit census data between 1880 and 1930, see Smith and Smith 1994) reveal that, in all documented populations, the sex ratios of the children were clearly shifted in favour of boys (on average 173 boys vs 100 girls), while among adults the ratio turns out slightly in favour of women (92 vs 100), but is balanced in principle. Thus, in the course of ontogeny from childhood to adult age, a substantial shift in the sexual proportion takes place on average. It can not be excluded that possible artefacts in the census data caused this general trend. Schrire and Steiger (1974) even assumed that through differences in marriage age between the sexes – young-married girls are the rule in Inuit societies – these individuals were recorded as women and not as girls in the censuses and that in this way a skewed sex ratio was introduced through the method of data acquisition. Furthermore, it is confirmed that the mortality rate of men lies significantly above that of women and can be explained largely by hunting accidents or subsequent infection, but also by killing offences. Finally, there is concurrent evidence from different ethnographic sources that there was considerable competition among men over women of marriage age and that it was particularly pronounced in those communities where childhood sex ratios were strongly skewed in favour of boys. Therefore, when female individuals were generally exposed to higher mortality at the time of birth or during childhood than boys, then at the same time the sex ratio in the adult population was subject to substantial pressure on numerical parity. Adjustments of the census data to model life tables moreover showed that, only on the assumption of low female marriage age and a clear reduction of female infants, i.e. infanticide, can agreement with existing models become possible in these computations. An average rate of 20.9±15.1 % girl infanticides can thus explain a large part of the variance in Inuit sex ratios, although it can certainly not be considered the only factor (Smith and Smith 1994). If the data are taken for granted, how then can behaviour be explained that leads to a loss of about one-fifth of all girls born?

Three interpretative models compete with one another, whose explanatory value of manipulations of sexual proportions is meaningful regardless of the conditions found among Inuit: population regulation as a long-term strategy, adjustment of the sex ratio at adult age and optimisation of parental invest-ment in offspring in the sense of rational economic decisions.

Increased female mortality affects the growth possibilities of a population more strongly than high mortality rates of men in the population. Under con-ditions of natural fertility, therefore, a delimitation of total population size would be achieved very effectively by a reduction in the number of girls or women, an explanation suggested for high infanticide rates for girls among Amazonian Native Americans (see Harris 1974; Divale and Harris 1976).

In the circumpolar habitat with its often extreme and unpredictable fluctu-ations of resource availability (Freeman 1988; Savelle and McCartney 1988; Smith 1991), it cannot be readily expected that it would be an ecological necessity for Inuit to delimitate total population size. On the contrary, peri-odic resource losses and weather changes are reported more frequently (see Smith and Smith 1994), which, as a consequence, could entail the killing of children. Therefore, Inuit populations would more frequently have to face the situation of increasing their total population again after such breakdowns, rather than stabilising its density. However, indications of long-term adjust-ments to the resource situation by a reduction in population size are not observed and, given the habitat conditions, are probably not expected either. But if some Inuit would preferentially do without raising girls, this could very likely lead to a relaxation of the resource situation on the population level, where many (also many free-riders) would benefit, while the costs would have to be borne by a few individuals only. Yet, this possibility is improbable, since there is no selection level for such group-supportive behaviour (see above), not least because each individual who does not adhere to the social agreement but instead continues to reproduce, would have a higher fitness gain in the long term. Since the Inuit foraging mode of subsistence requires high group mobility, the possibility of resources being monopolised by certain families is also void, because they would adapt their group size according to the given abundance. The population dynamics of Inuit therefore seems to be rather characterised by variable and often high fertility rates (figures reported vary from 5.0 to 8.0 live births; Smith and Smith 1994) and highly oscillating death rates that have to be understood as probabilistic responses to local resource situations, not as long-term (planned) strategies, in order to cause a balance between population and resources.

The second explanatory model, which suggests a balanced sex ratio in the adult population as a reason for infanticide, can first of all be supported by census data, which reveal clear connections between a sex proportion skewed in favour of boys during childhood and approximate parity at the adult age (Smith and Smith 1994). A functionalistic explanation is quickly found: girl infanticide would adjust the increased mortality of young men and, while this would result in a balanced sex ratio, the sustainability of the population is maintained by an approximately equal number of surviving women and men of marriageable age. However, on the one hand, this is not compatible with ethnographic reports on high competition over women, particularly in those communities exhibiting high rates of infanticide. On the other hand, again, is it hard to recognise on what level such behaviour would be selected if the sex ratio alone was considered the adaptive trait, whereas the mechanism of its emergence was not.

According to theoretical predictions about the distribution of parental investment among offspring (Fisher 1930), it follows that natural selection promotes those strategies by which parental fitness is maximised, i.e. it favours investment into the sex which yields the highest fitness gain for parents. If costs for both sexes were alike, this would show in a balanced number of sons and daughters. If costs differ, the prediction would be that parents would invest more strongly into the 'cheaper' sex. In sum, on the population level, sex proportions would therefore be promoted that would lead to balanced parental investments into girls or boys. To the extent by which the mechanisms of resource allocation are culturally inherited, selection would favour higher investments into the more frequent and more inexpensive sex, until a balance is reached on the population level. Parental fitness is thus determined by the optimal sex ratio of the offspring, i.e. the respective strategies chosen are affected by frequencies of one or the other sex. If there is a difference between the sexes in the costs to be invested, then selection would promote an unbalanced sex ratio.

Can this model be transferred to the conditions found among Inuit? It is true that the parental investment is shifted in favour of boys, but the survival rate of boys is particularly high in those communities where the highest rates of female infanticide can be inferred from census data. Differential parental investment in favour of boys was thus exercised by a manipulation of the infant sex ratio, corresponding in numbers to approximately 1.5–2.0 times the additional investment expenditure for boys before they attained social maturity (Smith and Smith 1994; Smith 1995b). In fact, boys are thus the more expensive children and the higher investment is not compatible with and cannot be explained by Fisher's predictions.

Such an imbalance only makes sense for the Inuit if it results in an advantage for their life-support system, i.e. if the higher investment pays off, such that parental fitness is increased (see Clutton-Brock 1991). Interestingly enough, in the few reported statements that the Inuit themselves issued about the reason for female infanticide, they say that boys when they are grown up would one day contribute more to the cost of living than girls. The ultimate reason for a higher investment in boys would thus not only be an increase in direct fitness for the parents or the male descendant, respectively, but also an improved resource supply and thus an increased probability of survival for parents and siblings, known as the primary helper-at-the-nest situation.

Under the given ecological circumstances a strategy by which the sex ratio is changed in favour of boys would thus provide more predictable advantages in terms of resource security. Since girls on average marry five years earlier than boys, the higher and also longer investment in boys pays off through a higher marginal fitness return. It also appears necessary, since the training of boys required to become successful hunters is time-intensive and, in view of resource constraints, it is an indispensable investment into the future survivability of the family. Correlations between the average yearly temperature and the extent of manipulation of the sex ratio in Inuit (expressed as boy/girl ratio; Fig. 5.8; see Irwin 1989) show that the colder the habitat, the more significant men become as providers for the population. Under these circumstances, it is conceivable that a selection of cultural strategies which take into consideration the facultative preferential killing of girls as a function of the ecological conditions works towards a maximisation of fitness, as the successful raising of more offspring can be secured, in comparison to a strategy which does not entail this behaviour. In those habitats, where there is higher mortality of men as a result of high-risk subsistence tasks, parents who practise female infanticide will thus compensate for the likely future loss of men by aiming at the optimum condition of a sex ratio of their children that is shifted in favour of boys. An optimisation of child number thus leads to a maximisation of life reproductive success. At the same time, such a rational decision would relax a specifically strained resource situation, so that an economic option is selected as an adaptive proximate response to the weighing of one's own and the offspring's survival levels, resulting in ultimate fitness gain.

What may represent a favourable personal life-history strategy (e.g. Hill and Kaplan 1999) in terms of direct or inclusive fitness at the level of the family, hence the level of individual reproductive decisions, can accumulate to equilibrium on the population level. Thus, on the basis of individual selection, adjustments of population densities below carrying capacity may also be conceivable. It remains unclear however how long the time-intervals between

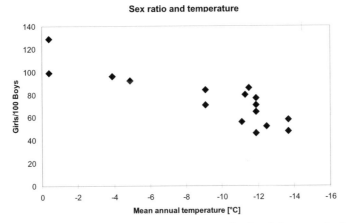

Fig. 5.8. Correlation of mean annual temperature and sub-adult sex ratio for 16 Inuit communities. The colder the environment, the more the sex ratio is biased towards boys (data from Irwin 1989)

periodic resource minima or other tangible constraints of the habitat have to be for selection to be still possible at the individual level. For this connection, Moore (1983) suggested a model that links cultural carrying capacity to the notion that it represents the period of time between two minima which would still allow an organism to adjust. In its longest temporal expansion, this space would correspond to the life span, while the shortest space would be given by the time that has to elapse before descendants can pay back the energy invested into them, either through their own reproduction or as helpers-at-the-nest. Assuming that the figures for humans are between five and 50 years (Moore 1983) and given an average of 20 years for males to reach full social maturity, then parents who would practise preferential female infanticide would very likely be able to take advantage of a stabilised supply situation for themselves and their descendants. The selection of cultural traits, which lead to an improvement of resource allocation, would then result from the preference for certain adaptive behavioural patterns that are being learned as cultural characteristics and prove to have effects that promote reproduction (see Irons 1996). Biological reproduction thus profits from social reproduction and an increase in social resilience.

Yet, the selective advantage of certain behaviour must not necessarily be connected with the stability of a population in its habitat. If stability is the result of selection towards certain behaviour, then it was selected rather despite than because of this effect (Bates and Lees 1979). Stability will thus result as an epiphenomenon of processes in which reproducing individuals by making use of given resources compete for their inclusive fitness, which they attempt to maximise within the existing ecological and social possibilities.

This selectionist view does not represent a refusal to ecological principles. On the contrary[Only the denomination of specific ecological factors makes possible the definition of that framework within which the development, maintenance or change of certain behavioural patterns take place. When preferential female infanticide is an expression of individual economic decisions of family planning (for a discussion of the economy of familial decisions, see Becker 1991b), then it can thus be understood as a context-dependent response to given environmental circumstances. In the case of Inuit, infanticide will have always originated from a current situation and the necessity to adjust population size in view of scarce resources without this being the option of choice in any case. In the end, it is the consequence of a specific social and reproductive context that leads to behaviours which reflect a reaction to coping with environmental constraints.]

In the same vein as census records are being used to help detect cases of population regulation in line with non-uniform parental investment or, more generally, differential resource allocation as part of life history strategies, it has been attempted to use the skeletal record to help identify infanticide in past societies (e.g. Mays 1993). Typical peaks of infant mortality around neonatal, or rather full-term gestational, age were suggested as evidence for this practice in various socio-cultural and temporal contexts. Yet, as the above example demonstrates, infanticide if practised at all is clearly not about the random disposal of offspring, but is instead related to cost/benefit constellations that crucially take into account the sex of the newborn. Those few examples, where preferential female infanticide has been examined using human remains are worth noting: either because they are reporting on a very specific case (e.g. for a late Roman brothel, see Faerman et al. 1998), or because they fail to demonstrate such a pattern. It has repeatedly been advocated that agrarian societies would be prone to preferential female infanticide should circumstances require it, because male offspring would have been considered the more valuable resource with regard to subsistence tasks and thus the provision of essential resources. In the case of a medieval community from Aegerten, canton Bern, molecular analysis did not support assumptions of lack of parental investment and subsequent death of baby girls (Lassen et al. 2000), nor can such evidence be provided from the Anglo-Saxon community of Raunds, Gloucestershire (Fig. 5.9). Rather, sex-related mortality patterns suggest death rates in line with clinical data on (male) excess late-foetal and peri-natal mortality and spontaneous abortions. Whilst there can be no doubt about the validity of pursuing the question of infanticide in the past, the examples serve to show that empirical data do not always support a general assumption. Infanticide is most often associated with situations of economic hardship, i.e. when the tapping of energy and material flows is compromised by a mismatch between existing resources and the number of people competing for them. And then it is socially sanctioned. Yet, one must not assume that costly individual fitness decisions and unconscious energetic calculations in

Raunds foetal and infant mortality

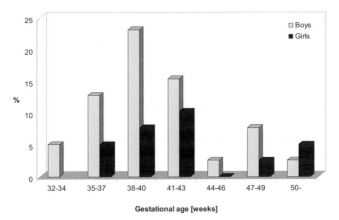

Fig. 5.9. Pre-term and peri-natal individuals in the Anglo-Saxon community of Raunds display a mortality profile whose sex ratio is inconspicuous in modern clinical terms. Even though the assumption of infanticide in past societies is generally justified, the empirical evidence needs to be carefully examined before premature conclusions are drawn. Age estimations from McGovern (2003)

order to sacrifice reproductive options are easily done or taken lightly. After all, it requires a rationale that links individual, ultimately fitness-related, decisions with community level issues of social resilience.

5.2.4 Subsistence, Change and Fertility

Modes of subsistence in human communities not only differ in the techniques of food production or the extraction of nutrients from the environment, but also in the associated settlement and activity patterns (see Chap. 2). A mode of food procurement which requires mobility and flexible group composition, due to resources being more scattered and not concentrated in certain patches, will on average only be able to sustain smaller numbers of individuals (e.g. p. 34 in Harris and Ross 1987) than a mode of subsistence which operates a stationary way of food production, lays in supply stocks and aims at the long-term stability of the resource basis. The implied population increase is reflected in the constantly (and in the long run exponentially) rising world population, which essentially results from improved and more effective possibilities of food production in the course of history (see Sect. 5.1). Trivially enough, however, a global or even regional rise in population size is only possible if the conditions of reproduction and thus the framework for reproductive decisions of individuals change accordingly. In Chap. 3, it was shown that the successful change in subsistence modes with the transition to agriculture is regularly connected with, yet not triggered by, population growth. If a general connection between subsistence and the possibilities of population increase can be derived from this, one should also be able to explain subsistence change as a reproduction strategy by which, through a change in the availability of food resources, improved reproductive conditions are made possible. How closely thus is the connection between production and reproduction?

5.2.4.1 Subsistence Change as Reproductive Strategy?

Today, the Mukogodo (Mutundu 1999) live as pastoral nomads in southern Kenya. Until the 1920s or 1930s, they were foragers, whose subsistence was based on animals and collected plants as well as honey, which they harvested from natural and manufactured beehives. The transition from a foraging to a pastoral mode of subsistence appeared during a time when the Mukogodo went through a culture change induced by the traditional pastoral societies in their neighbourhood. Cronk (1989) interpreted this subsistence change as a result of changing individual reproductive decisions and strategies, which finally led to profound social and societal changes.

As the Mukogodo increasingly kept cattle, sheep and goats, they shifted their settlement places from caves to favourable pasture grounds and began to build houses there. Thus, their contact with pastoral societies like the Maasai, Samburu or Digirri strengthened and marriage relations developed between these societies and the Mukogodo. Indeed, the reason for their beginning a pastoral subsistence lay in the fact that Mukogodo received cattle as bridewealth for women who married into the neighbouring pastoral societies. Livestock began to regularly replace beehives as traditional bridewealth. In reverse, once they began to keep livestock, Mukogodo men were forced to increase their herd stocks in order to be able themselves to pay for the bridewealth of their future wives. The value of beehives as bridewealth plummeted and, instead, it became necessary to procure and increase cattle stocks. With the old bridewealth regulation, the chances to be able to marry at all declined, whilst after cattle had been established as bridewealth, there were obviously clearly better possibilities to find a woman, both within their own and among the neighbouring communities. From a certain point, the availability of new resources was decisive for one's own reproductive possibilities and promoted decisions about cultural strategies, which served for the fulfilment of biological goals. The transition to pastoralism was thus a consequence of reproductive decisions.

The situation was aggravated by the fact that Mukogodo women outmigrated in higher numbers from their own community through marriage than women married in from neighbouring societies. In addition, foreign women were more expensive for Mukogodo men than their own women were to foreign men. However, by marrying out of their own community, new alliances could be formed, which would be important for mutual support in times of hardship – a security strategy commonly found with pastoral societies (see Sect. 2.1). Altogether this led to a situation in which safeguarding one's own reproduction and survival could only be achieved by means of high bridewealth, which in turn could only be accommodated through the procurement of livestock (Fig. 5.10).

If in this case subsistence change can be explained as a marriage strategy, does this equally apply to a parental strategy, which as a consequence of change leads to increased fertility and probability of survival of the offspring? From the available data (Cronk 1989) it does not follow compellingly that the transition would mean a more stable or better food supply in pastoral subsistence. Also, being foragers, the Mukogodo would have vast areas at their disposal during a local exhaustion of food resources that could be exploited. Nevertheless, it can be assumed that, in the process of adopting a pastoral subsistence, obvious advantages resulted from the sources of food now additionally available, even if foraged food resources were not in short supply.

It seems to have been of greater importance, however, that through the change in subsistence, foraging as a parallel food acquisition strategy or as a

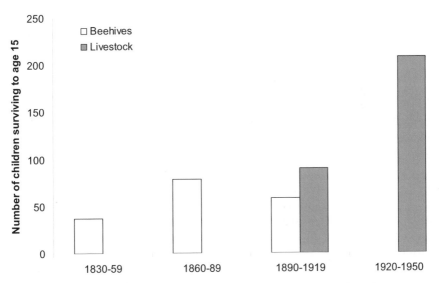

Fig. 5.10. Shifting amounts of beehives and livestock as bridewealth and consequences for marriage opportunities and reproductive success among the Mukogodo (modified from Cronk 1989)

possibility to return to the old traditions was in fact no longer available. As Mukogodo entered into marriage relations with their pastoral neighbours and adopted their customs and ways of life and finally also their language, they became tied into a cultural system in which foraging was regarded as socially stigmatising. Subsistence change developed from individual decisions arising as a response to changing social and reproductive ties within one's own community and between one's own and other communities, which in turn was the new pre-condition for the achievement of reproductive decisions. Subsistence change seems to have been no intended strategy here but the result of a cumulative effect of individual decisions that had effects on the group level. Once the process of the resource shift was set in course, only the new subsistence basis proved to be a suitable possibility to accomplish the avowed social goals of Mukogodo, i.e. in this case more women for the men and more children for the women (Cronk 1989). This, however quantified, may have introduced an element of peer pressure as well that is likely to have further promoted the adoption of the new subsistence mode.

This example has implications with regard to the question as to what extent the accomplishment of reproductive interests affects or accelerates the change in the subsistence basis altogether. Apart from the advantages of having food stocks as well as stationary and (in yield) relatively predictable food resources, it appears plausible to assume that the transition to a producing subsistence mode was also, or perhaps rather, the result of improved reproductive possibilities. If this applies, then regular connections between determinants of population development, such as food acquisition and fertility, should be observable and available for interpretation within an evolutionary framework.

[handwritten margin note: Note here that this is a secondary transition not a primary one.]

5.2.4.2 Subsistence and Differences in Fertility

The question of causal connections between subsistence mode and demographic structures of human populations was recently discussed widely and with controversial outcomes. Taking cross-cultural studies of societies subdivided by their patterns of subsistence management (see Sect. 2.1), but independently of their respective habitats, no differences were initially found in investigations of total fertility rates in traditional societies of foragers, horticulturalists and pastoralists as well as agriculturalists – admittedly broad categories (e.g. Campbell and Wood 1988). Other studies found higher fertility rates in agricultural and pastoral societies, as opposed to foragers (Hewlett 1991; Bentley et al. 1993; see also Cohen 1994). More recent cross-cultural investigations, employing more sophisticated designs, arrived at more differentiated results, supporting the assumption of differential fertility as a function of subsistence mode. Sellen and Mace (1997), for example, considered that populations usually use more than one subsistence technique for food acquisition, but also considered the fact that a possible correspondence of subsistence activities may be the result of cultural-historical connections between populations and thus were not independent of each other (Galton's problem, see Mace and Pagel 1994). Accordingly, a higher average number of children is positively correlated with an increasing reliance of the subsistence on agriculture and/or livestock farming (Table 5.1). The enhancement of any other subsistence mode, however, does not lead to a change in average fertility. Furthermore, it can be shown that a 10 % increase in the subsistence reliance on agriculture results in 0.4 more children (or births) per woman on average. While in permanently sedentary populations an increase in the amount of agriculture does not affect fertility, non-permanently sedentary populations do show such a correlation. Here, the higher the relative contribution of agriculture to the subsistence, the higher is the average fertility.

There is a significant 'fertility leap' in agricultural societies (Fig. 5.11)[5]. It thus shows that the reliability of food resources, i.e. their controlled cultivation and the possibility for supply stocks, are a pre-requisite for increased fertility and thus, at least potentially, for higher population densities. Variances within a given subsistence category may be high in individual cases, though. !Kung and Hadza, for example, differ substantially in their fertility rates and the growth potential of their communities (see Sect. 5.2.1). These differences, however, exist as a function of the different habitat conditions

---

[5] The leap becomes evident even when standard deviations in the average fertility rate are considered (re-calculated to be 1.1–1.6).

Table 5.1. Change in average total fertility rate depending on the percentage of various modes of subsistence in pre-industrial societies (modified from Sellen and Mace 1997)

Prevailing mode of subsistence	Number of societies (n)	Hunting/ gathering/ fishing (%)	Animal husbandry (%)	Agri- culture (%)	Fertility (number of children)
Hunting	4	100	0	0	5.2
Gathering	5	100	0	0	5.3
Fishing	8	95	0	5	5.2
Pastoralism	16	19	66	15	5.6
Incipient agriculture	3	60	5	35	6.1
Extensive agriculture	17	26	14	60	6.5
Intensive agriculture	16	17	25	59	6.6
Foragers	18	100	0	0	5.3
No foragers	51	22	33	45	6.1
No agriculture	34	57	33	10	5.4
Agriculture	35	24	19	59	6.3

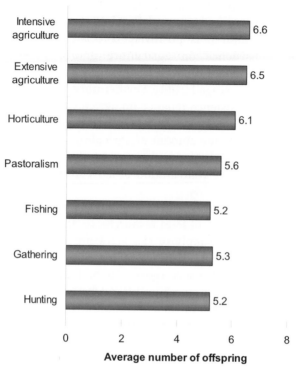

Fig. 5.11. Increase in total fertility with increasing amounts of food production (data from Sellen and Mace 1997)

they live in and the resulting possibilities of resource utilisation. This may well entail including children in the procurement of subsistence to significantly different extents, thus relieving mothers from arduous subsistence tasks and in sum making increased fertility possible. If individual fertility as a response to ecological basic conditions, e.g. resource availability, is a product of natural selection, then also patterns of population distribution and composition should correspond with extrinsic conditions of this kind. Under conditions of natural fertility it is to be expected then that subsistence mode and fertility are inter-dependent and do not vary just coincidentally with ecological conditions. Exactly this expectation is supported by the cross-cultural data.

Chapters 2 and 4 explored the correlations between the modes of production in human communities; and their social organisation was discussed. Results indicated that an increase in social complexity is generally coupled with an intensified dependence of the subsistence on produced food. Moreover, in a sample of 24 pre-industrial/traditional societies, cross-cultural data about characteristics of their social structure, such as hierarchical complexity or the way in which conflicts are settled showed close correlative connections with the degree of polygynous marriage patterns (Betzig 1982; see also Hill 1984). As societies depart increasingly further from the organisational level of the Family group or Local group, there is an increasing tendency towards polygyny at the top of the social order, i.e. among high-status men. Thus, if individuals occupy positions of power and increased influence within a community, these will eventually be used for the accomplishment of one's own reproductive interests.

If Betzig's (1982) model is taken further and connected with data about the dependence of these societies on agriculture and/or livestock farming (for details, see Murdock 1981), a highly significant correlation (Spearman, $P<0.1$) can be found between the mode of subsistence and the degree of polygyny. A combination of both results has two consequences. On the population level, the correlation of modes of production and degree of polygyny of high-ranking individuals can or will lead to increased fertility and thus supports the general trend shown above, according to which the transition to a producing mode of subsistence will result in an increased probability of more surviving offspring in a population. Production and reproduction are thus interconnected by the factors of increased resource predictability and reliability. On an individual level – with nevertheless numeric consequences on the population level in the long run – the same principle applies, except that an increase in control over means of production and other resources allows a conversion into improved possibilities of reproduction through more power and influence – assuming that polygyny represents a good proxy for reproductive chances.

Subsistence change thus possesses an intrinsic component through the prospect of altogether more favourable reproductive possibilities, whose effects can be demonstrated by altered total fertility rates on the population level. At the same time, new constellations for reproductive conditions arise within the population as a result of changes of the social organisation that are connected with the subsistence change. Apart from proximate causes, which can be due to improved energetic utilisation of resources or simply more secure living conditions, it is to be assumed that the option of fitness maximisation (resulting from an increase in the control and access possibilities of resources) played a role as an ultimate cause. To the extent in which social differentiation leads to an asymmetry in conditions of power and influence, it is to be expected that these have an effect on individual fitness. The gradual adoption of new subsistence techniques and the observation of changed yields and living conditions may have likewise acted as an intrinsic driver for the conversion, if accompanied by differential reproductive advantages and if it led to a continuation and solidification of the new subsistence mode with those individuals whose adoption of a new cultural strategy was, as it were, recompensed with increased fitness.

Subsistence change towards a form of subsistence management built on production and property changes the traditional social structures. New cultural needs and culture goals arise, which are achieved with varying success by the individuals of a population and eventueally become stablished. If success and fitness are intertwined in any way, one might be able to expect parallel strategies by which both goals are being pursued.

5.2.5 Cultural Goals and Reproduction

The formulation of neo-Darwinist hypotheses about culture as an extra-somatic mechanism for the conversion of reproductive success go back to Irons (1979, 1990). It implies that humans who consciously pursue certain culturally defined goals would implement a proximate strategy, by which in turn they would accomplish unconscious reproductive goals. It seems as if beyond the biological imperative of fitness maximising reproduction (Markl 1983) other, primarily non-biological goals are being pursued which add to the individual reproductive decisions humans make in the course of their lifetime. If status, prestige and power are definable as cultural goals, which for an individual are just as important to achieve and aspire as genetic continuity, then one should be able to examine whether the two goals of biological and cultural propagation are coupled and whether the achievement of the biological goal is improved or facilitated by the help of culturally defined default conditions and *vice versa*. Successful means to accomplish and meet aspired cultural standards can be expected to form a highly adaptive set of behaviours co-selected with the strategies by which the biological goal is achieved. If this applies, then differing patterns of culture and social organisation are the result of selection, i.e. evolutionary processes, during which the behaviours which became generally accepted and modified again and again were those which promoted the reproductive interests of humans as effectively as possible under given environmental conditions.

It is without a doubt difficult to find the appropriate direct reference to biological success resulting from the many possible and also contradictory human goals that arise from quite a range of different needs. One area of cultural goals, however, the propagation of immaterial capital in the form of honour, prestige and power, can be converted into wealth or possession and then seems to serve as the basis for high reproductive success (Casimir and Rao 1995). For various pre-industrial societies, such correlations between influence, resource availability and access to marriage or sexual partners were ascertained and demonstrated to lead to above average fitness, e.g. among Aché (Kaplan and Hill 1985), Yanomamö (Chagnon 1979), Yomut Turkmen (Irons 1979), Ifaluk (Turke and Betzig 1985), Mukogodo (Cronk 1991b), Mormons (Faux and Miller 1984; Josephson 2002) and white Americans (Essock-Vitale 1984). Whether this also applies in the same way to industrial or post-industrial societies has been questioned however (Vining 1986; Pérusse 1993)[6]. Furthermore, not every society displaying some sort of social stratification can necessarily be unambiguously integrated into a pattern of corresponding cultural and biological goals. Regardless, the general situation can be described as one in which above all high-ranking individuals within a community can be expected to have more children and also be able to offer these children a bioculturally adaptive environment, e.g. through better nutrition, better care and provision of better opportunities, and thus increase their chances of survival. This implies that humans, at least most of them, compete with others for the maximisation of their fitness and that they achieve this by satisfying their cultural needs and accomplishing their cultural goals. It will be decisive for the relative success achieved as to the extent that cultural default values embodied within the society were successfully converted into biological success, how close one has come to the culturally set goal. This should be measurable then by differential reproduction rates (see Borgerhoff-Mulder 1987a, b; Cronk 1991a). The question about close correlations between production and reproduction thus has to be raised again after discussing the connections with subsistence change, however now in a new guise, i.e. as the result of advanced steering through differentially distributed and accessible resources of land, livestock, money, other commodities or the sanctioned and aspired access to more than one marriage partner within a population.

As can be shown in at least one of the following examples, the size of a family alone is not necessarily a good indicator for the resources a family has at its disposal for the supply of its offspring. The basic pattern, according to which large possessions and high reproduction go together, is decisively modified as a function of ecological and social conditions, but in sum nevertheless leads to recognisable patterns of differential reproduction. These influences of extrinsic conditions shape behavioural responses that can be construed in retrospect as changeable adaptive mechanisms by which optimisation is achieved through the parallel pursuit of cultural and biological strategies. It may well be of even greater importance than pure total numbers, how many descendants of which sex actually survive or have the opportunity to reproduce and thus contribute to and benefit the total fitness of individuals. Which strategy is implemented in the end and how successful it is, is rather a ques-

---

6   It would have to be carefully examined, though, whether the overall decline in fertility is the result of impaired overall conditions that have a negative impact on reproductive decisions, or whether genetic fitness gain has been replaced by some other currency that appears to be a more worthwhile aspiration. The latter has recently been suggested in the context of high-status social competition, where the decision about the trade-off between more offspring and more status tends to be resolved in favour of status gain (status anxiety), which nevertheless is a worthwhile long-term fitness strategy enhancing inclusive fitness (Boone and Kessler 1999).

tion of status and possessions, but also how individual reproductive behaviour can help retain or increase these privileges.

Three case studies will briefly demonstrate what responses to specific socio-ecological conditions in different habitats and subsistence modes may be found for the accomplishment of cultural and reproductive goals and how this correspondence of emic and etic attributes is embedded in the socio-cultural fabric.

The Kipsigis live in the Kericho district of southwest Kenya. Until approximately 1930, they were predominantly pastoralists, keeping cattle, goat and sheep. After the arrival of Europeans at the beginning of the twentieth century and under the influence of colonial economic interests, Kipsigis subsistence changed gradually into a mixed mode of production, in which eventually maize was to become the main and most important cultivated crop. This change led to individual land ownership and an alteration in Kipsigis political economy. Herd animals, however, continued to be an important economic and socio-cultural component in the life of the Kipsigis, which was derived from their pastoral past. It can be only assumed whether the transition to a mixed subsistence was promoted also by improved possibilities of reproduction, but there is some likelihood for it, since the cultivation of maize as a cash crop led to new sources of income which then again were converted into property values. This can be regarded as a further clue to the importance of reproductive consequences connected with subsistence change (see Sect. 5.2.3). Herds and land are inherited along the paternal line: rich Kipsigis who yield a surplus of maize would sell it and invest the profit into the acquisition of cattle and land, but also use it to provide medical supplies and education for their children (Manners 1967).

Weddings are initiated and the necessary bridewealth negotiated by the parents of both partners. During the 1980s, it amounted to about six cows and six goats as well as some money (Borgerhoff-Mulder 1987a). A bridewealth of this dimension represents a substantial investment, since this number of animals corresponds to approximately a third of the average herd stock of a family. While the father is financing the first marriage, every further marriage has to be paid for by the groom and it usually takes about ten years for this to become economically feasible.

The area of land available to a man proves a suitable unit to measure and quantify cultural success in Kipsigis society. Independent of their individual age, culturally successful men are richer and better off socially when they own larger areas of land. Also, self-perception within Kipsigis society regards efforts to increase material possessions in terms of land and herds as a universal cultural goal. Thus, both the emic and etic aspect of the culture variable, i.e. the outside and inside views, are congruent in this case. The correlation between a man's wealth and the number of his surviving offspring is significant. The life-time reproductive success of Kipsigis men is the larger, the more land they possess. More land means a higher surplus from the maize harvest, which in turn facilitates the purchase of land and cattle, so that the amount of disposable commodities increases which is needed for the preparation of further marriages.

Indeed, the rate of polygyny of Kipsigis men is the principal reason for substantial reproductive variance and thus a function of differential wealth within the society. Neither the average marriage age of men, better material endowment of the children, nor any other variable leads to more descendants in richer individuals than the number of spouses. But land ownership is a central variable as well which women or their parents consider in their decisions over giving consent to a marriage (Borgerhoff-Mulder 1990, 1992). Since available food resources vary with the size of the available area of land, preferably those men will be selected who, due to more resources, are expected to be able to provide improved possibilities for raising more offspring.

Also among the Bakkarwal, a pastoral Muslim society in the north of India, material possession and an enlargement of herds are an avowed and emically embodied cultural goal, as well as a suitable means to attain power and reputation (Casimir and Rao 1995). In fact, this is one of the conditions in order to achieve membership in one of three status groups among socially high-standing and prestigious Bakkarwal men. It is only men from the group of the *Lambardar* who are entitled to hold an institutionalised position, which is normally hereditary and expressed through looking after political functions. *Lambardar* are the most powerful men of their community. Their power correlates impressively with the possibilities of enhancing possessions and prestige across generations in terms of above-average reproductive success.

Overall and valued against all other men, *Lambardar* not only have more children, but also more sons, more brothers and more livestock (Fig. 5.12). With that, they live up to ideals engrained within this patrilineal community, which favours the ability to influence and exert control over possessions, relational networks and profit orientation. This is further reflected in the observation that the average number of surviving children in Bakkarwal households can be described as a function of herd size, i.e. the bigger the herd, the more children will be able to survive. Among all households, again, the *Lambardar* possess the largest herds compared with all other families. Thus, the average number of both children and boys can be matched along the gradations of power and possession in Bakkarwal society.

In a society where families compete among themselves for resources and where this competition is dependent on the cohesion of family networks, particularly those of brothers, not only a higher number of children, but also a larger number of sons is adaptive. Sons of *Lambardar* can expect one day to inherit office, reputation and possessions from their fathers. Not only is the assumption of functions and status a cross-generational contract, as it were, an engrained cultural goal among these families, but at the same time it is an educational aspiration with consequences for the socialisation of their children, a privilege not granted or available to other status groups due to exclu-

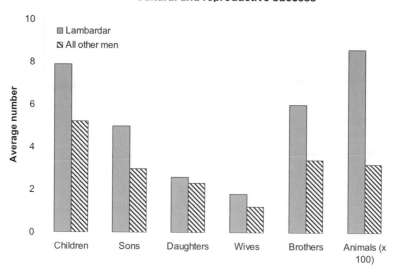

Fig. 5.12. Cultural and reproductive success of *Lambardar* families over their peers from other social groups among the Bakkarwal (modified from Casimir and Rao 1995).

sively assigned and non-hereditary rank positions within the *Lambardar* community (Casimir and Rao 1992).

Even though it cannot be pinpointed in detail why boys have an increased chance of survival as opposed to girls, the empirical result is nevertheless in agreement with predictions by Trivers and Willard (1973), according to which parents should move away from a uniform distribution of parental investment if the reproductive chances are unevenly distributed between the sexes and they are expected to invest more in that sex bearing the better chances. Socially higher-ranking individuals should therefore invest more strongly in sons, in particular if power, wealth and reproductive possibilities represent an entity that is socially embedded and passed on patrilineally. Among the Bakkarwal, cultural and behavioural strategies employed to accomplish these goals are in agreement with success that is measurable in the currency of ultimate biological interests of reproduction.

The Krummhörn, a region situated in the extreme northwest of Germany, in East Frisia, is said to represent one of the altogether rare ecological cases of a population approximately stabilised at the carrying capacity of its habitat during the eighteenth and nineteenth centuries. Delimitated by the North Sea to the West, the Dollart to the South and inhospitable swampland to the East, there were no possibilities for the population to expand, neither spatially nor numerically, exacerbated by the fact that the whole area had seen early and complete settlement of all available inhabitable patches (Engel 1990). The rich and fertile marsh soils allowed for a productive mixed subsistence based on agricultural farming and animal husbandry and, as a consequence, the transition from a subsistent to a market- and capital-orientated agriculture was realised early. Social differentiation within the population was pronounced. Access to and control of land and means of production were dominated by a stratum of rich landlords (big farmers), contrasted by relatively poor smallholders and/or wage labourers in the agricultural sector and by craftsmen and merchants in the service sector, who accomplished different amounts of wealth, depending on their profession. The mode of inheritance was peculiar in that the youngest son was the heir to his parents' farm (*ultimogeniture*) and who had to disburse his siblings, whereby twice as much had to be paid to brothers than to sisters.

The social structure of the population was determined in a remarkably rigid way by access to resources, in this case to land. Only beyond a certain plot size [25 Grasen (9.25 ha) of possession, or 50 Grasen of lease] land ownership made possible the pursuance of political or other social functions. Land was the crucial determinant for the acquisition of social prestige, influence and rights in this farmers' aristocracy; and differences in social standing between large farmers and agricultural workers were pronounced in various material and non-material ways. The solidification of the respective class-conscious attitudes was favoured by the Krummhörn being Calvinist protestant – again in an enclave situation within mainly Lutheran congregations to the East (Klindworth and Voland 1995). An ethics of social differentiation was a firm constituent of the cultural environment and facilitated differences in power and influence.

The reproductive conversion of these differences shows up in an average higher long-term fitness of big farmers as opposed to smallholders and landless workers. For example, figures for a certain marriage cohort (1720–1750) reveal that already ten years after marriage there is an increased fitness gain, which after 100 years amounts to a staggering difference in fitness of nearly twice the size of the population mean. Big farmer families did not only have more surviving children, but they were also able to provide better marriage chances for their children, while in half of the cases for the other social groups emigration from the Krummhörn remained the only option (Voland 1990). Increased marital fertility and more effective possibilities of social placement of offspring thus crucially contributed to the emergence of fitness advantages.

The situation in the Krummhörn has been compared with conditions of a population in a saturated habitat pursuing a *k*-strategy of reproduction, where too many descendants can lead to aggravations in resource supply and/or to reduced possibilities of opening up equally good chances of reproduction to all offspring. Therefore, a favourable social and ecological placement of children is considered the suitable strategy in order to optimally equip offspring with fitness-relevant resources and/or suited marriage partners and thus to secure long-term reproductive success. Under these conditions, socio-cultural strategies would aim primarily at safeguarding or, if possible, also at raising one's own status (e.g. in the case of smallholders) and would pursue the accomplishment of culturally defined goals, while at the same time improving the starting conditions for a maximisation of reproductive goals. Biological adaptation is thus converted with cultural means.

However, this situation presented big farmers with a problem. Under conditions in which more wealth helps yield more reproductive success, possessions can either be distributed among many, then less-rich, or few, but richer, descendants (Rogers 1990). But, since both land owners and landless people lived under the same ecological conditions, it should be possible to attribute the recognisable differences in reproduction and survival rates to differential possibilities of resource distribution (Voland and Dunbar 1995). This should then also hold true for the higher mortality of boys among the children of big farmers as, in a growth-limited situation like the one experienced in the Krummhörn, this group has to constantly compete for sustained favourable reproductive conditions in a habitat that is at the limit of its carrying capacity. As a consequence of the ultimogeniture inheritance regulation, the probability of survival and thus the reproductive options for a male descendant were the smaller, the more older brothers there were – i.e. fellow competitors for the position of heir but also those members of the family who had to be disbursed later and with double the sum for girls at that. Under these circumstances, sons were those descendants whose education caused more investment and costs and thus ran counter to the strategy of concentrating resources known as local resource competition (Cronk 1991 c). There are indications that, in the families of big farmers, this obviously led to manipulation of the sex ratio, as parental care was differentially distributed. This observation, however, does not apply to the lower classes, where due to the lack of possible resource allocation an opportunistic reproduction strategy prevailed, which at least offered some descendants the chance of social ascent and thus potentially improved reproductive possibilities. On the whole, for both girls and boys of big farmers, the probability of marriage diminished with birth order. Accordingly, ratios for unmarried children who emigrated without having married were high (Voland and Dunbar 1995). Yet, despite these reproductive restrictions, the strategy of choice under the given socioecological conditions, big farmers still achieved increased reproductive success compared with the landless families (Klindworth and Voland 1995).

These case studies demonstrate that mode of production and reproductive possibilities are linked through a positive feedback loop. In pre-industrial societies with different economic systems, as well as in different habitats, a culturally developed and socially embodied goal can be quantified economically in its effects on the biology of populations. More power and reputation can be translated into larger possession and wealth and this, in turn, is the condition and basis for accomplishing above-average reproductive success, whereby the number and/or the sex ratio of offspring in each case reinforces and perpetuates the existing system. In the long run, an extension of the resource basis (Kipsigis, Bakkarwal) or its stabilisation (Krummhörn) is a function of ecological conditions, i.e. part of the ecological resilience, and depends (at least in the cases presented here) on whether the habitat is open in principle or is closed, as in the case of the Krummhörn. The fundamental strategy however seems to be congruent. Resources are concentrated and held in concentration. This makes them predictable for those who are in control and allows these resources to be incorporated as cultural standards, as a calculable factor of active property adherence, for example through inheritance regulations. The predictability of the use and availability of resources is key to an on-average increased fertility and the maximisation of biological success in terms of inclusive fitness, which arises from the optimum conversion of a cultural strategy. Cultural and biological success come close, at least for those who make the rules.

5.3 Conclusion: Subsistence and Population Structure

The specific composition of human populations develops from the interrelationship between the ecological conditions of the habitat and their cultural mediation, which eventually have an impact on the size and density in which human communities can secure long-term survival in a given locality. This chapter deals with a central aspect of this interaction, i.e. the effects of different strategies and options of resource use on population structure and on its possibilities for development by considering influences of the mode of production, their change and their organisation within a society. Following on from this, questions are raised about how the use and availability of resources are connected with the necessity and possible mechanisms of regulation of population size and density. Central functional areas of life-support dealt with in the preceding chapters can be shown to become manifest and visible in the size and composition of populations[7].

With regard to numerical increase and development of populations, it is useful to differentiate between the two basic patterns of exponential and logistic growth. In line with ecological assumptions, exponential growth is sustainable until the ultimate carrying capacity of the habitat is reached, i.e. until the resource capacity is exhausted and prevents the numeric rise in population numbers. Usually, this process is therefore temporally limited. It turns into logistic growth, which leads a population into equilibrium at or

---

[7] Population is used here in the meaning of a complex integration of the genetic, demographic and social factors which arise from the responses of individuals to given and changing ecological basic conditions on a supra-individual level.

below the capacity limit, often realised by an oscillation around the carrying capacity.

The case most frequently found in human populations, however, is logistic growth with variable capacity limits, which corresponds to a combination of both basic patterns of growth. First, the technological possibilities available at a certain point in time allow an exponential rise in population size, which is subsequently slowed down by the delimitation of resources. Then, technological innovations provide an extended resource basis or improved possibilities for their utilisation and consequently trigger new growth expansion. The increase in world population, for example, representing exponential growth on a global scale, can be interpreted by way of this series of connected events of logistic growth. The carrying capacity of a habitat thus corresponds to that population size which can be regarded as the optimal and/or adaptive density of a population at a given point in time and in a given resource situation. A population so adjusted occupies a stability domain.

Exponential growth is a characteristic inherent to all organisms, but its accomplishment in terms of reproductive success is limited by natural selection. What can be observed at the level of the population as delimitation of resource availability caused by external factors is triggered within the population at the individual level through variable adaptation to the possibilities of resource use and the accomplishment of resource-related reproductive strategies. The regulation of population size or density is influenced by extrinsic and intrinsic processes, which were discussed in connection with use and availability of resources.

The effects of certain disease profiles were addressed in their significance as external density regulations. Subsequently, intrinsic mechanisms of population regulation were dealt with which, although permitting a systemic interpretation, can equally well or even better be regarded as the result of individual strategies. The question, at which level regulation of population actually takes place as a reaction to the ecological conditions of the habitat, was further explored taking infanticide as an example. Connections between subsistence change and fertility differences followed, to demonstrate the meaning of reproductive decisions for the accomplishment and maintenance of subsistence strategies. Finally, effects of differential access to resources within populations were considered in their consequences for group-typical reproductive possibilities.

The propagation of pathogens and the resulting distribution patterns of certain diseases work as an external element of population regulation. They are on the one hand connected with subsistence modes, but can on the other hand affect patterns of social organisation in a human community. The practice of certain modes of production leads to different density distributions. A foraging subsistence generally supports only small groups and the communities are dispersed throughout the habitat in only slightly aggregated distributions. Producing subsistence modes entail increasing intensification of food production and thus allow sustaining higher population densities and larger local populations. This is reflected by the spectrum of diseases, which are typically connected with certain subsistence modes; and predictions can be made as to the epidemiology of these diseases not only in extant groups but also in (pre-)historic populations.

Those infectious diseases which are persistent even under conditions of low population densities and cause population-effective consequences can be expected to exhibit high rates of infestation and not or only to a small extent induce immunity. Furthermore, endemic pathogens have to be expected that survive inside their hosts for an unlimited time and are also capable of reactivation (e.g. chickenpox, hepatitis) or slow infection (e.g. tuberculosis). An increase in population size in the course of stationary production and use of food resources leads to a change in the spectrum of pathogens and new paths of transmission. In some cases, only the transformation of landscape associated with a certain subsistence mode creates the conditions for a certain pathogenic organism to become endemic, for example in the case of malaria. Diseases can evolve into density-dependent regulators of population size, as soon as sufficiently high densities of the host population are reached that can sustain a sufficient number of infected individuals, which in the case of infectious diseases can require host populations of several hundred thousand individuals. Mobility favours the propagation of diseases also without the presence of large population concentrations, both in terms of mobility between larger settlements or cities (e.g. in the case of the Black Death) or as the result of migration (e.g. during colonisation events).

With the transition to agriculture, arguably the most far-reaching change in human subsistence, skeletal assemblages reveal a set of changes in disease frequencies, which in relation to the lifestyle of foragers are rated as a sign of generally degrading living conditions. Degenerative skeletal changes are regarded as an indicator of, among other things, the more intense workload accompanying agriculture, but they do not seem to change in frequency as a result of the conversion to food production. However, there is an increase in skeletal markers of infectious disease and the effects of possible synergies between infections and nutritional deficiencies or stress indicators in general. Geographical variability, however, is high and dependent on the respective abundance of resources, by which the transition could be nutritionally buffered. At the same time, however, it also shows that the prevalence of diseases varies with social inequality within populations, as higher-ranking individuals are usually less affected than those that belong to the lower social strata. An improved access to food resources thus seems to go with a diminished disease load.

At least on the level of biomes, distribution patterns of certain diseases can be shown to be correlated with certain characteristics of social organisation in human communities and to have an effect on reproductive possibilities. In tropical areas, where the diversity and prevalence of infectious and parasitic diseases is higher than in other geographical latitudes, marriage strategies are observed that are geared towards increasing the genetic variability of off-spring. Significantly more societies than elsewhere show high frequencies of polygynous marriages, which create a genetic diversity that helps scatter broadly the risk of low adaptability to high pathogen stress and thus increase the probability of more descendants surviving. Adaptive socio-cultural strate-gies of partner choice thus become a mechanism by which the specific dis-favour of a habitat can be moderated and converted into reproductive advan-tages.

Among the intrinsic factors that can lead to changes in population densi-ties, the ones of particular interest under system-ecological aspects are those that are connected with ascribed characteristics of self-regulation by which populations can keep their size and density below the carrying capacity of the habitat. It needs to be carefully examined, though, whether what appears to be self-regulation is in fact fictitious because its observation only results from the choice of an unsuitable viewpoint of those mechanisms and behaviours within a community which in sum lead to equilibrium between human popu-lation and resource supply.

The Tsembaga of the highlands of New Guinea, however, seem to represent such a case of autochthonous homoeostasis. It has been postulated (Rappa-port 1968) that by means of a cyclical operational sequence, where times of truce and peace alternate with phases of violent conflicts between local popu-lations, the distribution of land and thus the adjustment of population size to available resources is regulated by performing a ritual. This ritual is accom-plished whenever the number of pigs kept as livestock gets out of hand and indicates that too much of the arable crops from horticultural production has to be spent on feeding the pigs, so that the human population, having likewise grown in numbers, is left with too little food. On the occasion of a specifically arranged celebration, the pig stock is decimated by ritual slaughtering and, in outbursts of inter-group warfare following the ritual, the redistribution of land is accomplished. This would lead to stabilising access to land, the one resource scarcest under conditions of shifting cultivation, until an inevitably developing new imbalance is adjusted again by a ritual.

But is it actually this highly complex cyclical operational sequence, a cul-tural trait, that has been selected as an adaptive means of re-establishing homoeostasis in the system? As simulation experiments revealed, this is highly unlikely. Only under extremely stable environmental conditions could such a ritual cycle fulfil the function of stabilising regulation for the Tsembaga population, but hardly ever under the real conditions of relatively strong fluc-tuations of resource supply in the habitat. It is true, not only on the local level, but also on the regional level that the population seemingly aims at equilib-rium, although it is actually in a state of constant dis-equilibrium or a non-equilibrium pulsing state (Barrett and Odum 2000) which nevertheless tends towards balance in the long run. Homoeostasis becomes visible on the scale of a large time-window, but not however within the comprehensible duration of human generations.

Nevertheless, the ritual has evolved and does exist and, in spite of all justified criticism of its interpretation, it might still possess a function for the adjustment of the people/land ratio. The flexible, kinship-related adjustment of land is known to be a general principle among the peoples of the New Guinean highlands (e.g. Wohldt 2004), yet it does not necessarily have to be accompanied by complex rituals. If one assumes that the redistribution of land is understood as a mechanism by which resources are made accessible and available again to the advantage of members of certain groups of relatives, such as lineages or clans, this would lead to differential resource allocation within the population. Those individuals who benefit from the re-shuffling would thus be expected to increase their reproductive success through improved means of providing resources for their offspring.

The subsistence of the Rendille, a society of pastoralists in Kenya, largely depends on breeding and keeping camels. It was observed that the growth rate of the human population is lower than that of their most important resource, which has led to the assumption that, again, this would be a case in point for autochthonous balancing of a population below the carrying capacity (e.g. Sato 1980). The mechanism described to be responsible for this is based on the subdivision of the population into age sets that are connected with specific marriage procedures and synchronised with the generation length of the camels. Certain cultural prescriptions force daughters of fathers who belong to a certain age set (*sepaade*) to marry with a substantial time-delay, compared with other women. Likewise, there are prescriptions that also lead to a clear delay of marriage for young men. Heritage regulations benefit firstborn sons, so that the later-born siblings frequently out-migrate to neighbouring societies, in order to found their own existence there. Thus, women are also eventually coerced into exogamous marriage if they do not engage in polygynous marriage. Due to their higher age-at-marriage, *sepaade* women have fewer children than other women and moreover run the risk of having no male descendant who will inherit the herd. Men, in turn, try to compensate for this by marrying several women (Roth 1993).

The reason why this reproductive self-restraint of *sepaade* women is kept in place cannot be easily understood, particularly since it only reinforces a trend that characterises the Rendille anyway, i.e. the altogether small and clearly resource-oriented growth rate of the population. On the contrary, for *sepaade* women this system holds the risk of being maladaptive, since as a collateral effect of their husbands' polygyny and in the case that they do not give birth to a son, they must fear losing out in the process of resource allocation to their offspring. Even the firstborn sons are in danger of diminishing recognisable selective or fitness advantage, since their brothers would have to out-migrate and their descendant could only indirectly contribute to the firstborn son's inclusive fitness. However, there is still likely to be the advantage of retaining a more favourable distribution of resources for their future own descendants. Thus, at least their chances of increasing their life reproductive success under the given conditions are protected. Even the out-migrating males would have a higher probability of reproduction outside their native group, which likewise applies to women who leave the local population.

It follows that, whilst for individuals or families among the Rendille, even a low growth rate can be regarded as contributing to reproductive optimisation, only the tradition of the *sepaade* cannot be consistently explained at present, at least not in behavioural ecological terms. There seems to be, however, some individual benefit for poorer men who are allowed to marry *sepaade* women by performing bride service rather than having to pay bridewealth (Roth 1999). What looks like a sophisticated adjustment to the special environmental conditions of an arid habitat on the level of the total population, cannot be completely explained, but at least in the majority of aspects it can be satisfyingly explained by intrinsic behavioural adaptations in line with evolutionary biological and/or ecological assumptions about the optimisation of life reproductive success. Sacrificing fertility by *sepaade* women fits in with notions about individual, sex-specific power relationships which centre on the control of reproduction (Roth 2001). The existence of highly complex cultural strategies therefore need not be explained in the sense of self-regulating mechanisms by which survival is made possible for a whole society, but it can be regarded as the result of cumulative individual strategies that find their expression on the generic level of the population. Socio-cultural agreement secures what is individually adaptive and serves the increase in genetic fitness under the given ecological conditions. Rendille are reported to have long been aware of the conflicting goals for individuals, who may lose out on reproductive options, and the community, which benefits from an alignment of human and camel reproductive cycles. Marriage restrictions for *sepaade* women have their roots in past traditions of raids and territorial conflicts. When it became obvious that the circumstances had changed and the individual and communal benefit of this tradition was lost, this led to its demise (Roth 2001).

!Kung have been the prime example of populations to essentially adjust their size to the resources available in the habitat by maintaining long birth intervals. Their fertility is clearly below those of other foraging societies. Keeping a mean interval of approximately four years is accomplished by a combination of several mechanisms. Among those are an extended post-partum and lactation amenorrhoea as a result of frequent and continued breastfeeding, as well as a seasonal birth maximum triggered by the seasonally variable resource supply and thus a varying relative amount of body fat, as well as a delimitation of the carrying load for the individual !Kung woman, who has to carry both the weight of the collected food and a child during the food quest. Only those women would optimise their weight/distance ratio, who carry one child at a time, i.e. who can accordingly return larger quantities of food to the camp from their gathering trips. At the same time also, the number of surviving children is highest with those women who raise only one child every four years (Blurton Jones 1986). Even though this is a relatively long birth-interval period under conditions of natural fertility, it is by no means a 'waste' of reproductive capacities, but an optimisation of the reproductive possibilities available to a !Kung woman under the given habitat conditions.

However, this pattern of pronounced resource-limited reproduction is indeed only characteristic for !Kung from the Dobe area studied by Lee (1979) and represents a specific adaptation to the refugial conditions of their habitat. Other foragers, even other !Kung communities, under more favourable ecological conditions accordingly find and also use such different conditions for reproduction, which is reflected in the higher fertility and higher growth rates of their population.

It thus becomes clear that also, and particularly in this case, a 'wise' implementation of autochthonous or group-steered homoeostatic regulation of population size must not be assumed, but that instead adjustments of population densities below the carrying capacity have to be viewed as the result of cumulative individual reproductive decisions, which in sum can lead to a stabilisation of a certain population density, despite the fact that the idea of guided group selection has been reintroduced in the discussion of adaptive reproductive behaviour patterns (Boehm 1996; Wilson and Sober 1994). The reported statement of a !Kung woman that she would gladly have more children, if she could only nourish them (Marshall 1976), shows that there is a clear connection with limiting resource conditions also in the individual self-perception and linked to the decision for or against a further child.

With this in mind, the question whether population regulation is a myth or reality consequently follows on, as it could be shown that explanations for the adjustment of equilibria on the population level neither compellingly nor consistent with theory have anything to do with self-regulation mechanisms. Growth appears as the evolutionary principle that is never abandoned, but which is applied only in quasi-rational weighing by the individual in view of existing resource options. Postulating the advantage of group selection, however, is not in agreement with the principles of natural selection, but is probably rather a restriction of reproduction, if this means an improvement for personal or inclusive fitness for ecological reasons.

Does the selection of behaviour go as far as to allow also apparent counter-selective events like the killing of offspring to be regarded as an adaptive strategy? Inuit communities are a known, although controversially discussed, example for preferential infanticide, a behaviour that occurs in rising frequency: the more adverse the climatic condition of the habitat, thus the more uncertain the resource situation. In an Inuit society, the major responsibility for food procurement lies with the male hunters, who clearly suffer from increased risk of accidents and deaths. A successful and experienced hunter is highly esteemed, but casualties among the hunters are an equally big blow to the community. Intentionally manipulating the sex ratio of children in favour of boys, however, eventually leads to a balanced sex ratio in adults, conducive to securing food supply for the families. Although parental investment into boys is more costly, because they have to be trained longer and marry later than girls, this additional expenditure disburses itself and pays off with a more predictable resource security. By applying selective control, parents decide on further investment into offspring and optimise the number of children they will successfully be able to raise under the given conditions. A surviving boy will later in his life make a substantial contribution to the family's subsistence, as it were facilitate its continuation; and this will eventually benefit the parents as well, so that their possibilities for further reproduction are improved, thus allowing a maximisation of their life history strategy.

Regulation of population takes place, however, at the individual level; and in each case it takes place in a specific social and reproductive context, in which certain behaviours are selected as responses to given constraints of the environment – and socially sanctioned. The myth only exists in the assumption of auto-regulative mechanisms at the population level for the selection of which there are no conclusive explanations – it is no myth however that there are regulative adjustments by which individuals or families steer the number or the sex of their offspring in a resource-oriented fashion as a reaction to existing or expected environmental conditions. But probably rather as the ultimate reason, which suggests that far-reaching extrapolations into the past should be treated with caution.

Subsistence change is frequently accompanied by population increase, even if it cannot be sufficiently clarified in each individual case whether this is caused by new subsistence technology or whether it is rising population numbers as a sign of optimised habitat use that leads to a change in the economy. This question can now be put into a reproductive context.

The Mukogodo of southern Kenya went through profound culture change at the beginning of the twentieth century, in the process of which their subsistence changed from a foraging to a pastoral mode. Contact with neighbouring pastoral societies saw marriage relations develop that, however, made it increasingly harder for Mukogodo to participate, as the bridewealth had to be paid in cattle rather than bee hives. To nevertheless accomplish marriage, Mukogodo were in need of herds. From a certain point in time onwards, it was thus necessary to convert their economy towards generating and maintaining a new resource, in order to be able to found a family at all. At the same time, this conversion meant that not only had the old way of life to be abolished, but a return was practically excluded due to the close integration into social and economic relations with the neighbouring communities. The gradual change of the subsistence basis started with individual decisions aimed at changing the circumstances of marriage and reproductive options and finally spread within the entire population. Yet, the supply of new food resources also provided a desired side-effect with regard to subsistence security.

Such a process, where cultural contact induces a change in both social organisation and subsistence and which is conclusively congruent with the reproductive interests of individuals, offers a general explanatory pattern of ultimate causes of why humans would give up their traditional subsistence mode (see Chap. 3). Indeed, at a generic level, clear differences in average fertility between the basic forms of human modes of subsistence can be found, showing that with increasing control of cultivation and availability of food resources reproductive possibilities are expanding. In pre-industrial societies – or those before the demographic transition – rising fertility, but not necessarily a lowered mortality rate, triggers population growth. In addition to ecological causes or to motives developing from increasing social differentiation, the accomplishment of reproductive interests in the sense of personal maximisation of fitness is suggested as a further mainspring for the conversion of subsistence modes. The success of such a strategy would then also be accompanied by better resource security, which in turn would increase the chances of improving the probability of survival of the offspring. The linkage of socioecological basic conditions with reproductive options is thus likely to lead to subsistence change. But what effects might social differentiation within a population have on reproduction?

In societies where incorporeal categories of power, status and influence find their expression in larger possessions and wealth, in which reaching certain social positions is thus embedded and existent as a culturally defined goal that can be converted into material goods, it may be expected that better possibilities of access to resources are correlated with reproductive success. There is repeated and world-wide evidence for such connections between cultural success and reproductive possibilities in pre-industrial societies, in various habitats and with different modes of subsistence. While for the Kipsigis from Kenya a direct connection between ownership of land and herds of livestock and life reproductive success can be shown, the north Indian Bakkarwal and particularly the northwest German people of the Krummhörn, in addition to the reproductive significance of resource allocation, exemplify that the sex ratio of surviving offspring is important, rather than numbers of children. The probability that sons of big farmers would survive in the Krummhörn, for example, was higher, the fewer brothers survived and had to be compensated because they were not entitled to inherit.

Behind the maximisation of reproduction is the principle of accumulation or the concentration of resources, which is steered via the socially unequal availability of resources. In many pre-industrial societies, the varying distribution of material goods identifies culturally defined goals and this, in turn, is the condition for an optimal allocation of resources which can be invested in one's own descendants. Cultural and reproductive success is mutually reinforcing and dependent.

In specific cases, the concentration of resources and the trial to maximise individual genetic fitness has been carried to extremes. Historically reliable sources report culturally sanctioned incest (Van den Berghe and Mesher 1980) practiced among royal ruling families with established polygyny and patrilineal inheritance, the most notable examples being Inca rulers in Peru, Hawaiian clan chiefs or the Monomotapa kings of historic Zimbabwe. However, close scrutiny reveals that such a strategy is only found on the condition that the social exclusivity of the ruler is unquestioned; and this entails the possibility of begetting descendants who are genetically as closely related as possible – at the high risk and price of hereditary genetic damage (e.g. Vogel and Motulsky 1997). Depending upon requirement and possibilities, however, the principle is being softened and incest replaced by any other form of close inbreeding. What is culturally permitted does not necessarily have to be always executed, though: it may rather be diverted into symbolic or ritualised behaviours. The cultural norm is flexible, as long as only the ruler possesses the privilege to select for himself the strategy which, both genetically and regarding the concentration of resources, yields the most profit.

This chapter attempted to unite the areas of subsistence modes, subsistence change and the differential use of food resources in the context of possible reproductive consequences. It becomes evident that an interpretation informed by systems ecological theory allows these areas to be beneficially linked with behavioural–ecological assumptions.

The development of adaptive strategies of food acquisition can be explained on the one hand as effective biocultural adjustment to the ecological conditions of a given habitat, by which human communities co-ordinate matter and energy on a long-term basis and are able to achieve remarkable efficiencies (see Chap. 2). On the other hand, ecological success can also be based on how and whether populations succeed in adjusting densities and sizes to the resource options of the habitat. Individual reproductive decisions can be identified, which eventually lead to equilibria between resources and population on the population level. The systemic characteristic of regulation is the result of cumulative events on the individual level. Stability domains emerge from continued individual attempts to reconcile production and reproduction.

Looked at from a general point of view, this has produced strategies of habitat use, which resulted in improved predictability and control of resources and led to increased fertility. The transition of subsistence modes, for instance the change from foraging to farming, is thus connected with an average increase in the number of offspring. Whatever may have been the cause of subsistence transition (see Chap. 3), it can be described by its reproductive consequences where fitness gain provides an ultimate explanatory framework.

Finally, to the extent in which decisions about the availability and use of resources are being increasingly made with reference to social differentiation within a population, the possibilities of the concentration and accumulation of material goods are connected with differential reproductive success. At least in pre-industrial times, social differences that can purposefully affect the flows of matter and energy do not only have consequences for nutritional quality and the allocation of other resources. They are the conditions which eventually find their expression in the social variance of fitness maximisation.

6 Synthesis – Towards a Biocultural Human Ecology

6.1 Introduction

Human ecological footprints cannot be overlooked. Humans shape, and have shaped, the appearance and composition of entire ecosystems by the trademark of their ecological 'role', the intentional interference with flows of matter and energy. In the habitats they utilise, the niches occupied by humans are broad by any standards and not only in comparison with other species. Their shaping of these niches is facilitated by biological/physiological adaptations, yet accomplished primarily in terms of socio-cultural achievements. Like no other species, humans have emancipated themselves from the genetic fixations and reaction norms of their basic biological equipment; and they have developed and extended their adaptability by non-genetically coded means in the course of their history. Humans are thus also culturally a polytypic species. And this is notwithstanding the observation that human cultural and ethnic diversity is strongly linked with a latitudinal gradient and seems to form biogeographical patterns (Cashdan 2001; Collard and Foley 2002). The transformation and transmission of cultural information significantly contributed to an adjustment in the ecological circumstances of the respective habitats which was both flexible and enduring. By means of specific strategies of environmental use, humans found successful solutions as responses to stresses and limiting factors in their habitats, which enabled them to settle in areas quite remote from their places of origin in tropical Africa and which left no biome unoccupied. Cultural mediation of natural conditions often led to a re-shaping of landscapes and thus to alterations and adjustments of the living conditions, which in turn could subsequently impact upon biological processes and characteristics, for example as reflected in demographic parameters. Cultural responses are dynamically interrelated with biological presuppositions and consequences of human action; and they are intertwined in a complex of biocultural adaptations. Human ecology is thus about the biological outcomes of cultural strategies.

To accommodate this notion, human populations are preferably viewed from a holistic perspective denoting their interaction with abiotic and other biotic components of the system, also referred to as interrelations with the ecospace and the genetic biospace, respectively (cp. Lawrence 2003), with which they are connected through structural and functional relationships. Whilst the biosphere is finite, human ecosystems are not: they are open and subject to external influences even beyond the immediate boundaries of the system. Human populations form units which utilise and control flows of matter and energy in a given habitat in time and space, based on biological and especially cultural adaptations; and they make particular use of the ecological category of information, which extends far beyond the idea of a genetic blueprint and indeed forms the key component to facilitate survival through culturally embedded knowledge, belief systems and institutions. Since human populations are subject to general ecological principles, they can be analysed from a systems-theoretical point of view. Within the system, effects of changes in certain components can therefore be analysed with respect to their consequences for human populations; and effects of human action can in turn be considered as alterations of these components with respect to their consequences for the whole system. Human populations thus become available to an analysis of their connections and interrelations with other components of ecosystems.

The systems-ecology approach is beneficially complemented by a cross-cultural assessment of relationships between adaptive solutions of subsistence procurement in human communities and given natural conditions; and as such it pays homage to early attempts of this kind within Cultural Ecology. The framework of a systems-ecology analysis allows changes within the system and effects of relationships between components to be uncovered on the population level. However, higher-order properties of this kind, which become visible on the level of communities, are the cumulative result of individual decisions, driven by the aim to accomplish proximate tasks and targets or, in fact, driven by ultimate fitness-related goals. In order to consider such processes in an appropriate way, the classic concept of systems ecology needs to be extended by the concepts of behavioural ecology and evolutionary biology and to incorporate life-history considerations. The inclusion of such approaches, which take into account the long-term effects of reproductive interests and fitness and shorter term trade-offs of differential energy allocations, provides a connection between the individual level and the resulting consequences of these decisions on a population level.

Biocultural adaptations encompass reactions, processes, strategies and factors that can be observed as responses of human populations to given ecological conditions, which in turn react on the biology of populations, i.e. their potential for survival and reproduction as well as their spatial distribution. Therefore, and for obvious reasons, there has been an emphasis on analyses revolving around the issues of how humans are able to meet requirements or challenges of resource availability, procurement and sustainability, an approach that has also been adopted for this study. These issues are not only connected with the most fundamental principles of how humans manage to make their living, but they constitute and permeate through major functional aspects of living and experience in human communities, which allows a demonstration of not only the adaptability and success of biocultural strategies of coping with the environment, but also the consequences that result from possible failures. These are timeless issues, as well, and are thus particularly suited to support an approach which decidedly adopts the inclusion of historic conditions, which acknowledges that human ecosystems develop along a time-line and can only be understood if their current shape and form is considered against previous individual and collective decisions. In fact, human ecosystems provide us with a tool kit of long-term natural experiments which preserve the most varied and sophisticated record and memory – however encoded – of human/environment interaction. This property builds a link with present-day issues and future challenges. In particular, the aspect of resource sustainability provides epistemological continuity in times of growing concern about the disjunction between patterns of environmental consumption and the disempowerment of regulative structures (Gare 2000). Resources, their provision and meaning, their biological functions, social significance and cultural connotations are thus at the heart of an integrative, biocultural human ecology.

The main questions asked here to unravel these connections are briefly summarised. How do human communities cope with various ecological constraints in their habitats and solve the problem of ensuring energetic and caloric survival? The spectrum of mechanisms presented here can only provide a coarse cross-sectional account of the variability and diversity of subsistence strategies. Yet, it reveals an amazing breadth of small-scale solutions developed in attempts to tackle complex dynamics, both spatial and temporal, and uncertainty and surprise as the ever-present challenges and properties of habitats. And despite the necessity for tailored local responses, different levels of depth of interference with the environment, usually subsumed under distinct subsistence techniques (e.g. Sutton and Anderson 2004), show common features in resource appropriation strategies which are shared across subsistence categories for the implementation of complex socio-cultural structures that govern resource procurement. Increasing societal complexity, typically at the levels of local groups and regional polities, is connected with differences in individual possibilities to exert influence over the utilisation and control of resources. Apart from hierarchical structures of ownership, for example, of land and means of production, differential access to resources has measurable consequences for nutrition and health, where status-related dietary options are not only reflected in cultural sophistication of cuisine but also in more tangible differences of dietary quality and access to animal protein. Past societies can provide significant insights into this immediate expression of complex resource-related behaviour and the issue can be followed through time up to the present day. Human subsistence strategies, their very existence and duration in time, tend to be seen as the result of adjustments or some kind of equilibrium state between the population and other components of the ecosystem. Provided this is the outcome of a long-term adaptive process to prevailing natural conditions and resource supplies, however non-linear, fluctuating or stable they are, what has caused people to give up a certain mode of production? Which extrinsic or intrinsic factors were, or in fact are, responsible for subsistence change and where is the advantage or benefit of transferring a proven subsistence strategy into a new stability domain at the (calculated?) risk of a temporal disturbance of the equilibrium state? Strategies of flexible and efficient co-ordination of flows of matter and energy under various ecological constraints, differentials in individual resource control and access and ways to re-adjust temporary disturbances of equilibrium states eventually have to be analysed in terms of their effects and consequences for the size and composition of populations. These are the biological outcomes of cultural strategies, the pursuit of individual goals and collective interests, the interface between production and reproduction.

6.2 Subsistence and Social Structure

Natural habitats are characterised by various and different ecological constraints which generate biogeographical distribution patterns. They are subject to ubiquitous change across all scales and levels of organisation, which may result not only in environmental variability but also in temporary abundance and non-linear conditions of resource supply. This tendency towards instability constitutes the framework against which human communities have developed coping mechanisms for a dynamically robust persistence in the presence of fluctuation. Modes of production and food procurement strategies can thus be expected to allow for an analysis of biocultural adaptations, since means and ways to provide not only a sufficient but also a nutritionally balanced or adequate supply of food is among the most basic requirements that have to be fulfilled for long-term human settlement in a given habitat. Compared with other species, humans are generalists as regards their physiological properties. Even though they are usually placed at the level of third-order consumers within their food-webs, they are not specialised in terms of their nutritional physiology and, being omnivorous organisms, they are able to extract nutrients from a large number of plant and animal food sources. Without any doubt, this is a suitable requirement for a successful adaptation to various habitats providing diverse food items. In view of the fact, however, that humans managed to settle in any of the earth's biomes, from the circumpolar regions to savannahs, deserts and the tropics, the question arises whether this prerequisite was truly necessary, but just not sufficient. It shows that humans developed quite different strategies of long-term survival and still do so, depending on the ecological conditions they are faced to cope with. These strategies of food procurement can be correlated with limiting factors characteristic of the respective habitat, e.g. in terms of climate, productivity, vegetation and species diversity or soil conditions. The savannah or the steppe environment, for example, is not suited for agricultural use due to unpredictable rainfall and its cyclical and often lasting periods of draught. It is, however, suited for any subsistence strategy in which animals take the task unsolvable to humans of converting sparse vegetation into food. This strategy is typical of pastoralists, whose survival insurance is 'living food' in terms of their herds. Other biomes could provide comparable examples.

Such strategies of food procurement are far from being developed and maintained, because of certain biological properties of human beings. Strategies are adjusted to certain general set-ups and preferably aim to be effective for the long term. To be able to adapt to the conditions of a habitat, to survive and to procure for a secure nutritional basis in spite of limiting factors and habitat constraints, additional mechanisms are required to establish and accomplish these strategies within the community. They need to be firmly embedded into social agreements, rules and political institutions, which in turn sets the frame for a co-ordination of the subsistence activities within the community. These flanking mechanisms themselves become part of the strategies; and cultural continuity as part of the social organisation of food procurement allows coping with change and variation as it enhances resistance and adds to the resilience of human communities by allowing transitions among stable states. To push the analogy further: it enhances the adaptive capacity of the biocultural system (cp. Gunderson 2000).

The variability of human subsistence strategies reflects the multitude of ecological default values and local resource supplies and thus leads to a large variety of regional solutions, which are optimised as adjustments to typical habitat conditions. The diversity of a habitat can be utilised through the division of labour and space, like in foragers and pastoralists, in terms of the mobility, fission and fusion of groups (case studies of !Kung, Inuit, Karimojong). Other strategies aim at the inclusion of micro-habitats into subsistence activities, for example by exploiting resource islands in the Andean highlands (case study of Nuñoa) or by the small-scale parcelling-out of productive land in temperate areas (case study of Wiltshire).

In biomes characterised by cyclically fluctuating or constantly extreme climatic conditions, strategies are established by the community that ensure an even distribution of essential food resources among the population and thereby spread and reduce subsistence risks for individuals. Such a behaviour can either be regularly restricted to times of scarcity and need (e.g. Karimojong) or anchored as a general principle (e.g. !Kung, Inuit). The social consensus relies on the principle of mutuality and hence ensures advantages beneficial to individuals in the long run. Mechanisms of physiological optimisation, such as coca-chewing in the Andean highlands, are found to be maintained by cultural safeguarding in terms of networks of barter and exchange (Nuñoa), or they are the result of co-evolving genetic and cultural traits, such as lactose tolerance and its high prevalence in herders. The general principle appears to be that limiting factors, constraints and unpredictability can be compensated for by establishing mechanisms that function as a buffer against the inherent dynamics of environmental variability.

Human populations prefer or choose to concentrate their subsistence activities on resources which are energetically efficient and reliable in the long term (e.g. !Kung, Tsembaga, Karimojong). However, the exploitation of alternative sources of food, or rather, in the first place, active knowledge of their occurrence and abundance serves as a back-up mechanism. They provide potential sources of food supplements as essential parts of the overall subsistence strategy in order to cope with the dynamics of resource availability, for example in times of long-lasting draught (!Kung, Karimojong), provided the habitat conditions allow for it. With the exception of extreme habitats, e.g. the circumpolar regions, which practically do not allow alternatives to the given nutritional pattern, human populations are in principle able to flexibly adjust their trophic relation within food-webs to the respective ecological circumstances of the habitat and thus enhance local stability. This denotes ability in the first place. Whether it can be accomplished smoothly or on what scale this is conceivable at all, however, certainly depends on the extent to which food preferences or prescriptions have become part of cultural identities and cosmologies, i.e. how deeply they are engrained in the social fabric of lifestyles associated with certain subsistence strategies.

Food procurement strategies aim at optimising energy yield in relation to energy input. With increasing intensification of food production, costs will rise for the invested work effort. Therefore, the energetic efficiency of the various modes of production is directed towards different factors, in order to keep the energy balance consistent or positive, in other words to optimise energy return on investment. Foragers, for example, pursue strategies optimising yield per unit of time by taking into account resource abundance at certain patches. Such inter-relations can be impressively demonstrated by modelling optimal foraging behaviour. Yet, there are no simple and direct connections between a certain mode of production and the respective time budget, since even within one subsistence category the time spent for food procurement differs considerably depending on the respective habitat conditions, i.e. the patchiness of available resources and their variable abundance. Cultivators and farmers necessarily show a higher investment into production due to the continued work effort that has to be invested into plots of arable land. They must thus aim to optimise the efficiency per unit of cultivated land and increase the yields, for example through a suitable combination or sequence of crops or even fallows. Pastoralists, on the other hand, can optimise the efficiency and energetic return of the major resource, their herds, and invest in subsistence strategies aiming at local ecological stability. Attempts to quantify the efficiency of resource extraction strategies have a long tradition, based on the notion that energy is a universally applicable currency for measuring the adaptiveness of such strategies. While dismissed at some stage as the equivalent of a caloric obsession behind attempts to understand how human communities manage to adjust to uncertainty and unpredictability, the idea has regained momentum through the concept of energy gain, or energy return on investment (Tainter et al. 2003), an idea strikingly similar to the notion of total energetic efficiency, as it relates the amount of energy extracted to the amount of energy necessary to tap the source and manufacture and refine a product.

The efficiency of food production in a given habitat is not bound to subsistence economy and may be widely influenced and governed by socio-political conditions (cp. Sect. 2.2.7). This has, in the past, led to a dissociation of production from subsistence economy (for example, Wiltshire), but it can be observed wherever coping mechanisms are disrupted and resilience is lowered, i.e. when opportunities of culturally established mechanisms to buffer, self-organise and learn are restricted (cp. Trosper 2003).

Comparing the different modes of production reveals that, at face value, energetic efficiency is increased with growing investment of work, time and energy into food production. The total efficiency of the subsistence system, however, decreases both in absolute terms and compared with the energy expended. The intensification of production through technological advance increases thermodynamic inefficiency (Pimentel et al. 1973; see Stepp et al. 2003) as more and more complex sequences of energy exchange between populations and habitats and of energy extraction are required. This aggravates a balancing of energy and material flows. Yet, energy is the only convertible reference unit that allows a cross-comparison of different subsistence systems. Hence, what leads to the disclosure of general rules of energy use on the macro-level, or even predictions about settlement and social organisation from energy gain patterns (see Tainter et al. 2003), can be more beneficially analysed on the micro-level. By focusing on a family or household as the basic unit of both production and reproduction, it is possible to successfully assess the interference with matter and energy flows which result from individual decisions. Variances in the efficiency of managing resources on the micro-level can then be analysed with respect to the effect on differential probabilities of survival; and thus energy use can eventually be evaluated in terms of reproductive consequences. While attempts to assess energy flows on a population level are doomed to fail, because not every aspect and facet of human life and behaviour can be converted into energy currency, the micro-level offers a genuine opportunity to link life-history parameters and effects of energy allocation with measurable fitness outcomes. Cultural strategies of resource procurement are thus reflected in fitness returns.

A survey of social systems reveals that there is a close correlation between modes of production and the complexity of societal structures, i.e. between resource utilisation and social organisation (e.g. Johnson and Earle 2000). Typically, some kind of control over resource availability within a community becomes a characteristic trait in those societies displaying food production and/or institutionalised social differentiation, as it occurs on the level of local groups and, especially, in societies organised as regional polities. Thus, the development of the very socio-political structures that within a community allow for an increase in the differential utilisation and regulation of flows of energy and matter in a given habitat leads to an unequal distribution of societal conditions of resource availability. The natural conditions of the habitat, then, are usually of minor importance.

The implications of social inequality for the asymmetrical access to resources are a commonly known feature and often at the heart of debates over resource sustainability today (and in the past) and ways to remedy its consequences (Gare 2000; Janssen and Scheffer 2004). The phenomenon as such, however, is equally accessible or can be reconstructed for past populations (even of those historical periods with scarce or almost no written record), by fully utilising the archaeological record and analytical techniques of archaeological science. The reconstruction of subsistence patterns and the variability of dietary regimes is one of the salient examples of the potential that lies in applying an ecological framework to populations and communities of the past and thus to provide a bridge capable of linking the long history of human/environment interaction with the current environmental debate, i.e. to find common ecological determinants. For a number of quite different socio-cultural, geographical and temporal settings, from the Mississippian city of Cahokia in North America to Iron Age communities of Bohemia and early medieval societies in south-western Germany, it can meanwhile be demonstrated that unequal opportunities of availability and utilisation of food resources within populations is reflected in social differences in dietary behaviour. This body of evidence will continue to grow and provide empirical support for an apparently universal behavioural pattern associated with ranked and hierarchically organised societies, i.e. the occurrence of socially governed and determined resource control and thus the opportunity of differential resource utilisation. Power relations leading to improved resource control provide tangible benefits; and dietary differences expressed through status-dependent differences in the availability of food resources reflect more than just sophisticated cuisine. The overall pattern reveals the obviously better access either to animal protein in general or, in fact, to resources whose production requires a higher energy input and maintenance but which give improved health, likely as a result of better living conditions and nutritional supply. Such differentials translate into different levels in the social fabric. The immediate effects of status-related resource control not only result in gender differences, but characterise certain age and family groups of different ranks.

Detecting pronounced status-related dietary patterns in children only testifies to the unequally distributed options of resource utilisation between families. As these are the core units of both production and reproduction, children directly reflect the outcomes of social control over resources and their allocations. Past societies thus become accessible in principle to interpretations in terms of behavioural ecology, consistent with the notions of life-history trade-offs.

The application of scientific methods to archaeological questions has an important role here, as it enables a detailed insight into the subtleties of past life-styles and the social and economic organisation which will eventually allow an evaluation of their repercussions on human-ecological relationships. Analytical investigations of human remains continue to make an impressive contribution to furnishing ecological issues with empirical data; and the topics covered range from subsistence to movement and migration, population dynamics, palaeogenetics and kinship (Ambrose and Katzenberg 2000; Kaestle and Horsburgh 2002; Bentley et al. 2003; Fuller et al. 2003; Hummel 2003; Montgomery et al. 2005), i.e. topics of immediate significance to fundamental processes and relations in human ecosystems. Their increasing incorporation into a truly human-ecological, i.e. biocultural, context will add a new quality to the understanding of our past and thus facilitate a unified interpretive framework that effortlessly links history and presence. +

6.3 Transitions

Human populations have gone through a number of subsistence transitions, most notably those that independently and at different times and locations replaced the foraging mode with the production and control of food resources. While in hindsight the benefits seem to be obvious, subsistence change is a complex, disruptive and convoluted process, which one would not expect to happen very often in the first place. Subsistence strategies, the combination of technological solutions for environmental exploitation and their implementation into social agreements and cultural institutions are a means of lowering ecological resilience, i.e. reducing the time necessary for the system to return to a stable state after disruption (Gunderson 2000; Trosper 2003). By implication, this enhances social resilience, i.e. the ability of communities to cope with external stresses (Adger 2000). Taken together, established ways of social and cultural mediation which made a certain mode of production viable and sustainable under given ecological conditions, may be called adaptive due to their very, often long-term, existence, as they offer considerable buffer capacity against various kinds of perturbation. Such combined strategies have likely evolved and their maintenance been accomplished through decision-making that was influenced by the level of prior investment, which should increase the hesitation towards pursuing change (sunk-cost effect; Janssen and Scheffer 2004). Why then do we find subsistence change, anyway, and why is a hitherto successful subsistence strategy being abandoned and a new one implemented? Major events of this kind are known to be connected with powerful extrinsic causes, for example climatic changes, or intrinsic developments, such as socio-political dynamics within a population, which inevitably require new and perhaps more productive modes of production and which allow the redefinition of power relations and the means of resource control. In other words subsistence strategies change when ecological and/or socio-cultural circumstances are being substantially altered. Subsistence change thus corresponds to a departure from equilibrium between the human population and other system components and eventually results in the establishment of a new stability domain, yet at the cost of exceeding resilience capacities and structural change due to disruption.

A heuristic transition model can help in understanding the processual dynamics of subsistence change by subdividing the process into three phases: (1) a precursor phase, as it were the lead time within which people can get acquainted with a new mode of production, (2) a phase of adoption of a new subsistence technique or rather the substitution of the previous mode and (3) a consolidation phase during which the system establishes itself in a new stability domain. Major transition events, for example the establishment of farming in the Levant, the large-scale introduction of maize into traditional subsistence systems of North America, or the advent of the Neolithic in Southern Scandinavia, all have an important feature in common: either the resource and its cultivation and management (cereals, maize) or the whole subsistence kit was known to the human communities over very long periods of time. This acquaintance became, as it were, part of the resilience and resistance these communities had developed for many generations and allowed them to test a new subsistence technology and to prepare and implement a gradual change in the socio-cultural framework. The actual transition, first of all irrespective of its cause, was then a matter of a few generations only, in any case a short time compared with the lead phase. A longest possible precursor phase and a short substitution phase appear to represent the ideal conditions under which the re-instalment of a temporarily disrupted equilibrium is accomplished. This, for example, is particularly important in the transition to subsistent agriculture, because food production at an intermediate level of implementation (ca. 10–50 %) is obviously not stable and thus rarely to be found in human communities in general. Thus, the yield of agricultural production has to be so big that fluctuations in resource supply and energetic uncertainties in the transition to the new mode of subsistence can be relatively quickly compensated. In other words: the higher the flux of nutrient-cycling and energy from the new subsistence mode, the more quickly will the effects of the perturbation be 'flushed out' (p. 858 in Begon et al. 2003), i.e. the quicker the disturbed flow of energy and resources is channelled into a stable supply, the

quicker can subsistence change and the return to a new stability domain be achieved. Rapid subsistence change thus appears to be an energetic necessity. But the issue to remember here is the importance of a long period of availability and the possibility to prepare change in order to cater for its successful implementation.

The extent of social disruption, as it were the amount of strain that social resilience has to suffer, appears to be traumatic, irrespective of whether subsistence change is largely driven by intrinsic or extrinsic factors. It is always connected with a re-organisation of social and power relations and may, as with the spreading of agriculture into Central and Western Europe, be bound to empirically verifiable demic and genetic diffusion, as well as the wholesale import of a new technology, together with the connected social and political structures.

In all cases, sedentariness is already established as a crucial prerequisite for the production of locally bound food resources, before the transition to a new mode of subsistence takes place. Such developments, being part of an adaptively relevant environment (Irons 1998) and eventually pre-adaptive to subsistence change, are connected with accompanying socio-cultural change. Strategies of food procurement are always and by necessity embedded into an ecological and social structure, so that changing an existing mode of production also affects alterations in central parts of societal organisation, such as property rights or division of labour. The increasing social differentiation and complexity evolve in parallel with the establishment of an agricultural subsistence. In the case of southern Scandinavian Mesolithic populations, this is even regarded by some as the driving motivation for an adoption of the new technology. Subsistence change, therefore, facilitates a new definition of matter and energy flows in the course of the establishment of a food procurement strategy, but the novel strategy also facilitates the achievement of new culturally defined goals, such as property or status, and leads to a level of resource control hitherto not possible. Especially the second aspect suggests that subsistence change can also be related to individual and fitness-oriented decisions within populations, which in turn affect differential resource allocation.

6.4 Resources and Fitness

The distribution and dispersal pattern of human populations generally follows principles of resource abundance and availability. On a global scale, this is mirrored by a clear relationship of ethnic and cultural diversity with habitat diversity and a latitudinal gradient. In this respect, humans are just like any other animal, as biogeographical rules seem to apply. The question remains, though, what kind of effects strategies of food procurement, conditions and course of subsistence change and mechanisms of regulating access to resources have on populations as biological and ecosystemic units. How are the distribution pattern, size and composition of human populations affected by responses to the constraints of natural conditions and what are the reproductive consequences of cultural strategies?

Such patterns are dependent on the carrying capacity of a given habitat at a given time. [Humans] are peculiar in this respect, as they [not only display population dynamics against the ecological capacity of the habitat but are able themselves to alter and interfere with it through superimposing a cultural carrying capacity.] This carrying capacity is variable and takes the form of different stable states related to environmental and cultural properties; and it is determined both by the resource situation and by behaviour options. Being resource-related, the capacity is finite in the first place: it restricts human population growth and leads to a pulsing state around the sustainable limit. Frequently, however, capacity limits are extended through technological innovations and concomitant societal adjustments. In fact, this is the major force that facilitated population growth in general, so that the expansion of human population sizes and densities can be characterised as logistical growth with variable capacity limits – provided environmental conditions allow for it. Human populations, however, are known to resort to alternative strategies, in particular under conditions of environmental unpredictability, viz. to develop and socially implement mechanism of regulation as adjustments of the resource/population ratio. The element of social implementation cannot be overemphasised. Depriving people of their reproductive options or, in fact, actively adjusting sex ratios, needs to be and is firmly embedded in the socio-cultural fabric of human communities: it is part of belief systems and is thus part of human ecosystem properties.

While it would seem obvious that one is essentially looking at individual reproductive decisions, it has frequently been assumed that populations as entities would be capable of autochthonous homoeostatic regulation, i.e. that they would pursue an adjustment of their size and composition to the carrying capacity with the help of cultural strategies. Yet, whatever example has been examined to support this assumption, it can be shown that [mechanisms of density regulation, which may be viewed as being beneficial to the group, actually turn out as the sum of fitness-oriented strategies of individual or kin selection, whose effect then only appears as a stabilisation of the population referring to the foreseeable resource supply.] The complexity of socio-cultural sublimation is intriguing, be it the cyclical ritual of decimating pig stocks in an attempt to initiate the re-distribution of land, the limiting resource of the Tsembaga Maring, the subsistence farmers in the highlands of New Guinea, the complex regulations of reproductive self-restriction enabling the Rendille (pastoralists of Kenya) to synchronise the procreation cycles of humans and camels, or the foraging !Kung of the Kalahari desert, who seem to use relatively long birth intervals as a self-regulating mechanism of population control. Thus, even though group selectionist interpretations, which would explain certain behavioural norms as beneficial and advantageous to the community and thus postulate the group as the unit of adaptive change, have experienced a revival in the guise of guided group selection (Roth 2001), concepts of behavioural ecology suffice to explain such behaviour. This does not diminish the system's ecological explanatory value of culturally shaped behavioural strategies, but only leads the way to an empirically testable and theoretically consistent level of analysis. Population regulation takes place, even in the extreme terms of the at first sight seemingly counter-selective behaviour of infanticide. Decisive, however, is the cultural context, in which individuals make reproductive decisions as an often economy-driven response to ecological circumstances of their habitat and which are embedded in socially sanctioned and widely supported rules on the community level.

Reproduction is so firmly culturally mediated that it seems to be much less a matter of biological possibility than of social desirability and cultural opportunity. Empirical examples, such as the case of the Mukogodo, are rare, but there is good reason to assume that reproduction may well have been among the drivers for subsistence change in general, as a new subsistence mode would offer improved possibilities to find marriage partners and to reproduce, first individually, then as a collectively adopted motive. Overall, even when considering the large variances within any one subsistence category, there is an increase in the total fertility of human populations, the more the subsistence is based on the production of food. An alteration in the conditions of production and an increasingly higher predictability of resource utilisation affects the possibilities of reproduction and thereby the size or density of a population.

Provided that a change in mode of production also corresponds with increasing differences in the utilisation and availability of resources within a society, it may be expected that this should result in socially different reproductive success. The accumulation of property and wealth is part of the cultural goals in many societies and as such can be expected to lead to an improved allocation of resources to the offspring and reproductive advantage for those who are able to accomplish these cultural goals more effectively than others. For both historical and recent populations under various ecological and socio-cultural conditions status, the varying ability to exert control over resources has been demonstrated to correlate with reproductive success. Moreover, not only the number of offspring, but also the sex ratio may be regulated in the course of adaptive strategies of resource utilisation in such a way that the concentration and amplification of resources and a long-term fitness advantage will be achieved under the constraints of given environments. Whether this is part of reproductive strategies sensitive to fertility risks (see Winterhalder and Leslie 2002) remains to be seen, yet it would be reasonable to assume, as an improved resource allocation would be one way to buffer against the inherent difficulty of unpredictable lifetime fertility.

In the end, patterns of utilisation and management of material and energy flows can be characterised in effect as strategies typical of certain populations in certain ecological settings. At the same time, decisions made within a community on the individual or family level can be identified and analysed as those mechanisms finally leading to certain patterns of distribution, composition and population size under given ecological conditions. Biocultural strategies of producing, exploiting and controlling resources inform the fragile and subtle interplay of environmental properties and individual decision-making that constitute the compatibility of production and reproduction.

6.5 Epilogue

The emphasis of this book is on resources, the tangible objects of our relationship with natural and cultural environments. Their availability, provision and maintenance permeate all aspects of our lives: from the appropriation of nature as a part of life-support strategies to the possibility of their use and allocation as a means of accomplishing individual goals and collective interests. Concerns about their overexploitation seem to be a relatively recent phenomenon, albeit poignant in the face of either ever-increasing energetic input or diminishing returns. Yet, sustainability does not appear to have been a powerful driver in the past either. Collapses or situations on the brink of failure are a regular part of our history, for example the crisis in fourteenth-century Europe, during a time of social and ecological disruption and instability, was only 'preventively checked' as it were by the wildfire spread of the bubonic plague.

In general, humans tend to favour short-term gain over long-term stability and equilibrium. This implies a built-in propensity to cause breaches of resistance capacities and resilience in ecosystems. And it poses a hierarchical dilemma (Gare 2000), because overcoming one constraint, for example exploiting a new food source, generates a new, higher-order constraint, e.g. potentially allowing population growth and subsequent depletion of resources. But it is one of the hallmark features that humans have developed ways to not only cope with constraints but to overcome them, thereby keeping the dynamics of energy extraction, growth and environmental degradation in motion. Even patterns and strategies of population regulation and reproductive restraint may not exist or have existed for what we would like them to be: efforts towards sustainability. Rather, they are very pragmatic individual decisions with possible group level consequences made for a multitude of proximate reasons, which serve ultimate purposes. While the effects may indeed be conducive to a more considerate use of resources, the causes are not. And whenever circumstances allow people will seek to optimise reproductive and cultural success, resources will not be spared.

The past and present cultural and ethnic diversity of human communities reflect the breadth of local, small-scale, solutions that people have established to facilitate their survival and maintenance of stability domains. Subsistence systems that have been sustained for extended periods of time have managed processes at multiple scales and thus increased resilience. Overall, these solutions have been extremely successful; and this is why they have elicited interest recently. Not only are myths and oral traditions known to contain encoded environmental messages (e.g. Wright and Dirks 1973). Especially what is termed traditional ecological knowledge has increasingly become attractive to scholars concerned with environmental management (see Martinez 2000), as it encompasses low-level strategies which enable the buffer capacity of systems to be increased against the impact of larger or unpredictable perturbations (Gunderson 2000).

This book has attempted to demonstrate the importance of understanding some of the peculiarities and idiosyncrasies involved in such biocultural strategies. The human condition is composed of culturally mediated biology; and this inherently dual existence is key to the way in which humans have organised their living in the past and continue to do so today. The multitude of solutions, of varying duration and success, that human communities have come up with provides the reservoir from which we can enhance our understanding of current socio-ecological systems. Deciphering this information from the heritage and remains of past populations and preserving the still-existing diversity is more than a quaint academic drive or an outcome of antiquarian and ethnographic curiosity. It provides the empirical evidence of human adaptability, which is likely to hold insights and answers that allow a more comprehensive appreciation of the biocultural ecological context of our existence.

References

Abel W (1967) Geschichte der deutschen Landwirtschaft vom frühen Mittelalter bis zum 19. Jahrhundert, 2. Aufl. Ulmer, Stuttgart

Abel T, Stepp JR (2003) A new ecosystems ecology for Anthropology. Conserv Ecol 7:12 (http://www.consecol.og/vol7/iss/3/art12)

Adair LS (1987) Nutrition in the reproductive years. In: Johnston FE (ed) Nutritional anthropology. Liss, New York, pp 119–154

Adger WN (2000) Social and ecological resilience: are they related? Prog Hum Geogr 24:347–364

Agnew C, Anderson E (1992) Water resources in the arid realm. Routledge, London

Alexander RD (1979) Darwinism and human affairs. University of Washington, Seattle

Alland A (1975) Adaptation. Annu Rev Anthropol 4:59–73

Allen CJ (1981) To be Quechua: the symbolism of coca chewing in highland Peru. Am Ethnol 8:157–171

Ambrose SH (1993) Isotopic analysis of paleodiets: methodological and interpretive considerations. In: Sandford MK (ed) Investigations of ancient human tissue. Chemical analyses in anthropology. Gordon & Breach, Langhorne, pp 59–130

Ambrose SH, Katzenberg MA (eds) (2000) Geochemical approaches to palaeodietary analysis. Kluwer/Plenum, London

Ambrose SH, Norr L (1993) Experimental evidence for the relationship of the carbon isotope ratios of whole diet and dietary protein to those of bone collagen and carbonate. In: Lambert JB, Grupe G (eds) Prehistoric human bone: archaeology at the molecular level. Springer, Berlin Heidelberg New York, pp 1–38

Ambrose SH, Buikstra J, Krueger HW (2003) Status and gender differences in diet at mound 72, Cahokia, revealed by isotopic analysis of bone. J Anthropol Archaeol 22:217–226

Ammerman AJ, Cavalli-Sforza LL (1984) The neolithic transition and the genetics of populations in Europe. Princeton University, Princeton

Anderson RM, May RM (eds) (1982) Population biology of infectious diseases. Springer, Berlin Heidelberg New York

Arias P (1999) The origins of the Neolithic along the Atlantic coast of continental Europe. J World Prehist 13:403–464

Armelagos GJ (1990) Health and disease in prehistoric populations in transition. In: Swedlund AC, Armelagos GJ (eds) Disease in populations in transition. Anthropological and epidemiological perspectives. Bergin & Garvey, New York, pp 127–144

Arneborg J, Heinemeier J, Lynnerup N, Nielsen HL, Rud N, Sveinbjörnsdóttir AE (1999) Change of diet of the Greenland Vikings determined from stable carbon isotope analysis and ^{14}C dating of their bones. Radiocarbon 41:157–186

Aufderheide AC, Rodriguez-Martín C (1998) The Cambridge encyclopaedia of human palaeopathology. Cambridge University, Cambridge

Baales M (2001) From lithics to spatial and social organisation: interpreting the lithic distribution and raw material composition at the final Palaeolithic site of Kettig (Central Rhineland, Germany). J Archaeol Sci 28:127–141

Bailey RC, Jenike MR, Ellison PT, Bentley GR, Harrigan AM, Peacock NR (1992) The ecology of birth seasonality among agriculturalists in central Africa. J Biosocial Sci 24:393–412

Baker PT (1988) Human ecology and human adaptability. In: Harrison GA, Tanner JM, Pilbeam DR, Baker PT (eds) Human biology. Oxford University, Oxford, pp 439–547

Baker PT, Mazess RB (1963) Calcium: unusual sources in the highland Peruvian diet. Science 142:1466–1467

Balikci A (1970) The Netsilik Eskimo. Natural History, Garden City, N.Y.

Bargatzky T (1997) Ethnologie. Eine Einführung in die Wissenschaft von den urproduktiven Gesellschaften. Buske, Hamburg

Barnard A (2000) History and theory in anthropology. Cambridge University, Cambridge

Baron H, Hummel S, Herrmann B (1996) *Mycobacterium tuberculosis* complex DNA in ancient human bones. J Archaeol Sci 23:667–671

Barrett GW, Odum EP (2000) The twenty-first century: the world at carrying capacity. BioScience 50:363–368

Bar-Yosef O (1989) Introduction to section on archaeology. In: Hershkovitz I (ed) People and culture in change. BAR Int Ser 508:1–5

Bar-Yosef O (1998) The Natufian culture in the event, threshold to the origins of agriculture. Evol Anthropol 6:159–177

Bar-Yosef O (2002) The upper Palaeolithic revolution. Annu Rev Anthropol 31:363–393

Bar-Yosef O, Belfer-Cohen A (1989a) The levantine "PPNB" interaction sphere. In: Hershkovitz I (ed) People and culture in change. BAR Int Ser 508:59–72

Bar-Yosef O, Belfer-Cohen A (1989b) The origins of sedentism and farming communities in the Levant. J World Prehist 3:447–498

Bar-Yosef O, Belfer-Cohen A (1992) From foraging to farming in the Mediterranean Levant. In: Gebauer AB, Price TD (eds) Transitions to agriculture in prehistory. Prehistory, Madison, pp 21–48

Bar-Yosef O, Meadows RH (1995) The origins of agriculture in the Near East. In: Price TD, Gebauer AB (eds) Last hunters – first farmers. New perspectives on the prehistoric transition to agriculture. School of American Research, Santa Fe, pp 39–94

Bates DG, Lees SH (1979) The myth of population regulation. In: Chagnon NA, Irons W (eds) Evolutionary biology and human social behaviour. Duxbury, North Scituate, pp 273–289

Bayliss-Smith TP (1974) Constraints on population growth: the case of the Polynesian outlier atolls in the precontact period. Hum Ecol 2:259–295

Bayliss-Smith TP (1982) The ecology of agricultural systems. Cambridge University, Cambridge

Beall CM, Strohl KP, Gothe B, Brittenham GM, Barragan M, Vargas E (1992) Respiratory and hematological adaptations of young and older Aymara men native to 3600 m. Am J Hum Biol 4:17–26

Becker C (1991a) Erste Ergebnisse zu den Tierknochen aus Tall Seh Hamad – Die Funde aus Raum A des Gebäudes P. In: Kühne H (ed) Die rezente Umwelt von Tall Seh Hamad und Daten zur Umweltrekonsrtuktion der assyrischen Stadt Dur-Katlimmu. (Berichte der Ausgrabung Tall Seh Hamad/Dur-Katlimmu 1) Reimer, Berlin, pp 117–132

Becker CD, Ostrom E (1995) Human ecology and resource sustainability: the importance of institutional diversity. Annu Rev Ecol Syst 26:113–133

Becker GS (1991b) A treatise of the family. Harvard University, Cambridge, Mass.

Begon M, Harper JL, Townsend L (2003) Ecology: individuals, populations and communities, 3rd edn. Blackwell Scientific, Oxford

Belfer-Cohen A, Schepartz LA, Arensburg B (1999) New biological data for the Natufian populations in Israel. In: Bar-Yosef O, Vall FR (eds) The Natufian culture in the Levant. (Archaeological series 1) International Monographs in Prehistory, Ann Arbor, pp 411–424

Bender DA (1993) Introduction to nutrition and metabolism. University College, London

Bender MM, Baerreis DA, Steventon RL (1981) Further light on carbon isotopes and Hopewell agriculture. Am Antiq 46:346–353

Benecke N (1994) Der Mensch und seine Haustiere. Die Geschichte einer jahrtausendealten Beziehung. Theiss, Stuttgart

Bennett JW (1990) Ecosystems, environmentalism, resource conservation, and anthropological research. In: Moran EF (ed) The ecosystem approach in anthropology. From concept to practice. University of Michigan, Ann Arbor, pp 435–457

Bennike P (1993) The people. In: Hvass S, Storgaard B (eds) Digging into the past: 25 years of Danish archaeology. Universitetsforlag, Århus, pp 34–39

Bentley GR (1985) Hunter-gatherer energetics and fertility: A reassessment of the !Kung San. Hum Ecol 13:79–109

Bentley GR, Jasienska G, Goldberg T (1993) Is the fertility of agriculturalists higher than that of nonagriculturalists? Curr Anthropol 34:778–785

Bentley RA, Krause R, Price TD, Kaufmann B (2003) Human mobility at the Neolithic settlement of Vaihingen, Germany: evidence from strontium isotope analysis. Archaeometry 45:471–486

Berkes F, Jolly D (2001) Adapting to climate change: social-ecological resilience in a Canadian western arctic community. Conserv Ecol 5:18–36

Betzig L (1982) Despotism and differential reproduction: a cross-cultural correlation of conflict asymmetry, hierarchy, and degree of polygyny. Ethol Sociobiol 3:209–221

Binford LR (1968) Methodological considerations of the archaeological use of ethnographic data. In: Lee RB & DeVore I (eds) Man the hunter. Aldine, Chicago, pp 268–273

Bird R (1999) Cooperation and conflict: the behavioural ecology of the sexual division of labour. Evol Anthropol 8:65–75

Blakely RI, Beck L (1981) Trace elements, nutritional status, and social stratification at Etowah, Georgia. In: Cantwell A-M, Griffin J, Rothschild N (eds) The research potential of anthropological museum collections. Ann NY Acad Sci 376:417–431

Blumler MA (1996) Ecology, evolutionary theory and agricultural origins. In: Harris DR (ed) The origins and spread of agriculture and pastoralism in Eurasia. University College, London, pp 25–50

Blurton Jones N (1986) Bushman birth spacing: A test for optimal interbirth intervals. Ethology and Sociobiology 7: 91-105

Blurton Jones N, Sibley RM (1978) Testing adaptiveness of culturally determined behaviour: do bushmen women maximize their reproductive success by spacing births widely and foraging seldom? In: Reynolds V, Blurton Jones N (eds) Human behaviour and adaptation. Taylor and Francis, London, pp 135–157

Blurton Jones N, Smith LC, O'Connell JF, Hawkes K, Kamuzora CL (1992) Demography of the Hadza, an increasing and high density population of savanna foragers. Am J Phys Anthropol 89:159–181

Bocherens H (1997) Isotopic biogeochemistry as a marker of Neanderthal diet. Anthropol Anz 55:101–120

Bocquet-Appel J-P (1985) Small populations: demography and paleoanthropological inferences. J Hum Evol 14:683–691

Boehm C (1996) Emergency decisions, cultural selection mechanisms and group selection. Current Anthropology 37: 763-793

Böhmer-Bauer K (1990) Nahrung, Weltbild und Gesellschaft. Ernährung und Nahrungsregeln der Massai als Spiegel der gesellschaftlichen Ordnung. Breitenbach, Saarbrücken

Bogin BA (2001) The growth of humanity. Wiley–Liss, New York

Bollig M, Casimir MJ (1993) Pastorale Nomaden. In: Schweizer T, Schweizer M, Kokot W (eds) Handbuch der Ethnologie. Reimer, Berlin, pp 521-559

Bollig M, Schulte A (1999) Environmental change and pastoral perceptions: degradation and indigenous knowledge in two African pastoral communities. Hum Ecol 27:493–514

Bongaarts J (1980) Does malnutrition affect fecundity? A summary of evidence. Science 208:564–569

Bongaarts J (1982) Malnutrition and fertility. Science 215:1273–1274

Bonsall C, Macklin MG, Anderson DE, Payton RW (2002) Climate change and the adoption of agriculture in north-west Europe. Eur J Archaeol 5:7–21

Boockmann H (1988) Einführung in die Geschichte des Mittelalters, 4 Aufl. Beck, Munich

Boone JL, Kessler KL (1999) More status or more children? Social status, fertility reduction, and long-term fitness. Evol Hum Behav 20:257–277

Borgerhoff Mulder M (1987a) On cultural and reproductive success: Kipsigis evidence. Am Anthropol 89:617–634

Borgerhoff Mulder M (1987b) Adaptation and evolutionary approaches to anthropology. Man 22:25–41

Borgerhoff Mulder M (1990) Kipsigis women´s preferences for wealthy men: evidence for female choice in mammals? Behav Ecol Sociobiol 27:255–264

Borgerhoff Mulder M (1992) Women's strategies in polygynous marriage. Kipsigis, Datoga, and other East African cases. Hum Nat 3:45–70

Boserup E (1965) The conditions of agricultural growth: the economics of agrarian change under population pressure. Aldine, Chicago

Bosinski G (1981) Gönnersdorf. Eiszeitjäger am Mittelrhein. Veroeff Landesmus Koblenz 7:9–120

Bosinski G (1988) Upper and final paleolithic settlement patterns in the Rhineland, West Germany. In: Dibble HL, Montet-White A (eds) Upper pleistocene prehistory of western Eurasia. University of Pennsylvania, Philadelphia, pp 375–386

Bosinski G (1989) Die große Zeit der Eiszeitjäger. Jahrb RGZM 34:1–139

Bosinski G (1990) Homo sapiens. L'Histoire des chasseurs du paléolithique supérieur en Europe (40000-10000 avant J.-C.). Errance, Paris

Bourdieu P (1997) Die feinen Unterschiede. Kritik der gesellschaftlichen Urteilskraft (Translation from the French original, 1979, 'La distinction. Critique sociale du jugement'). Suhrkamp, Frankfurt am Main

Bradley C, Moore CC, Burton ML, White DR (1990) A cross-cultural historical analysis of subsistence change. Am Anthropol 92:447–457

Brightman R (1996) The sexual division of foraging labour: biology, taboo, and gender politics. Comp Stud Soc Hist 38:687–729

Bronson FH (1995) Seasonal variation in human reproduction: environmental factors. Q Rev Biol 70:141–164

Brookfield HC, Brown P (1963) Struggle for land: agriculture and group territories among the Chimbu of the New Guinea highlands. Oxford University, Melbourne

Brothwell DR, Pollard AM (eds) (2001) Handbook of archaeological science. Wiley, Chichester

Brown AB, Blakely RL (1985) Biocultural adaptation as reflected in trace element distribution. J Hum Evol 14:461–468

Brush SB (1975) The concept of carrying capacity for systems of shifting cultivation. Am Anthropol 77:799–811

Buikstra JE, Milner GR (1991) Isotopic and archaeological interpretations of diet in the Central Mississippi Valley. J Archaeol Sci 18:319–329

Buikstra JE, Autry W, Breitburg E, Eisenberg L, Merwe N van der (1988) Diet and health in the Nashville Basin: human adaptation and maize agriculture in middle Tennessee. In: Kennedy BV, LeMoine GM (eds) Diet and subsistence: current archaeological perspectives. University of Calgary Archaeological Association, Calgary, pp 243–259

Burstein J, Grimprel E, Lukehart S, Norgard M, Radolf J (1991) Sensitive detection of *Treponema pallidum* by using the polymerase chain reaction. J Clin Microbiol 29:62–69

Burton JH, Price TD (1990) Paledietary applications of barium values in bone. In: Pernicka E, Wagner GA (eds) Archaeometry '90. Birkhäuser, Basel, pp 787–795

Burton JH, Price TD (2000) The use and abuse of trace elements for palaeodietary research. In: Ambrose SH, Katzenberg MA (eds) Biogeochemical approaches to palaeodietary analysis. Kluwer/Plenum, London, pp 159–171

Burton JH, Wright LE (1995) Nonlinearity in the relationsship between bone Sr/Ca and diet: Paleodietary implications. Am J Phys Anthropol 96:273–282

Burton JH, Price TD, Middleton WD (1999) Correlation of bone Ba/Ca and Sr/Ca due to biological purification of calcium. J Archaeol Sci 26:609–616

Burton JH, Price TD, Cahue L, Wright LE (2003) The use of barium and strontium abundances in human skeletal tissues to determine their geographic origin. Int J Osteoarchaeol 13:88–95

Butler C (1989) The plastered skulls of Ain Ghazal: preliminary findings. In: Hershkovitz I (ed) People and culture in change. BAR Int Ser 508:141–145

Butzer KW (1976) Early hydraulic civilization in Egypt. A study in cultural ecology. University of Chicago, Chicago

Butzer KW (1980) Civilizations: organisms or systems? Am Sci 68:517–523

Butzer KW (1982) Archaeology as human ecology. Cambridge University, New York

Campbell K, Wood J (1988) Fertility in traditional societies. In: Diggory P, Teper S, Potts M (eds) Natural human fertility: social and biological mechanisms. Macmillan, London, pp 39–69

Cane S (1996) Australian Aboriginal subsistence in the Western Desert. In: Bates DG, Lees SH (eds) Case studies in human ecology. Plenum, London, pp 17–53

Carey HC (1837) Principles of political economy. Philadelphia

Caro TM, Borgerhoff Mulder M (1987) The problem of adaptation in the study of human behaviour. Ethol Sociobiol 8:61–72

Cashdan E (2001) Ethnic diversity and its environmental determinants: effects of climate, pathogens, and habitat diversity. Am Anthropol 103:968–991

Casimir MJ (1990) Energieproduktion, Energieverbrauch und Energieflüsse in einer Talschaft des westlichen Himalaya. Saeculum 42:246–261

Casimir MJ (1991) Flocks and food: a biocultural approach to the study of pastoral foodways. Böhlau, Cologne

Casimir MJ (1993) Gegenstandsbereiche der Kulturökologie. In: Schweizer T, Schweizer M, Kokot W (eds) Handbuch der Ethnologie. Reimer, Berlin, pp 215–239

Casimir MJ, Rao A (1992) Kulturziele und Fortpflanzungsunterschiede. Aspekte der Beziehung zwischen Macht, Besitz und Reproduktion bei den nomadischen Bakkarwal im westlichen Himalaya. In: Voland E (ed) Fortpflanzung: Natur und Kultur im Wechselspiel. Suhrkamp, Frankfurt am Main, pp 270–289

Casimir MJ, Rao A (1995) Prestige, possessions, and progeny. Cultural goals and reproductive success among the Bakkarwal. Hum Nat 6:241–272

Catton WR, Dunlap RE (1980) A new ecological paradigm for post-exuberant sociology. Am Behav Sci 24:15–47

Cavalli-Sforza LL, Menozzi P, Piazza A (1994) The history and geography of human genes. Princeton University, Princeton, N.J.

Chagnon NA (1979) Is reproductive sucess equal in egalitarian societies? In: Chagnon NA, Irons W (eds) Evolutionary biology and human social behavior: an anthropological perspective. Duxbury, North Scituate, pp 374–401

Chapman M (1990) The social definition of want. In: Chapman M, Macbeth H (eds) Food for humanity. Oxford Polytechnic, Oxford, pp 26–33

Chapman M (1980) Infanticide and fertility among Eskimos: a computer simulation. Am J Phys Anthropol 53:317–327

Chapman J, Crites GD (1987) Evidence for early maize (*Zea mays*) from the Icehouse Bottom Site, Tennessee. Am Antiq 52:352–354

Charnov EL (1976) Optimal foraging, the marginal value theorem. Theor Pop Biol 9:129–136

Childe VG (1928) The most ancient East: the oriental prelude to European history. Paul/Trench/Trubner, London

Churchill SE, Smith FH (2000) Makers of the early Aurignacian of Europe. Yearb Phys Anthropol 43:61–115

Cleuziou S (1981) Oman peninsula in the early second millennium BC. In: Härtel H (ed) South Asian archaeology 1979. Reimer, Berlin, pp 279–293

Cleuziou S (1984) Oman peninsula and its relations eastwards during the 3rd millennium BC. In: Lal B, Gupta SP (eds) Frontiers of the Indus civilization. (Sir Mortimer Wheeler memorial volume.) New Delhi, pp 371–394

Clutton-Brock TH (1991) The evolution of parental care. Princeton University, Princeton

Cohen JE (1995) Population growth and earth's human carrying capacity. Science 269:341–346

Cohen MN (1975) Population pressure and the origins of agriculture: an archaeological example from the coast of Peru. In: Polgar S (ed) Population, ecology, and social evolution. Mouton, The Hague, pp 79–121

Cohen MN (1989) Health and the rise of civilization. Yale Unversity, New Haven

Cohen MN (1994) Demographic expansion: causes and consequences. In: Ingold T (ed) Companion encyclopedia of anthropology. Routledge, London, pp 265–296

Cohen MN, Armelagos GJ (eds) (1984) Paleopathology at the origins of agriculture. Academic, Orlando

COHMAP (1988) Climatic changes of the last 18,000 years: observations and model simulations. Science 241:1043–1051

Coleman E (1976) Infanticide in the early Middle Ages. In: Stuard SM (ed) Women in mediaeval society. University of Pennsylvania, Philadelphia, pp 47–70

Colinvaux PA (1982) Towards a theory of history. Fitness, niche and clutch of Homo sapiens. J Ecol 70:393–412

Collard IF, Foley RA (2002) Latitudinal patterns and environmental determinants of recent human cultural diversity: do humans follow biogeographical rules? Evol Ecol Res 4:371–383

Colledge S, Conolly J, Shennan S (2004) Archaeobotanical evidence for the spread of farming in the eastern Mediterranean. Curr Anthropol 45 [Suppl]:S35–S58

Condon RG, Scaglion R (1982) The ecology of human birth seasonality. Hum Ecol 10:495–511

Constantini L (1978/79) Identification of two rows barley and early domesticated sorghum. Archaeol UAE 2/3:70–71

Cook E (1971) The energy flow in an industrial society. Sci Am 225:134–144

Cook GC (1978) Did persistence of intestinal lactase into adult life originate on the Arabian peninsula? Man 13:418–427

Cooper A, Poinar HN (2000). Ancient DNA: do it right or not at all. Science 289:1139

Coppolillo PB (2000) The landscape ecology of pastoral herding: spatial analysis of land use and livestock production in East Africa. Hum Ecol 28:527–560

Coughenour MB, Ellis JE, Swift DM, Coppock DL, Galvin K, McCabe JT, Hart TC (1985) Energy extraction and use in a nomadic pastoral ecosystem. Science 230:619–625

Crognier E, Baali A, Hilali M-K (2001) Do "helpers at the nest" increase their parents' reproductive success? Am J Hum Biol 13:365–373

Cronk L (1989) From hunters to herders: subsistence change as a reproductive strategy among the Mukogodo. Curr Anthropol 30:224–234

Cronk L (1991a) Human behavioral ecology. Annu Rev Anthropol 20:25–53

Cronk L (1991b) Wealth, status, and reproductive success among the Mukogodo of Kenya. Am Anthropol 93:345–360

Cronk L (1991c) Preferential parental investment in daughters over sons. Hum Nat 2:387–417

Crosby AW (1986) Ecological imperialism: the biological expansion of Europe, 900-1900. Cambridge University, Cambridge

Crosby AW (2003) The Columbian exchange. Biological and cultural consequences of 1492. Praeger, London

Curio E (1988) Gerontophagy and the usefulness of Trivers' parental investment concept. Ethology 79:78–80

Däcke F-O (2005) Das alamannische Gräberfeld von Kirchheim unter Teck. Die Ausgrabungen von 1970. (Marburger Studien zur Vor- und Frühgeschichte.) Hitzeroth, Marburg (in press)

Daly M, Wilson M (1984) A sociobiological analysis of human infanticide. In: Hausfater G, Bluffer Hrdy S (eds) Infanticide. Aldine, New York, pp 487–502

Damas D (1972) Central Eskimo systems of food sharing. Ethnology 11:220–240

Davis S (1987) The archaeology of the animals. Batsford, London

Delvoye P, Demaegd M, Delogne-Desnoek J, Robyn C (1977) The infuence of the frequency of nursing and of previous lactation experience on serum prolactin in lactating mothers. J Biosoc Sci 9:447–451

Denell R (1985) The hunter-gatherer/agricultural frontier in prehistoric temperate Europe: identity and interaction. In: Green S, Perlman SM (eds) The archaeology of frontiers and boundaries. Academic, New York, pp 113–140

Dewey KG (1997) Energy and protein requirements during lactation. Annu Rev Nutr 17:19–36

Dickemann M (1979) Female infanticide and the reproductive strategies of stratified human societies. In: Chagnon N, Irons W (eds) Evolutionary biology and human social organization. Duxbury, North Scituate, pp 321–367

Diesendorf M, Hamilton C (1997) Human ecology, human economy. Ideas for an ecologically sustainable future. Allen & Unwin, St Leonards

Dietz T, Ostrom E, Stern PC (2003) The struggle to govern the commons. Science 302:1907–1912

Dirlmeier U (1978) Untersuchungen zu Einkommensverhältnissen und Lebenshaltungskosten in oberdeutschen Städten des Spätmittelalters. (Philosophisch-Historische Klasse Jahrgang 1978.) Abhandlungen der Heidelberger Akademie der Wissenschaften, Heidelberg

Divale WT, Harris M (1976) Population, warfare, and the male supremacist complex. Am Anthropol 78:521–538

Dobson A (1994) People and disease. In: Jones S, Martin R, Pilbeam D (eds) The Cambridge encyclopaedia of human evolution. Cambridge University, Cambridge, pp 411–420

Dongus H (1961) Naturräumliche Gliederung Deutschlands Blatt 171. Institut für Landeskunde, Bad Godesberg

Dongus H (1991) Naturräumliche Gliederung Deutschlands Blatt 187. Bundesforschungs-institut für Landeskunde und Raumordnung, Bad Godesberg

Dostal W (1985) Egalität und Klassengesellschaft in Südarabien. (Wiener Beiträge zur Kulturgeschichte und Linguistik 20.) Berger & Söhne, Vienna

Draper HH (1977) The aboriginal Eskimo diet. Am Anthropol 79: 309–316

Dubreuil L (2004) Long-term trends in Natufian subsistence: a use–wear analysis of ground stone tools. J Archaeol Sci 31:1613–1629

Dugdale AE, Payne PR (1987) Modelling seasonal changes in energy balance. In: Taylor TG, Jenkins NK (eds) Proceedings of the 13th international congress of nutrition. Libbey, London, pp 141–144

Dumond DE (1972) Population growth and political centralization. In: Spooner B (ed) Population growth. Anthropological implications. MIT, Cambridge, Mass., pp 286–310

Dunlap RE (2002) Paradigms, theories , and environmental sociology. In: Dunlap RE, Buttel FH, Dickens P, Gijswijt A (eds) Sociological theory and the environment.Rowman & Littlefield, Lanham, pp 329–350

Durham WH (1991) Coevolution. Genes, culture and human diversity. Stanford University, Stanford

Durham WH (1992) Applications of evolutionary culture theory. Annu Rev Anthropol 21:331–355

Dyson-Hudson N (1980) Strategies of resource exploitation among East African savanna pastoralists. In: Harris DR (ed) Human ecology in savannah environments. Academic, London, pp 171–184

Dyson-Hudson R (1972) Pastoralism: Self image and behavioral reality. In: Irons W, Dyson-Hudson N (eds) Perspectives on nomadism. Brill, Leiden, pp 30–47

Dyson-Hudson R, Dyson-Hudson N (1969) Subsistence herding in Uganda. Sci Am 220:76–89

Dyson-Hudson R, Dyson-Hudson N (1970) The food production system of a semi-nomadic society: the Karimojong, Uganda. In: McLoughlin PFM (ed) African food production systems. Johns Hopkins, Baltimore, pp 92–123

Earle T (1994) Political domination and social evolution. In: Ingold T (ed) Companion encyclopaedia of anthropology. Humanity, culture and social life. Routledge, London, pp 940–961

Edens C (1986) Bahrain and the Arabian Gulf during the 2nd millennium BC: urban crisis and colonialism. In: Al-Khalifa SH, Rice M (eds) Bahrain through the ages; the archaeology. KPI, London

Edward JB, Benfer RA (1993) The effects of diagenesis on the Paloma skeletal material. In: Sandford M (ed) Investigations of ancient human tissue. Chemical analyses in anthropology. Gordon & Breach, Langhorne, Pa., pp 183–268

Elias N (1997) Über den Prozeß der Zivilisation. Soziogenetische und psychogenetische Untersuchungen. Suhrkamp, Frankfurt am Main

Elias R, Hirao Y, Patterson CC (1982) The circumvention of natural biopurification of calcium along nutrient pathways by atmospheric input of industrial lead. Geochim Cosmochim Acta 46:2561–2580

Ellen R (1982) Environment, subsistence and system. The ecology of small-scale social formations. Cambridge University, Cambridge

Ellen R (1990) Trade, environment, and the reproduction of local systems in the Moluccas. In: Moran EF (ed) The ecosystem approach in anthropology. From concept to practice. University of Michigan, Ann Arbor, pp 191–227

Ellen R (1994) Modes of subsistence: hunting and gathering to agriculture and pastoralism. In: Ingold T (ed) Companion encyclopaedia of anthropology. Humanity, culture and social life. Routledge, London, pp 197-225

Ellenberg H (1955) Wuchsklimakarte Südwestdeutschlands. Reise- und Verkehrsverlag, Stuttgart

Ellis JE, Jennings CH, Swift DM (1979) A comparison of energy flow among the grazing animals of different societies. Hum Ecol 7:135–149

Ellison PT (1991) Reproductive ecology and human fertility. In: Macsie-Taylor CGN, Lasker GW (eds) Applications of biological anthropology to human affairs. Cambridge University, Cambridge, pp 14–54

Ellison PT (1994) Advances in human reproductive ecology. Annu Rev Anthropol 23:255–275

Ellison PT (1995) Understanding natural variation in human ovarian function. In: Dunbar RIM (ed) Human reproductive decisions. Biological and social perspectives. Macmillan, Basingstoke, pp 22–51

Engel C (1990) Reproduktionsstrategien im sozio-ökologischen Kontext. PhD thesis, University of Göttingen, Göttingen

Engelberg J, Boyarsky LL (1979) The noncybernetic nature of ecosystems. Am Nat 114:317–324

Ergenzinger PJ, Kühne H (1991) Ein regionales Bewässerungssystem am Habur. In: Kühne H (ed) Die rezente Umwelt von Tall Seh Hamad und Daten zur Umweltrekonsrtuktion der assyrischen Stadt Dur-Katlimmu. (Berichte der Ausgrabung Tall Seh Hamad/Dur-Katlimmu 1.) Reimer, Berlin, pp 163–190

Eshed V, Gopher A, Galili E, Hershkovitz I (2004a) Musculoskeletal stress markers in Natufian hunter-gatherers and Neolithic farmers in the Levant: the upper limb. Am J Phys Anthropol 123:303–315

Eshed V, Gopher A, Gage TB, Hershkovitz I (2004b) Has the transition to agriculture reshaped the demographic structure of prehistoric populations? New evidence from the Levant. Am J Phys Anthropol 124:315–239

Essock-Vitale SM (1984) The reproductive success of wealthy Americans. Ethol Sociobiol 5:45–49

Ewald PW (1994) Evolution of infectious disease. Oxford University, Oxford

Ezzo J (1994a) Putting the "Chemistry" into archaeological bone chemistry analysis: modeling potential paleodietary indicators. J Anthropol Archaeol 13:1–34

Ezzo J (1994b) Zinc as a paleodietary indicator: an issue of theoretical validity in bone-chemical analysis. Am Antiq 59:606–621

FAO (1980) Handbook on human nutritional requirements. FAO, Rome

Fabig A (1998) Die Rekonstruktion der Ernährungsgrundlage als Möglichkeit der Differenzierung zwischen berufs- und ernährungsbedingten Elementeinträgen in das Skelett am Beispiel einer Bergbaubevölkerung. MSc thesis, Unversity of Göttingen, Göttingen

Fabig A, Herrmann B (2002) Trace elements in buried human bone: intra-population variability of Sr/Ca and Ba/Ca ratios – diet or diagenesis? Naturwissenschaften 89:115–119

Faerman M, Kahila Bar-Gal G, Filon D, Greenblatt C, Stager L, Oppenheim A, Smith P (1998) Determining the sex of infanticide victims from the late Roman era through ancient DNA analysis. J Archaeol Sci 25:861–865

Faux SF, Miller HL (1984) Evolutionary speculations on the oligarchic development of Mormon polygyny. Ethol Sociobiol 5:15–31

Fenton A (1997) Prestige, hunger and charity: Aspects of status through food. In: Teuteberg HJ, Neumann G, Wierlacher A (eds) Essen und kulturelle Identität. Akademie, Berlin, pp 155–163

Fingerlin G (1974) Zur alamannischen Siedlungsgeschichte des 3.–7. Jahrhunderts. In: Hübner W (ed) Die Alamannen in der Frühzeit. Veroeff Alemannen-Inst Freiburg/Br 34:45–88

Fischer A (1982) Trade in Danubian shaft-hole axes and the introduction of Neolithic economy in Denmark. J Dan Archaeol 1:7–12

Fisher RA (1930) The genetical theory of natural selection. Clarendon, London

Flannery KV (1973) The origins of agriculture. Annu Rev Anthropol 2:271–310

Flatz G, Rotthauwe HW (1973) Lactose nutrition and natural selection. Lancet 2:76–77

Foin TC, Davis WG (1984) Ritual and self-regulation of the Tsembaga Maring ecosystem in the New Guinea Highlands. Hum Ecol 12:385–412

Foin TC, Davis WG (1987) Equilibrium und nonequilibrium models in ecological anthropology: an evaluation of 'stability' in Maring ecosystems in New Guinea. Am Anthropol 89:9–31

Fratkin E (1986) Stability and resilience in East African pastoralism: the Rendille and the Ariaal of Northern Kenya. Hum Ecol 14:269–286

Fratkin E (1998) Ariaal pastoralists of Kenya. Surviving drought and development in Africa's arid lands. Allyn and Bacon, London

Fratkin E, Galvin KA, Roth EA (eds) (1994) African pastoralist systems. An integrated approach. Rienner, Boulder

Frayer DW (1984) Biological and cultural change in the European late Pleistocene and early Holocene. In: Smith FH, Spencer F (eds) The origins of modern humans: a world survey of the fossil evidence. Liss, New York, pp 211–250

Freeman MMR (1971) A social and ecologic analysis of systematic female infanticide among the Nestilik Eskimo. Am Anthropol 73:1011–1018

Freeman MMR (1988) Tradition and change: problems and persistence in the Inuit diet. In: Garine I de, Harrison GA (eds) Coping with uncertainties in food supply. Clarendon, Oxford, pp 150–169

Freye H-A (1978) Kompendium der Humanökologie. Fischer, Jena

Fried MH (1967) The evolution of political society: an essay in political economy. Random House, New York

Frifelt K (1975) On prehistoric settlements and chronology of the Oman peninsula. East West 25:329–423

Fuller BT, Richards MP, Mays SA (2003). Stable carbon and nitrogen isotope variations in tooth dentine serial sections from Wharram Percy. J Archaeol Sci 30:1673–1684

Gage TB (1980) Optimal diet choice in Samoa. Am J Phys Anthropol 52:229

Galvin KA, Coppock DL, Leslie PW (1994) Diet, nutrition, and pastoral strategy. In: Fratkin E, Galvin KA, Roth EA (eds) African pastoralist systems. An integrated approach. Rienner, Boulder, pp 113–121

Gare AE (2000) Is it possible to create an ecologically sustainable world order: the implications of hierarchy theory for human ecology. Int J Sust Dev World Ecol 7:277–290

Garine, I de (1994) The diet and nutrition of human populations. In: Ingold T (ed) The companion encyclopaedia of anthropology. Humanity, culture and social life. Routledge, London, pp 226–264

Gaston KJ, Blackburn TM (2000) Pattern and process in macroecology. Blackwell Science, Oxford

Gerste B (1992) Fischstäbchen statt Lebertran. Sind moderne Inuit besser ernährt als ihre Vorfahren? (Kölner ethnologische Arbeitspapiere 4.) Holos, Bonn

Glaeser B (ed) (1989) Humanökologie. Grundlagen präventiver Umweltpolitik. Westdeutscher, Opladen

Glaeser B (1995) Environment, development, agriculture. Integrated policy through human ecology. University College, London

Glover E (1991) The molluscan fauna from Shimal, Ras al-Kaimah, United Arab Emirates. In: Schippmann, K, Herling A, Salles J-F (eds) Golf-Archäologie. Liedorf, Buch am Erlach, pp 205–220

Godin GJ (1972) A study of the energy expenditure of a small Eskimo population. (Physiology section report, IBP Human adaptability project, Igloolik, North West Territory.) University of Toronto, Toronto

Goldman N, Westhoff CF, Paul LE (1987) Variations in natural fertility: the effect of lactation and other determinants. Pop Stud 41:127–146

Golley FB (1993) A history of the ecosystem concept in ecology: more than the sum of the parts. Yale University, New Haven

Goodman AH, Brooke Thomas R, Swedlund AC, Armelagos GH (1988) Biocultural perspective and stress in prehistoric, historical and contemporary population research. Yearb Phys Anthropol 31:169–202

Goudie A (2000) The human impact on the natural environment, 5th edn. MIT, Cambridge, Mass.

Gradmann R (1902) Der Dinkel und die Alamannen. Wuerttemberg Jahrb Stat Landeskd 1901/1902:103–158

Greenblatt CL (ed) (1998) Digging for pathogens. Balaban, Rehovot

Grote K (ed) (1993) Die Abris im südlichen Leinebergland bei Göttingen. Archäologische Befunde zum Leben unter Felsschutzdächern in urgeschichtlicher Zeit. Veroeff urgeschicht Samml Landesmus Hannover 43

Groube L (1996) The impact of diseases upon the emergence of agriculture. In: Harris DR (ed) The origins and spread of agriculture and pastoralism in Eurasia. University College, London, pp 101–129

Grupe G (1986) Umwelt und Bevölkerungsentwicklung im Mittelalter. In: Herrmann B (ed) Mensch und Umwelt im Mittelalter. Deutsche Verlagsanstalt, Stuttgart, pp 24–34

Grupe G (1990) Sozialgruppenabhängiges Nahrungsverhalten im frühen Mittelalter am Beispiel der Skelettserie von Altenerding, Ldkr. Erding. Anthropol Anz 48:365–374

Grupe G, Schutkowski H (1989) Dietary shift during the 2nd millennium BC at prehistoric Shimal, Oman pensinula. Paleorient 15:77–84

Gunderson LH (2000) Ecological resilience – in theory and application. Annu Rev Ecol Syst 31:425–439

Gundlach C von (1986) Agrarinnovation und Bevölkerungsdynamik – aufgezeigt am Wandel der Dreifelderwirtschaft zur Fruchtwechselwirtschaft unter dem Einfluß der Kartoffeleinführung im 18. Jahrhundert. Eine Fallstudie im südwestdeutschen Raum. PhD thesis, University of Freiburg, Freiburg

Guting K (1987) Theorien über den Infantizid und ihre Darstellung am Beispiel der Netsilik-Eskimo. (Mundus Reihe Ethnologie 17.) Holos, Bonn

Hahn R (1993) Die menschlichen Skelettreste aus den Gräberfeldern von Neresheim und Kösingen, Ostalbkreis. In: Knaut M (ed) Die alamannischen Gräberfelder von Neresheim und Kösingen, Ostalbkreis. Forsch Ber Vor- Fruehgesch Baden-Wuerttemberg 48:357–428

Hamilton WD (1964) The genetical evolution of social behavior. J Theor Biol 7:1–52

Hamilton WD, Henderson PA, Moran N (1981) Fluctuation of environment and coevolved antagonist polymorphism as factors in the maintenance of sex. In: Alexander RD, Tinkle DW (eds) Natural selection and social behavior: recent research and theory. Chiron, New York, pp 363–381

Hardesty DL (1975) The niche concept: suggestions for its use in human ecology. Hum Ecol 3:71-85

Hardesty DL (1977) Ecological anthropology. Wiley, New York

Härke H (1990) "Warrior graves"? The background of the Anglo-Saxon weapon burial rite. Past Present 126:22–43

Härke H (1997) Early Anglo-Saxon social structure. In: Hines J (ed) The Anglo-Saxons from the Migration Period to the 8th century: an ethnographic perspective. Boydell, San Marino, pp 125–166

Harlan JR, Zohary D (1966) Distributions of wild wheat and barley. Science 153:1074–1080

Harner M (1977) The ecological basis of Aztec sacrifice. Am Ethnol 4:117–135

Harris DR (1973) The prehistory of tropical agriculture. In: Renfrew C (ed) The explanation of culture change: models in prehistory. Duckworth, London, pp 391–417

Harris DR (1996a) Introduction: themes and concepts in the study of early agriculture. In: Harris DR (ed) The origins and spread of agriculture and pastoralism in Eurasia. University College, London, pp 1–9

Harris DR (1996b) The origins and spread of agriculture and pastoralism in Eurasia: an overview. In: Harris DR (ed) The origins and spread of agriculture and pastoralism in Eurasia. University College, London, pp 552–573

Harris M (1966) The cultural ecology of India´s sacred cattle. Curr Anthropol 7:51–59

Harris M (1974) Cows, pigs, wars, and witches: the riddles of culture. Random House, New York

Harris M (1977) Cannibals and kings: the origins of culture. Random House, New York

Harris M (1989) Kulturanthropologie. Campus, Frankfurt am Main

Harris M (2001) Cultural materialism: the struggle for a science of culture. Altamira, Oxford

Harris M, Ross EB (1987) Death, sex, and fertility. Population regulation in preindustrial and developing societies. Columbia University, New York

Harrison GA (1988) Human genetics and variation. In: Harrison GA, Tanner JM, Pilbeam DR, Baker PT (eds) Human biology. An introduction to human evolution, variation, growth, and adaptability. Oxford University, Oxford, pp 147–336

Harrison GA, Waterlow JC (eds) (1990) Diet and disease in traditional and developing societies. Cambridge University, Cambridge

Harrison RG, Katzenberg MA (2003) Palaeodiet studies using stable carbon isotopes from bone apatite and collagen: examples from Southern Ontario and San Nicholas Island, California. J Anthropol Archaeol 22: 227–244

Hassan FA (1983) Earth resources and population: an archaeological perspective. In: Ortner DJ (ed) How humans adapt – a biocultural odyssey. Smithsonian Institution, Washington, D.C., pp 191-216

Hastorf CA (1990) The ecosystem model and longterm prehistoric change: an example from the Andes. In: Moran EF (ed) The ecosystem approach in anthropology. From concept to practice, pp 131-157. University of Michigan, Ann Arbor, pp 131–157

Hatch JW, Geidel RA (1985) Status-specific dietary variation in two world cultures. J Hum Evol 14:469–476

Hawkes K (1996) Behavioral ecology. In: Levinson D, Ember M (eds) Encyclopaedia of cultural anthropology. Holt, New York, pp 121–125

Hawkes K, Hill K, O'Conell J (1982) Why hunters gather: optimal foraging and the Ache of eastern Paraguay. Am Ethnol 9:379–398

Hawkes K, O'Connell JF, Blurton Jones NG (1995) Hadza children's foraging: juvenile dependency, social arrangements, and mobility among hunter-gatherers. Curr Anthropol 36:688–700

Haynes G (2002) The catastrophic extinction of North American mammoths and mastodons. World Archaeol 33:391–416

Heinrich J (2001) Cultural transmission and the diffusion of innovations: adoption dynamics indicate that biased cultural transmission is the predominate force in behavioural change. Curr Anthropol 103:992–1013

Henke W (1989) Biological distances in Late Pleistocene and Early Holocene human populations in Europe. In: Hershkovitz I (ed) People and culture in change. BAR Int Ser 508:541–563

Herhahn CL, Hill BJ (1998) Modeling agricultural production strategies in the northern Rio Grande valley, New Mexico. Hum Ecol 26:469–487

Herrmann B (1989) Umweltgeschichte. In: Herrmann B, Budde A (eds) Natur und Geschichte. Niedersächsiches Umweltministerium, Hannover, pp 145–153

Herrmann B (1994) (ed) Archäometrie. Naturwissenschaftliche Analyse von Sachüberresten. Springer,Berlin Heidelberg New York

Herrmann B (1996) Umweltgeschichte als Integration von Natur- und Geisteswissenschaften. In: Bayerl G (ed) Cottbuser Studien zur Geschichte der Technik, Arbeit und Umwelt 1. Waxmann, Münster, pp 21–30

Herrmann B (1997) „Nun blüht es von End' zu End' allüberall". Die Eindeichung des Nieder-Oderbruches 1747–1753. Cottbuser Studien zur Geschichte von Technik, Arbeit und Umwelt 4. Waxmann, Münster

Hertz O (1995) Ecology and living conditions in the Arctic: the Uummannamiut (in Danish). Ejlers, Copenhagen

Hewlett BS (1991) Demography and childcare in preindustrial societies. J Anthropol Res 47:1–37

Hill J (1984) Prestige and reproductive success in man. Ethol Sociobiol 5:77–95

Hill K (1993) Life history theory and evolutionary anthropology. Evol Anthropol 2:78–88

Hill K, Hurtado M (1996) Aché life history. The ecology and demography of a foraging people. Aldine de Gruyter, New York

Hill K, Kaplan H (1999) Life history traits in humans: theory and empirical studies. Annu Rev Anthropol 28:397–430

Hill K, Hawkes K, Hurtado M, Kaplan H (1984) Seasonal variance in the diet of Aché hunter-gatherers in Eastern Paraguay. Hum Ecol 12:101–135

Hill K, Kaplan H, Hawkes K, Hurtado M (1985) Men's time allocation to subsistence work among the Aché of Eastern Paraguay. Hum Ecol 13:29–47

Hill K, Kaplan H, Hawkes K, Hurtado AM (1986) Foraging decisions among Aché hunter-gatherers: new data and implications for optimal foraging models. Ethol Sociobiol 8:1–36

Holden C, Mace R (1997) Phylogenetic analysis of the evolution of lactose digestion in adults. Hum Biol 69:605–628

Hole F (1994) Origins of agriculture. In: Jones S, Martin R, Pilbeam D (eds) The Cambridge encyclopaedia of human evolution. Cambridge University, Cambridge, pp 373–379

Hole F (1996) The context of caprine domestication in the Zagros region. In: Harris DR (ed) The origins and spread of agriculture and pastoralism in Eurasia. University College, London, pp 263–281

Holling CS (1973) Resilience and stability of ecological systems. Annu Rev Ecol Syst 4:1–23

Hovers E (1989) Settlement and subsistence patterns in the Lower Jordan Valley from Epipaleolithic to Neolithic times. In: Hershkovitz I (ed) People and culture in change. BAR Int Ser 508:37–51

Howell N (2000) Demography of the Dobe !Kung, 2nd edn. De Gruyter, New York

Hummel S (2003) Ancient DNA typing. Springer, Berlin Heidelberg New York

Hunn ES, Williams NM (1982) Introduction. In: Williams NM, Hunn ES (eds) Resource managers: North American and Australian hunter-gatherers. Westview, Boulder, pp 1–16

Huntington E (1945) Mainsprings of civilisation. Wiley, New York

Hutchinson DL, Norr L (1994) Late prehistoric and early historic diet in Gulf Coast Florida. In: Larsen CS, Milner GR (eds) In the wake of contact. Biological responses to conquest. Wiley–Liss, New York, pp 9–20

Hutchinson DL, Larsen CS, Schoeninger MJ, Norr L (1998) Regional variation in the pattern of maize adoption and use in Florida and Georgia. Am Antiq 63:397–416

Irons W (1979) Cultural and biological success. In: Chagnon NA, Irons W (eds) Evolutionary biology and human behavior: an anthropological perspective. Duxbury, North Scituate, pp 257–272

Irons W (1990) Lets make our perspective broader rather than narrower. A comment on Turke's „Which humans behave adaptively, and why does it matter?" and on the so-called DA-DP debate. Ethol Sociobiol 11:361–374

Irons W (1996) Adaptation. In: Levinson D, Ember M (eds) Encyclopaedia of cultural anthropology. Holt, New York, pp 1–5

Irons W (1998) Adaptively relevant environments versus the environment of evolutionary adaptedness. Evol Anthropol 6:194–204

Irwin C (1989) The sociocultural biology of Netsilingmiut female infanticide. In: Rasa A, Vogel C, Voland E (eds) The sociobiology of sexual and reproductive strategies. Chapman & Hall, London, pp 234–264

Jackes M, Lubell D, Meiklejohn C (1997) Healthy but mortal: human biology and the first farmers of western Europe. Antiquity 71:639–658

Jacomet S, Kreuz A (1999) Archäobotanik. Aufgaben, Methoden und Ergebnisse vegetations- und agrargeschichtlicher Forschung. Ulmer, Stuttgart

Jäger H (1980) Bodennutzungssysteme (Feldsysteme) der Frühzeit. Abh Akad Wiss Goettingen, Phil-Hist Kl 3:197–228

Janssen MA, Scheffer M (2004) Overexploitation of renewable resources by ancient societies and the role of sunk-cost effects. Ecol Soc 9:6 (http://www.ecologyandsociety.org/vol9/iss1/art6)

Jennbert K (1985) Neolithisation – a Scanian perspective. J Dan Archaeol 4:196–197

Jensen J (1988) Wirtschaftsethnologie. In: Fischer H (ed) Ethnologie. Einführung und Überblick. Reimer, Berlin, pp 83–111

Jochim MA (1981) Strategies for survival. Cultural behavior in an ecological context. Academic, New York

Johnson AW, Earle T (2000) The evolution of human societies. From foraging group to agrarian state, 2nd edn. Stanford University, Stanford

Josephson SC (2002) Does polygyny reduce fertility? Am J Hum Biol 14:222–232

Jordan CF (1981) Do ecosystems exist? Am Nat 118:284–287

Kaplan H, Hill K (1985) Hunting ability and reproductive success among male Aché foragers: preliminary results. Curr Anthropol 26:131–133

Kaplan H, Hill K (1992) The evolutionary ecology of food acquisition. In: Smith EA, Winterhalder B (eds) Evolutionary ecology and human behavior. de Gruyter, New York, pp 167–201

Kaestle FA, Horsburgh KA (2002) Ancient DNA in anthropology: methods, applications, and ethics. Yearb Phys Anthropol 45:92–130

Katz SH, Hediger L, Valleroy LA (1974) Traditional maize processing techniques in the New World. Science 184:765–773

Katzenberg MA (1992) Changing diet and health in pre- and protohistoric Ontario. In: Huss-Ashmore R, Schall J, Hediger M (eds) Health and lifestyle change. MASCA Res Pap Sci Archaeol 9:23–31

Katzenberg MA (2000) Stable isotope analysis: A tool for studying past diet, demography, and life history. In: Katzenberg MA, Saunders SR (eds) Biological anthropology of the human skeleton. Wiley–Liss, New York, pp 305–327

Katzenberg MA, Schwarcz HP, Knyf M, Melbye FJ (1995) Stable isotope evidence for maize horticulture and paleodiet in southern Ontario, Canada. Am Antiq 60:335–350

Keegan WF (1986) The optimal foraging analysis of horticultural production. Am Anthropol 88:92–107

Keene AS (1985) Nutrition and economy: models for the study of prehistoric diet. In: Gilbert RI, Mielke JH (eds) The analysis of prehistoric diets. Academic, Orlando, pp 155–190

Keil G (1986) Seuchenzüge des Mittelalters. In: Herrmann B (ed) Mensch und Umwelt im Mittelalter. Deutsche Verlagsanstalt, Stuttgart, pp 109–128

Kellum B (1974) Infanticide in England in the later Middle Ages. Hist Child Q 1:357–388

Kemp WB (1971) The flow of energy in a hunting society. Sci Am 225:105–115

Kislev ME (1989) Pre-domesticated cereals in the Pre-Pottery Neolithic A period. In: Hershkovitz I (ed) People and culture in change. BAR Int Ser 508:147–151

Kislev ME, Nadel D, Carmi I (1992) Epi-Palaeolithic (19,000 BP) cereal and fruit diet at Ohalo II, Sea of Galilee, Israel. Rev Palaeobot Palynol 71:161–166

Klausing O (1967) Naturräumliche Gliederung Deutschlands Blatt 151. Bad Godesberg Bundesforschungsanstalt für Landeskunde und Raumordnung

Klein RG (1999) The human career. Human biological and cultural origins, 2nd edn. University of Chicago, Chicago

Klindworth H, Voland E (1995) How did the Krummhörn elite males achieve above-average reproductive success? Hum Nat 6:221–240

Knaut M (1993) Die alamannischen Gräberfelder von Neresheim und Kösingen. (Forschungen und Berichte zur Vor- und Frühgeschichte von Baden-Württemberg 48.) Theiss, Stuttgart

Koch, E (1998) Neolithic bog pots from Zealand, Møn, Lolland and Falster. Nordiske Fortidsminder, Copenhagen

Köhler U, Seitz S (1993) Agrargesellschaften. In: Schweizer T, Schweizer M u. Kokot W (eds) Handbuch der Ethnologie. Reimer, Berlin, pp 561–592

Kokabi M (1997) Fleisch für Lebende und Tote. Haustiere in Wirtschaft und Begräbniskult. In: Archäologisches Landesmuseum Baden-Württemberg (ed) Die Alamannen. Theiss, Stuttgart, pp 331–336

Konner M, Worthman C (1980) Nursing frequency, gonadal function, and birth spacing among !Kung hunter-gatherers. Science 207:788–791

Kraybill N (1977) Pre-agricultural tools for the preparation of foods in the old world. In: Reed CA (ed) Origins of agriculture. Mouton, The Hague, pp 485–521

Krebs JR (1978) Ecology. The experimental analysis of distribution and abundance. Blackwell, New York

Krebs JR, Davies NB (1991) Behavioral ecology. An evolutionary approach, 3rd edn. Blackwell Scientific, Oxford

Kühne H (Hrsg) (1991) Die rezente Umwelt von Tall Seh Hamad und Daten zur Umweltrekonsrtuktion der assyrischen Stadt Dur-Katlimmu. Berichte der Ausgrabung Tall Seh Hamad/Dur-Katlimmu 1. Reimer, Berlin

Lack D (1954) The natural regulation of animal numbers. Clarendon, Oxford

Lambert JB, Grupe G (eds) (1993) Prehistoric human bone – archaeology at the molecular level. Springer, Berlin Heidelberg New York

Landers J (1994) Reconstruction ancient populations. In: Jones S, Martin R, Pilbeam D (eds) The Cambridge encyclopedia of human evolution. Cambridge University, Cambridge, pp 402–405

Larsen CS (ed) (1990) The archaeology of Mission Santa Catalina de Guale: 2. Biocultural interpretation of a population in transition. Anthropol Pap Am Mus Nat Hist 68

Larsen CS (1997) Bioarchaeology. Interpreting behaviour from the human skeleton. Cambridge University, Cambridge

Larsen CS, Milner GR (eds) (1994) In the wake of contact. Biological responses to conquest. Wiley–Liss, New York

Larsen CS, Schoeninger MJ, Merwe NJ van der, Moore KM, Lee-Thorpe JA (1992) Carbon and nitrogen stable isotopic signatures of human dietary change in the Georgia Bight. Am J Phys Anthropol 89:197–214

Larsen CS, Griffin MC, Hutchinson DL, Noble VE, Norr L, Pastor RF, Ruff CB, Russell KF, Schoeninger MJ, Schultz M, Simpson SW, Teaford MF (2001) Frontiers of contact: bioarchaeology of Spanish Florida. J World Prehist 15:69–123

Larsson L (1990) The Mesolithic of Southern Scandinavia. J World Prehist 4:257–310

Lassen C, Hummel S, Herrmann B (2000) Molecular sex identification of stillborn and neonate individuals (Traufkinder) from the burial site Aegerten. Anthropol Anz 58:1–8

Laughlin WS, Harper AB (1976) Nutrition, evolution and adaptation. (Paper presented at a symposium on anthropological aspects of human nutrition.) School of American Research, Santa Fe, N.M.

Lawrence RJ (2003) Human ecology and its implications. Landscape Urban Plan 65:31–40

Layton R (1997) An introduction to theory in anthropology. Cambridge University, Cambridge

Leavitt GC (1990) Sociobiological explanations of incest avoidance: A critical review of evidential claims. Am Anthropol 92:971–993

Lee RB (1968) What hunters do for a living, or, how to make out on scarce resources. In: Lee RB, DeVore I (eds) Man the hunter. Aldine Atherton, Chicago, pp 30–48

Lee RB (1972) Population growth and the beginnings of sedentary life among the !Kung Bushmen. In: Spooner B (ed) Population growth: anthropological implications. MIT, Cambridge, Mass., pp 329–342

Lee RB (1976) !Kung spatial organization. In: Lee RB, DeVore I (eds) Kalahari Hunter-Gatherers. Studies of the !Kung San and their neighbors. Harvard University, Cambridge, Mass., pp 73–97

Lee RB (1979) The !Kung San: men, women, and work in a foraging society. Cambridge University, Cambridge

Lee RB (1980) Lactation, ovulation, infanticide, and women's work: a study of hunter-gatherer population regulation. In: Cohen MN, Malpass RS, Klein HG (eds) Biosocial mechanisms of population regulation. Yale University, New Haven, pp 321–348

Le Huray JD, Schutkowski H (2005) Diet and social status during the La Tène period in Bohemia – carbon and nitrogen stable isotope analysis of bone collagen from Kutná Hora-Karlov and Radovesice. J Anthropol Archaeol 24:135–147

Lee-Thorp J, Sponheimer M (2003) Three case studies used to reassess the reliability of fossil bone and enamel isotope signals for palaeodietary studies. J Anthropol Archaeol 22:208–216

Lees SH, Bates DG (1990) The ecology of cumulative change. In: Moran EF (ed) The ecosystem approach in anthropology. From concept to practice. University of Michigan, Ann Arbor, pp 247–277

Legge T (1996) The beginning of caprine domestication in southwest Asia. In: Harris DR (ed) The origins and spread of agriculture and pastoralism in Eurasia. University College, London, pp 238–262

Leonard WR, Thomas RB (1989) Biological responses to seasonal food stress in Highland Peru. Hum Biol 61:65–85

Leslie PW, Bindon JR, Baker PT (1984) Caloric requirements of human populations: A model. Hum Ecol 12:137–162

Libby WF, Berger R, Mead JF, Alexander GV, Ross JF (1964) Replacement rates for human tissue from atmospheric radiocarbon. Science 146:1170–1172

Little EA, Schoeninger MJ (1995) The Late Woodland diet of Nantucket Island and the problem of maize in coastal New England. Am Antiq 60:351–368

Little MA (1989) Human biology of African pastoralists. Yearb Phys Anthropol 32:215–247

Little MA (1995) Adaptation, adaptability, and multidisciplinary research. In: Boaz NT, Wolfe LD (eds) Biological anthropology: the state of the science. Oregon State University, Eugene, pp 121–147

Little MA, Morren GEB (1976) Ecology, energetics, and human variability. Brown, Dubuque

Little MA, Dyson-Hudson N, Dyson-Hudson R, Ellis JE, Galvin KA, Leslie PW, Swift DM (1990) Ecosystem approaches in human biology: Their history and a case study of the South Turkana Ecosystem Project. In: Moran EF (ed) The ecosystem approach in anthropology. From concept to practice. University of Michigan, Ann Arbor, pp 389–434

Long A, Benz BF, Donahue J, Jull A, Toolin L (1989) First direct AMS dates on early maize from Tehuan, Mexico. Radiocarbon 31:1035–1040

Low BS (1988) Pathogen stress and polygyny in humans. In: Betzig L, Borgerhoff Mulder M, Turke P (eds) Human reproductive behavior. A Darwinian perspective. Cambridge University, Cambridge, pp 115–127

Low BS (1990) Marriage systems and pathogen stress in human societies. Am Zool 30:325–339

Low BS (1993) Ecological demography: a synthetic focus in evolutionary anthropology. Evol Anthropol 1:177–187

Lunn PG, Austin S, Whitehead RG (1984) The effect of improved nutrition on plasma prolactin concentrations and postpartum infertility in lactating Gambian women. Am J Clin Nutr 39:227–235

Lutz W, Sanderson W, Scherbov S (2001) The end of world population growth. Nature 412:543–545

Lynott MJ, Boutton TW, Price JE, Nelson DE (1986) Stable carbon isotopic evidence for maize agriculture in southeast Missouri and northeast Arkansas. Am Antiq 51:51–65

Mace R, Pagel M (1994) The comparative method in anthropology. Curr Anthropol 35:549–564

Madsen T (1985) Comments on early agriculture in Scandinavia. Norw Archaeol Rev 18:91–93

Malinowski B (1960) A scientific theory of culture and other essays. Oxford University, London

Malthus T (1798) Essay on the principle of population (reprinted 1986). Pickering, London

Manners RA (1967) The Kipsigis of Kenya: Culture change in a „model" East African tribe. In: Steward JH (ed) Contemporary change in traditional societies, vol 1: introduction and African tribes. University of Illinois, Urbana, pp 205–359

Marieb E (1998) Human anatomy and physiology, 4th edn. Longman, Harlow

Markl H (1983) Wie unfrei ist der Mensch? Von der Natur in der Geschichte. In: Markl H (ed) Natur und Geschichte. Oldenbourg, Munich, pp 11–50

Marshall L (1976) The !Kung of the Nyae Nyae. Harvard University, Cambridge, Mass.

Martin PS (1967) Prehistoric overkill. In: Martin PS, Wright HE (eds) Pleistocene extinctions. Yale University, New Haven, pp 75–120

Martin PS, Klein RG (1984) Pleistocene extinctions. University of Arizona, Tuscon

Martinez D (2000) Traditional ecological knowledge, ecosystem science, and environmental management. Ecol Appl 10

Mascie-Taylor CGN (1993) The biological anthropology of disease. In: Mascie-Taylor CGN (ed) The anthropology of disease. Oxford University, Oxford, pp 1–72

Mays S (1993) Infanticide in Roman Britain. Antiquity 67:883–888

Mays S (1997) Carbon stable isotope ratios in medieval and later human skeletons from northern England. J Archeol Sci 24:561–567

Mazess RB, Mather W (1978) Biochemical variation: bone mineral content. In: Jamison PL, Zegura SL, Milan FA (eds) Eskimos of Northwestern Alaska. A biological perspective. (US/IBP synthesis series 8.) Dowden, Hutchinson & Ross, Stroudsburg, pp 134–138

McCorriston J, Hole F (1991) The ecology of seasonal stress and the origins of agriculture in the near east. Am Anthropol 93:46–69

McGovern CH (2003) Evdence for infanticide in Anglo-Saxon communities? MSc thesis, University of Bradford, Bradford

McGovern TH, Bigelow GF, Amorosi T, Russell D (1996) Northern islands, human error, and environmental degradation. In: Bates DG, Lees SH (eds) Case studies in human ecology. Plenum, London, pp 103–152

McMichael AJ (2004) Environmental and social influences on emerging infectious diseases: past, present and future. Philos Trans R Soc Lond Ser B 359:1049–1058

McNeil W (1976) Plagues and people. Penguin, London

Meggitt MJ (1965) The lineage system of the Mae Enga of New Guinea. Oliver & Boyd, Edinburgh

Meiklejohn C, Zvelebil M (1991) Health status of European populations at the agricultural transition and the implications for the adoption of farming. In: Bush H, Zvelebil M (eds) Health in past societies. Biocultural interpretations of human skeletal remains in archaeological contexts. BAR Int Ser 567:129–145

Meindel RS (1994) Human populations before agriculture. In: Jones S, Martin R, Pilbeam D (eds) The Cambridge encyclopedia of human evolution. Cambridge University, Cambridge, pp 406–410

Menell S (1986) Über die Zivilisierung der Eßlust. Z Soziol 15:405–421

Menell S, Murcott A, Otterloo AH van (1992) The sociology of food: eating, drinking and culture. Curr Sociol 40 (152 pp)

Mertz W (1981) The essential trace elements. Science 213:1332–1338

Middleton R (1962) Brother–sister and father–daughter marriage in ancient Egypt. Am Sociol Rev 27:603–611

Milner GR (1990) The late prehistoric Cahokia cultural system of the Mississippi valley: foundations, florescence, and fragmentation. J World Prehist 4:1–44

Minturn L, Stashak J (1982) Infanticide as a terminal abortion procedure. Behav Sci Res 17:70–90

Mithen SJ (1989) Modeling hunter-gatherer decision making: complementing optimal foraging theory. Hum Ecol 17:59–83

Mithen SJ (1990) Thoughtful foragers. A study of prehistoric decision making. Cambridge University, Cambridge

Molnar S (1983) Human variation, 2nd edn. Prentice-Hall, Englewood Cliffs, N.J.

Moore J (1983) Carrying capacity, cycles, and culture. J Hum Evol 12:505–514

Montgomery J, Evans JA, Powlesland D, Roberts CA (2005) Continuity or colonization in Anglo-Saxon England? Isotope evidence for mobility, subsistence practice and status at West Heslerton. Am J Phys Anthropol 126:123–138

Moran EF (1990) Ecosystem ecology in biology and anthropology: a critical assessment. In: Moran EF (ed) The ecosystem approach in anthropology. From concept to practice, pp 3–40. University of Michigan, Ann Arbor, pp 3–40

Moran EF (1996) Environmental anthropology. In: Levinson D, Ember M (eds) Encyclopedia of cultural anthropology. Holt, New York, pp 383–389

Moran EF (2000) Human adaptability. An introduction to ecological anthropology, 2nd edn. Westview, Boulder

Morton JD, Schwarcz HP (2004) Palaeodietary implications from stable isotope analysis of residues on prehistoric Ontario ceramics. J Archaeol Sci 31:503–517

Mosley WH (1979) The effects of nutrition on natural fertility. In: Menken JA, Leridon H (eds) Patterns and determinants of natural fertility. Ordina, Liège, pp 83–105

Munro ND (2004) Zooarchaeological measures of hunting pressure and occupation intensity in the Natufian – implications for agricultural origins. Curr Anthropol 45 [Suppl]:S5–S33

Murcott A (1990) Introducing an anthropological approach to familiar food practices. In: Chapman M, Macbeth H (eds) Food for humanity. Centre for the Science of Food and Nutrition, Oxford, pp 49–56

Murdock GP (1967) Ethnographic atlas. University of Pittsburgh, Pittsburgh

Murdock GP (1981) Atlas of world cultures. University of Pittsburgh, Pittsburgh

Murdock GP, White DR (1969) Standard cross-cultural sample. Ethnology 8:329–369

Mutundu KK (1999) Ethnohistoric archaeology of the Mukogodo of North-Central Kenya. Oxbow, Oxford

Nadel SF (1947) The Nuba. Oxford University, London

Nagaoka L (2002) Explaining subsistence change in southern New Zealand using foraging theory models. World Archaeol 34:84–102

Nentwig W (1995) Humanökologie. Fakten, Argumente, Ausblicke. Springer, Berlin Heidelberg New York

Netting RMcC (1981) Balancing on an Alp: ecological change and continuity in a Swiss mountain community. Cambridge University, Cambridge

Netting RMcC (1996) Cultural ecology. In: Levinson D, Ember M (eds) Encyclopedia of cultural anthropology. Holt, New York, pp 267–271

Nielsen PO (1987) The beginning of the Neolithic: assimilation or complex change? J Dan Archaeol 5:240–243

Noe-Nygaard N (1988) $\delta^{13}C$ values of dog bones reveal the nature of changes in man's food resources at the Mesolithic–Neolithic transition, Denmark. Isotope Geosci 73:87–96

Noss AJ, Hewlett BS (2001) The contexts of female hunting in Central Africa. Am Anthropol 103:1024–1040

O'Connor T (2000) The archaeology of animal bones. Sutton, Stroud

Odum EP (1983) Systems ecology: an introduction. Wiley, New York

Orchard J (1994) Third millennium oasis towns and environmental constraints on settlement in the Al-Hajar region. Part I: The Al-Hajar oasis towns. Iraq 56:63–100

Orchard J (1995) The origins of agricultural settlement in the Al-Hajar region. Iraq 57:145–158

Orlove BS (1980) Ecological anthropology. Annu Rev Anthropol 9:235–273

Ortner DJ (2003) Identification of pathological conditions in human skeletal remains. Academic, London

Ostrom E (1999) Coping with the tragedy of the commons. Annu Rev Polit Sci 2:493–535

Padberg B (1996) Die Oase aus Stein. Humanökologische Aspekte des Lebens in mittelalterlichen Städten. Akademie, Berlin

Panter-Brick C, Layton RH, Rowley-Conwy P (2001) Lines of inquiry. In: Panter-Brick C, Layton RH, Rowley-Conwy (eds) Hunter-Gatherers. An interdisciplinary perspective. Cambridge University, Cambridge, pp 1–11

Parker Pearson M (1993) The archaeology of death and burial. Sutton, Stroud

Patten BC, Odum EP (1981) The cybernetic nature of ecosystems. Am Nat 118:886–895

Pauketat T (2002) A fourth-generation synthesis of Cahokia and Mississippianization. Midcontinent J Archaeol 27:149–170

Pennington R, Harpending H (1988) Fitness and fertility among Kalahari !Kung. Am J Phys Anthropol 77:303–319

Pennington R (2001) Hunter-gatherer demography. In: Panter-Brick C, Layton RH, Rowley-Conwy (eds) Hunter-Gatherers. An interdisciplinary perspective. Cambridge University, Cambridge, pp 170–204

Peoples JG (1982) Individual or group advantage? A reinterpretation of the Maring ritual cycle. Curr Anthropol 23:291–310

Peregrine PN (1996) Ranked societies. In: Levinson D, Ember M (eds) Encyclopaedia of cultural anthropology. Holt, New York, pp 1057–1060

Pernicka E , Wagner GA (eds) Archaeometry '90. Karger, Basel

Persson P (1999) The inception of the Neolithic: studies on the introduction of agriculture in northern Europe (in Swedish). Kust till Kustböcker, Uppsala

Pérusse D (1993) Cultural and reproductive success in industrial societies: testing the relationship at the proximate and ultimate levels. Behav Brain Sci 16:267–322

Pfannhauser W (1988) Essentielle Spurenelemente in der Nahrung. Springer, Berlin Heidelberg New York

Picon-Reátegui E (1978) The food and nutrition of high-altitude populations. In: Baker PT (ed) The biology of high-altitude peoples. (US/IBP synthesis series 14.) Cambridge University,Cambridge, pp 219–149

Piepenbrink H, Schutkowski H (1987) Decomposition of skeletal remains in desert dry soil. A roentgenological study. Hum Evol 2:481–491

Pimentel D, Hurd EL, Belloti AC (1973) Food production and the energy crisis. Science 182:443–449

Plotnicov L (1996) Social stratification. In: Levinson D, Ember M (eds) Encyclopedia of cultural anthropology. Holt, New York, pp 1205–1210

Polet C, Katzenberg MA (2003) Reconstruction of the diet in a medieval monastic community from the coast of Belgium. J Archaeol Sci 30:525–533

Powell ML, Bridges PS, Mires AMW (1991) What mean these bones? Studies in southeastern bioarchaeology. University of Alabama, Tuscaloosa

Price TD (1996) The first farmers of southern Scandinavia. In: Harris DR (ed) The origins and spread of agriculture and pastoralism in Eurasia. University College, London, pp 346–362

Price TD (2000) The introduction of farming in northern Europe. In: Price TD (ed) Europe's first farmers. Cambridge University, Cambridge, pp 260–300

Price TD, Brinch Petersen E (1989) Ein Lagerplatz der Mittelsteinzeit in Dänemark. In: Siedlungen der Steinzeit. Spektrum, Heidelberg, pp 44–52

Price TD, Gebauer AB (eds) (1995) Last hunters – first farmers. New perspectives on the prehistoric transition to agriculture. School of American Research, Santa Fe

Privat KL, O'Connell TC, Richards MP (2002) Stable isotope analysis of human and faunal remains from the Anglo-Saxon cemetery at Berinsfield, Oxfordshire: dietary and social implications. J Archaeol Sci 29:779–790

Rafi A, Spiegelman M, Stanford J, Lemma E, Donoghue H, Zias J (1994) DNA of *Mycobacterium leprae* detected by PCR in ancient bone. Int J Osteoarchaeol 4:287–290

Rao A (1993) Zur Problematik der Wildbeuterkategorie. In: Schweizer T, Schweizer M, Kokot W (eds) Handbuch der Ethnologie. Reimer, Berlin, pp 491–520

Rappaport RA (1968) Pigs for the ancestors. Ritual in the ecology of a New Guinea people. Yale University, New Haven

Rappaport RA (1971) The flow of energy in an agricultural society. Sci Am 225:117–132

Rappaport R (1990) Ecosystems, populations and people. In: Moran EF (ed) The ecosystem approach in anthropology. From concept to practice. University of Michigan, Ann Arbor, pp 41–72

RDA (1991) Recommended dietary allowances. (Food and Nutrition Board, Commission on Life Sciences.) National Research Council, Washington, D.C.

Remmert H (1988) Energiebilanzen in kleinräumigen Siedlungsarealen. Saeculum 39:110–118

Richards M (2003) The Neolithic invasion of Europe. Annu Rev Anthropol 32:135–162

Richards M, Côrte-Real H, Forster P, Macaulay V, Wilkinson-Herbots H, Demaine A, Papiha S, Hedges R, Bandelt H-J, Sykes B (1996) Paleolithic and Neolithic lineages in the European mitochondrial gene pool. Am J Hum Genet 59:185–203

Richards M, Macaulay V, Hickey E, et al (2000) Tracing European founder lineages in the Near East mtDNA pool. Am J Hum Genet 67:1251–1276

Richards MP, Molleson TI, Vogel JC, Hedges REM (1998). Stable isotope analysis reveals variations in human diet at the Poundbury Camp Cemetery site. J Archaeol Sci 25:1247–1252

Richards MP, Price TD, Koch E (2003a) Mesolithic and Neolithic subsistence in Denmark: new stable isotope data. Curr Anthropol 44:288–295

Richards MP, Schulting R, Hedges REM (2003b) Sharp shift in diet at onset of Neolithic. Nature 425:366

Richards MP, Fuller BT, Sponheimer M, Robinson T, Ayliffe L (2003 c) Sulphure isotopes in palaeodietary studies: a review and results from a controlled feeding experiment. Int J Osteoarchaeol 13:37–45

Riches D (1976) The Netsilik Eskimo: a special case of selective female infanticide. Ethnology 13:351-361

Rindos D (1984) The origins of agriculture. An evolutionary perspective. Academic, Orlando

Roberts CA, Manchester K (1997) The archaeology of disease. Sutton, Stroud

Rogers AR (1990) Evolutionary economics of human reproduction. Ethol Sociobiol 11:479–495

Rogers AR (1992) Resources and population dynamics. In: Smith EA, Winterhalder B (eds) Evolutionary ecology and human behavior. de Gruyter, New York, pp 375–402

Rollefson GO, Köhler-Rollefson I (1989) The collapse of early neolithic settlements in the southern Levant. In: Hershkovitz I (ed) People and culture in change. BAR Int Ser 508:73–89

Rösch M (1997) Pflanzenreste aus alamannischen Siedlungen. In: Archäologisches Landesmuseum Baden-Württemberg (ed) Die Alamannen. Theiss, Stuttgart, pp 323–330

Rösch M, Jacomet S, Karg S (1992) The history of cereals in the region of the former Duchy of Swabia (Herzogtum Schwaben) from the Roman to the post-medieval period: results of archaeobotanical research. Veg Hist Archaeobot 1:193–231

Rosenzweig ML (1971) And replenish the earth. The evolution, consequences, and prevention of overpopulation. Harper & Row, New York

Rosetta L (1995) Nutrition, physical workloads and fertility regulation. In: Dunbar RIM (ed) Human reproductive decisions. Biological and social perspectives. Macmillan, Basingstoke, pp 52–75

Ross EB (ed) (1980) Beyond the myths of culture: essays in cultural materialism. Academic, New York

Ross EB (1986) Potatoes, population, and the Irish famine: the political economy of demographic change, In: Handwerker WP (ed) Culture and reproduction. An anthropological critique of demographic transition theory. Westview, Boulder, pp 196–220

Rosser ZH, Efgrafov O, Syrrou M, et al (2000) Y-Chromosomal diversity in Europe is clinal and influenced primarily by geography, rather than by language. Am J Hum Genet 67:1526–1543

Roth EA (1993) A reexamination of Rendille population regulation. Am Anthropol 95:597–611

Roth EA (1999) Proximate and distal variables in the demography of Rendille pastoralists. Hum Ecol 27:517–536

Roth EA (2001) Demise of the *Sepaade* tradition: cultural and biological explanations. Am Anthropol 103:1014–1023

Roth H, Theune C (1995) Das frühmittelalterliche Gräberfeld von Weingarten. (1 Katalog der Grabfunde.) Theiss, Stuttgart

Rowley-Conwy PA (1983) Sedentary hunters: the Ertebølle example. In: Bailey GN (ed) Hunter-gatherer economy in prehistory. Cambridge University, Cambridge, pp 111–126

Rowley-Conwy PA (1984) The laziness of the short-distance hunter: the origins of agriculture in western Denmark. J Anthropol Archaeol 3:300–324

Rowley-Conwy PA (1985) The origins of agriculture in Denmark: a review of some theories. J Dan Archaeol 4:188–195

Rowley-Conwy P (2004) How the West was lost. A reconsideration of agricultural origins in Britain, Ireland and Southern Scandinavia. Curr Anthropol 45 [Suppl]:S83–S113

Rozin P (1982) Human food selection: the interaction of biology, culture and individual experience. In: Barker LM (ed) The psychobiology of human food selection. AVI Publishing, Westport, pp 225–254

Rozin P (1987) Psychobiological perspectives on food preferences and avoidances. In: Harris M, Ross EB (eds) Food and evolution. Toward a theory of human food habits. Temple University, Philadelphia, pp 181–205

Sabo G (1991) Long term adaptations among arctic hunter-gatherers: a case study from southern Baffin Island. Garland, New York

Sahlins M (1968) Notes on the original affluent society. In: Lee RB, DeVore I (eds) Man the hunter. Aldine, Chicago, pp 85–89

Sandford MK (1993) Understanding the biogenetic–diagenetic continuum: interpreting elemental concentrations of archaeological bone. In: Sandford MK (ed) Investigations of ancient human tissue. Chemical analyses in anthropology. Gordon & Breach, Langhorne, Pa., pp 3–57

Sandford MK, Weaver DS (2000) Trace element research in anthropology: new perspective and challenges. In: Katzenberg MA, Saunders SR (eds) Biological anthropology of the human skeleton. Wiley–Liss, New York, pp 329–350

Sandström B (1989) Dietary pattern and zinc supply. In: Mills CF (ed) Zinc in human biology. Springer, Berlin Heidelberg New York

Sato S (1980) Pastoral movements and the subsistence unit of the Rendille of Northern Kenya: with special reference to camel ecology. Senri Ethol Stud 6:1–78

Savelle JM, McCartney AP (1988) Geographical and temporal variation in Thule Eskimo subsistence economy: a model. Res Econ Anthropol 10:21–7

Schiefenhövel W (1986) Populationsdynamische Homöostase bei den Eipo in West-Neuguinea. In: Kraus O (ed) Regulation, Manipulation und Explosion der Bevölkerungsdichte. (Veröffentlichungen der Joachim-Jungius-Gesellschaft der Wissenschaften Hamburg 55) Vandehoeck & Ruprecht, Göttingen, pp 53–72

Schiefenhövel W (1989) Reproduction and sex-ratio manipulation through preferential female infanticide among the Eipo, in the Highlands of West New Guinea. In: Rasa A, Vogel C, Voland E (eds) The sociobiology of sexual and reproductive strategies. Chapman & Hall, London, pp 170–193

Schild R (1984) Terminal palaeolithic of the north European plain: a review of lost chances, potentials and hopes. Adv World Archaeol 3:193–274

Schoeller DA (1999) Isotope fractionation: why aren't we what we eat? J Archaeol Sci 26:667–673

Schoeninger MJ (1979) Diet and status at Chalcatzingo: some empirical and technical aspects of strontium analysis. Am JPhys Anthropol 51:295–310

Schoeninger MJ (1981) The agricultural "revolution": its effect on human diet in prehistoric Iran and Israel. Paleorient 7:73–91

Schrire C, Steiger WL (1974) A matter of life and death: an investigation into the practice of female infanticide in the Arctic. Man 9:161–184

Schultz M (2001) Palaeohistopathology of bone: a new approach to the study of ancient diseases. Yearb Phys Anthropol 44:106–147

Schulze ED, Zwölfer H (eds) (1987) Potentials and limitations of ecosystem analysis. Springer, Berlin Heidelberg New York

Schutkowski H (1993a) Menschliche Skelettfunde des Mesolithikums vom Abri Bettenroder Berg IX bei Reinhausen, Ldkr. Göttingen. In: Grote K (ed) Die Abris im südlichen Leinebergland bei Göttingen. Archäologische Befunde zum Leben unter Felsschutzdächern in urgeschichtlicher Zeit. Veroeff Samml Landesmus Hannover 43:175–184

Schutkowski H (1993b) Sex determination of infant and juvenile skeletons. I. Morphognostic features. Am J Phys Anthropol 90:199–205

Schutkowski H (1995) What you are makes you eat different things – interrelations of diet, status, and sex in the early medieval population of Kirchheim unter Teck, FRG. Hum Evol 10:119–130

Schutkowski H (2000) Neighbours in different habitats – subsistence and social differentiation on the eastern Swabian Alb. Anthropol Anz 58:113–120

Schutkowski H (2002a) Modelling the ecology of subsistence change. (Paper presented at the 71st annual meeting of the American association of physical anthropologists, Buffalo, New York) Am J Phys Anthropol [Suppl] 34

Schutkowski H (2002b) Mines, meals and movement: a human ecological approach to the interface of 'history and biology'. In: Smith M (ed) Human biology and history. Taylor & Francis, London, pp 195–211

Schutkowski H (2005) Continuity, change, differentiation – reconstruction of subsistence patterns from human skeletal remains of Tall Seh Hamad, Syria. In: Kühne H (ed) Reports on the excavation at Tall Seh Hamad/Dur-Katlimmu. Reimer, Berlin (in press)

Schutkowski H, Herrmann B (1996) Geographical variation of subsistence strategies in early Mediaeval populations of southwestern Germany. J Archaeol Sci 23:823–831

Schutkowski H, Grupe G (1997) Zusammenhänge zwischen Cribra orbitalia, archäometrischen Befunden am Skelett und Habitatbedingungen. Anthropol Anz 55:155–166

Schutkowski H, Herrmann B (1999) Humanbiologische Aspekte des Landschaftsbildes. In: Gerken B, Görner M (eds) The development of European landscapes with large herbivores, history, models and perspectives. Nat- Kulturlandschaft 3:14–20

Schutkowski H, Wiedemann F, Bocherens H, Grupe G, Herrmann B (1999) Diet, status and decomposition at Weingarten. Trace element and isotope analyses on early mediaeval skeletal material. J Archaeol Sci 26:675–685

Schwarcz HP (2000) Some biochemical aspects of carbon isotopic palaeodiet studies. In: Ambrose SH, Katzenberg MA (eds) Biogeochemical approaches to palaedietary analysis. Kluwer/Plenum, New York, pp 189–209

Schwarcz HP, Schoeninger MJ (1991) Stable isotope analyses in human nutritional ecology. Yearb Phys Anthropol 34:283–321

Schwarcz HP, Melbye J, Katzenberg MA, Knyf M (1985) Stable isotopes in human skeletons of southern Ontario: reconstructing paleodiet. J Archaeol Sci 12:187–206

Scoones I (1999) New ecology and the social sciences: what prospects for a fruitful engagement? Annu Rev Anthropol 28:479–507

Scrimshaw SCM (1984) Infanticide in human populations: societal and individual concerns. In: Hausfater G, Bluffer Hrdy S (eds) Infanticide. Aldine, New York, pp 439–462

Sealy J (2001) Body tissue chemistry and palaeodiet. In: Brothwell DR, Pollard MA (eds) Handbook of archaeological science. Wiley, Chichester, pp 269–279

Seidl I, Tisdell CA (1999) Carrying capacity reconsidered: from Malthus' population theory to cultural carrying capacity. Ecol Econ 31:395–408

Sellen DW, Mace R (1997) Fertility and mode of subsistence: a phylogenetic analysis. Curr Anthropol 38:878–889

Service E (1962) Primitive social organization: an evolutionary perspective. Random House, New York

Shantzis SR, Behrens WW (1973) Population control mechanisms in a primitive agricultural society. In: Meadows DL, Meadows DH (eds) Towards global equilibrium. Wright–Allen, Cambridge, Mass., pp 257–288

Sherratt A (1980) The Cambridge encyclopaedia of archaeology. Crown, New York

Siegmund F (1998) Social structures and relations. In: Wood I (ed) Franks & Alamanni in the Merovingian Period: an ethnographic perspective. Boydell, San Marino, pp 177–199

Silberbauer GB (1972) The G/wi bushmen. In: Bicchieri MG (ed) Hunters and gatherers today. Holt, Rinehart & Winston, New York, pp 271–326

Sillen A, Lee-Thorp J (1991) Dietary change in the late Natufian. In: Bar-Yosef O, Vall FR (eds) The Natufian culture in the Levant. Int Monogr Prehist Archaeol Ser 1:399–410

Sillen A, Sealy J, Lee Thorp J, Horwitz LK, Merwe NJ van der (1989) Trace element and isotope research in progress: implications for near eastern archaeology. In: Hershkovitz I (ed) People and culture in change. BAR Int Ser 508:321–334

Simon JL (1977) The economics of population growth. Princeton University, Princeton

Simoons FJ (1969) Primary adult lactose intolerance and the milking habit: a problem in biological and cultural interrelations. I. Review of medical research. Am J Dig Dis 14:819–836

Simoons FJ (1970) Primary adult lactose intolerance and the milking habit: a problem in biological and cultural interrelations. II. A culture historical hypothesis. Am J Dig Dis 15:695–710

Simoons FJ (2001) Persistence of lactase activity among northern Europeans: a weighing of evidence for the calcium absorption hypothesis. Ecol Food Nutr 40:397–469

Smith BD (1989a) Origins of agriculture in eastern North America. Science 246:1566–1571

Smith BD (1995a) The emergence of agriculture. Scientific American Library, New York

Smith EA (1983) Anthropological applications of optimal foraging theory. Curr Anthropol 24:625–651

Smith EA (1991) Inujjuamiut foraging strategies. Evolutionary ecology of an arctic hunting economy. de Gruyter, New York

Smith EA (1992a) Human behavioral ecology: I. Evol Anthropol 1:20–25

Smith EA (1992b) Human behavioral ecology: II. Evol Anthropol 1:50–55

Smith EA (1995b) Inuit sex ratio: a correction. Curr Anthropol 36:658–659

Smith EA, Smith SA (1994) Inuit sex-ratio variation. Population control, ethnographic error, or parental manipulation? Curr Anthropol 35:595–624

Smith EA, Winterhalder B (eds) (1992) Evolutionary ecology and human behavior. de Gruyter, New York

Smith P (1989b) Paleonutrition and subsistence patterns in the Natufians. In: Hershkovitz I (ed) People and culture in change. BAR Int Ser 508:375–348

So JK (1980) Human biological adaptation to arctic and subarctic zones. Annu Rev Anthropol 9:63–82

Sokal RR, Oden NL, Wilson C (1991) Genetic evidence for the spread of agriculture in Europe by demic diffusion. Nature 351:143–145

Souci SW, Fachmann W, Kraut H (1981) Food composition and nutrition tables 1981/82. Wissenschaftliche Verlagsgesellschaft, Stuttgart

Southwick CH (1996) Global ecology in human perspective. Oxford University, New York

Spencer P (1973) Nomads in alliance: symbiosis and growth among the Rendille and Samburu of Kenya. Oxford University, London

Speth JD, Spielmann KA (1983) Energy source, protein metabolism, and hunter-gatherer subsistence strategies. J Anthropol Archaeol 2:1–31

Spielmann KA (1989) A review: dietary restrictions on hunter-gatherer women and the implications for fertility and infant mortality. Hum Ecol 17:321–345

Spielmann KA, Schoeninger MJ, Moore K (1990) Plains–Pueblo interdependence and human diet at Pecos Pueblo, New Mexico. Am Antiq 55:745–765

Spiro ME (1965) A typology of social structure and the patterning of social institutions: a cross-cultural study. Am Anthropol 67:1097–1119

Steckel RH (1995) Stature and the standard of living. J Econ Lit 33:1903–1940

Steckel RH, Rose JC (eds) (2002) The backbone of history: health and nutrition in the Western hemisphere. Cambridge University, New York

Stein Z, Susser M (1978) Famine and fertility. In: Mosley WH (ed) Nutrition and human reproduction. Plenum, New York, pp 11–28

Stepp JR, Jones EC, Pavao-Zuckerman M, Casagrande D, Zarger RK (2003) Remarkable properties of human ecosystems. Conserv Ecol 7:11 (http://www.consecol. org/vol7/iss3/art11)

Steuer H (1982) Frühgeschichtliche Sozialstrukturen in Mitteleuropa. Abhandlungen der Akademie der Wissenschaften in Göttingen. (Philologisch-Historische Klasse, 3 Folge, Nr 128.) Vandenhoek u. Ruprecht, Göttingen

Steuer H (1984) Die frühmittelalterliche Gesellschaftsstruktur im Spiegel der Grabfunde. In: Roth H, Wamers E (eds) Hessen im Frühmittelalter. Archäologie und Kunst. Thorbecke, Sigmaringen, pp 78–86

Steuer H (1987) Archäologie und die Erforschung der germanischen Sozialgeschichte des 5. bis 8. Jahrhunderts. In: Simon D (ed) Akten des 26. deutschen Rechtshistorikertages 1986. Stud Eur Rechtsgeschichte 30:443–453

Steuer H (1989) Archaeology and history: proposals on the social structure of the Merovingiam kingdom. In: Randsborg K (ed) The birth of Europe: archaeological and social development in the first millennium AD. L'Ermadi Bretschneider, Rome, pp 100–123

Steward JH (1949) Cultural causality and law: a trial formulation of the development of early civilizations. Am Anthropol 51:1–27

Steward JH (1955) Theory of culture change. University of Illinois, Urbana

Sullivan CH, Krueger HW (1981) Carbon isotope analysis in separate chemical phases in modern and fossil bone. Nature 292:333–335

Sutter RC (2003) Nonmetric subadult skeletal sexing traits: I. A blind test of the accuracy of eight previously proposed methods using prehistoric known-sex mummies from Northern Chile. J Forensic Sci 48:1–9

Sutton MQ, Anderson EN (2004) Introduction to cultural ecology. Berg, Oxford

Swedlund AC, Armelagos GJ (eds) (1990) Disease in populations in transition. Anthropological and epidemiological perspectives. Bergin & Garvey, New York

Tainter JA, Allen TFH, Little A, Hoekstra TW (2003) Resource transitions and energy gain: contexts of organisation. Conserv Ecol 7:4 (http://www.consecol.org/vol7/iss3/art4)

Tanaka J (1976) Subsistence ecology of the central Kalahari San. In: Lee RB, DeVore I (eds) Kalahari hunter-gatherers. Studies of the !Kung San and their neighbors. Harvard University, Cambridge, pp 98–119

Tansley AG (1935) The use and abuse of vegetational concepts and terms. Ecology 16:284–307

Tauber H (1981) ^{13}C evidence for dietary habits of prehistoric man in Denmark. Nature 292:332–333

Thomas RB (1976) Energy flow at high altitude. In: Baker PT, Little MA (eds) Man in the Andes. A multidisciplinary study of high-alititude Quechua. (US/IBP synthesis series.) Dowden, Hutchinson & Ross, Stroudsburg, pp 379–404

Tomasello M (1999) The human adaptation for culture. Annu Rev Anthropol 28:509–529

Tomczak PD (2003) Prehistoric diet and socio-economic relationships within the Osmore Valley of southern Peru. J Anthropol Archaeol 22:262–278

Torroni A, Bandelt H-J, Macaulay V, et al (2001) A signal, from human mtDNA, of post-glacial recolonisation in Europe. Am J Hum Genet 69:844–852

Trivers RL (1971) The evolution of reciprocal altruism. Q Rev Biol 46:35–57

Trivers RL (1972) Parental investment and sexual selection. In: Campbell B (ed) Sexual selection and the descent of man 1871–1971. Aldine, Chicago, pp 136–179

Trivers RL, Willard DE (1973) Natural selection of parental ability to vary the sex ratio of offspring. Science 179:90–92

Trömel M, Loose S (1995) Das Wachstum technischer Systeme. Naturwissenschaften 82:160–169

Trosper RL (2003) Resilience in pre-contact Pacific Northwest social ecological systems. Conserv Ecol 7:6 (http://www.consecol.org/vol7/iss3/art6)

Turke PW, Betzig LL (1985) Those who can do: wealth, status, and reproductive success on Ifaluk. Ethol Sociobiol 6:79–87

Uerpmann H-P (1990) Die Anfänge von Tierhaltung und Pflanzenbau. (Die ersten Bauern Bd 2) Schweizerisches Landesmuseum, Zurich, pp 27–37

Uerpmann H-P (1996) Animal domestication – accident or intention? In: Harris DR (ed) The origins and spread of agriculture and pastoralism in Eurasia. University College, London, pp 227–237

Ulijaszek SJ (1995) Human energetics in biological anthropology. Cambridge University, Cambridge

Ulijaszek SJ, Strickland SS (1993) Nutritional anthropology. Prospects and perspectives. Smith–Gordon, London

Valentová J (1991) A Celtic inhumation cemetery at Kutná Hora-Karlov. In: Archaeology in Bohemia 1986–1990. Institute of Archaeology, Prague, pp 221–224

Van den Berghe PL, Mesher GM (1980) Royal incest and inclusive fitness. Am Ethnol 7:300–317

Van Zeist W, Wasylikowa K, Behre K-E (eds) (1991) Progress in Old World paleoethnobotany. Balkema, Rotterdam

Vayda AP, Rappaport RA (1976) Ecology, cultural and noncultural. In: Richerson P, McEvoy J (eds) Human ecology. Duxbury, North Scituate

Vernier B (1984) Vom rechten Gebrauch der Verwandten und der Verwandtschaft: Die Zirkulation von Gütern, Arbeitskräften und Vornamen auf Karpathos (Griechenland). In: Medick H, Sabean S (eds) Emotionen und materielle Interessen – Sozialanthropologische und historische Beiträge zur Familienforschung. Vandenhoeck & Ruprecht, Göttingen, pp 55–110

Vining DR (1986) Social versus reproductive sucess: the central theoretical problem of human sociobiology. Behav Brain Sci 9:167–216

Vivelo FR (1981) Handbuch der Kulturanthropologie. Klett–Cotta, Stuttgart

Vlastos S (1986) Peasant protests and uprisings in Tokugawa Japan. University of California, Berkeley

Vogel C, Voland E (1987) Evolution und Kultur. In: Immelmann K, Scherer KR, Vogel C, Schmoock P (eds) Psychobiologie – Grundlagen des Verhaltens. Fischer, Stuttgart, pp 101–131

Vogel F, Motulsky AG (1997) Human genetics, 3rd edn. Springer Berlin Heidelberg New York

Vogel JC, Merwe N van der (1977) Isotopic evidence for early maize cultivation in New York State. Am Antiq 42:238–242

Vogt B, Franke-Vogt U (eds) (1987) Shimal 1985/1986. Excavations of the German archaeological mission to Ras al-Kahimah, UAE. Berlin Beitr Vorderen Orient 8

Voland E (1990) Differential reproductive success within the Krummhörn population (Germany, 18th and 19th centuries). Behav Ecol Sociobiol 26:65–72

Voland E (1998) Evolutionary ecology of human reproduction. Annu Rev Anthropol 27:347–374

Voland E, Dunbar RIM (1995) Resource competition and reproduction. The relationship between economic and parental strategies in the Krummhörn population (1720–1874). Hum Nat 6:33–49

Waldhauser J (1987) Keltische Gräberfelder in Böhmen: Dobra Voda, Letky sowie Radovesice, Stránce und Tuchomyšl. Ber Roemisch-german Komm 68:25–179

Walker PL, Hewlett BS (1990) Dental health, diet and social status among Central African foragers and farmers. Am Anthropol 92:383–398

Weidemann M (1982) Kulturgeschichte der Merowingerzeit nach den Werken Gregors von Tours. RGZM, Mainz

Weisgerber G (1983) Copper production during the third millennium BC in Oman and the question of Makan. J Oman Stud 6:269–276

Westman P, Sohlenius G (1999) Diatom stratigraphy in five offshore sediment cores from the northwestern Baltic proper implying large scale circulation changes during the last 8500 years. J Paleolimnol 22:53–69

White L (1943) Energy and the evolution of culture. Am Anthropol 45:335–356

White R (1983) The roots of dependency. University of Nebraska, Lincoln

WHO/FAO (1973) Energy and protein requirements. (FAO Food and Nutrition Series 7.) FAO, Rome

Wiechmann I, Grupe G (2005) Detection of *Yersinia pestis* DNA in two early medieval skeletal finds from Aschheim (Upper Bavaria, 6th century A.D.) Am J Phys Anthropol 126:48–55

Wilkes G (1989) Maize: domestication, racial evolution, and spread. In: Harris DR, Hillman GC (eds) Foraging and farming: the evolution of plant exploitation. Unwin Hyman, London, pp 440–455

Willerding U (1980) Anbaufrüchte der Eisenzeit und des frühen Mittelalters, ihre Anbauformen, Standortverhältnisse und Erntemethoden. Abh Akad Wiss Goettingen, Phil-Hist Kl 3:126–196

Willerding U, Hillebrecht M-L (1994) Paläo-Ethnobotanik – Fragestellung, Methoden, Ergebnisse. In: Herrmann B (ed) Archäometrie. Naturwissenschaftliche Analyse von Sachüberresten. Springer, Berlin Heidelberg New York, pp 137–152

Williams CT (1988) Alteration of chemical composition of fossil bones by soil processes and groundwater. In: Grupe G, Herrmann B (eds) Trace elements in environmental history. Springer, Berlin Heidelberg New York, pp 27–40

Williams GC (1966) Natural selection, the costs of reproduction, and a refinement of Lack's principle. Am Nat 100:687–690

Wilmsen EN, Durham D (1988) Food as a function of seasonal environment and social history. In: Garine I de, Harrison GA (eds) Coping with uncertainty in food supply. Clarendon, Oxford, pp 52–87

Wilson CS (1985) Nutritionally beneficial cultural practices. World Rev Nutr Diet 45:68–96

Wilson DS, Sober E (1994) Reintroducing group selection into the human behavioural sciences. Behav Brain Sci 19:777–787

Winterhalder B (1980) Environmental analysis in human evolution and adaptation research. Hum Ecol 8:125–170

Winterhalder B (1984) Reconsidering the ecosystem concept. Rev Anthropol 11:310–313

Winterhalder B (2001) The behavioural ecology of hunter-gatherers. In: Panter-Brick C, Layton RH, Rowley-Conwy (eds) Hunter-gatherers, an interdisciplinary perspective. Cambridge University, Cambridge, pp 12–38

Winterhalder B, Leslie P (2002) Risk-sensitive fertility. The variance compensation hypothesis. Evol Hum Behav 23:59–82

Winterhalder B, Larsen R, Thomas RB (1974) Dung as an essential resource in a highland Peruvian community. Hum Ecol 2:89–104

Wohldt PB (2004) Descent group composition and population pressure in a fringe Enga clan, Papua New Guinea. Hum Ecol 32:137–162

Wood JW (1998) A theory of preindustrial population dynamics. Curr Anthropol 39:99–135

Wood JW, Johnson P, Campbell KL (1985) Demographic and endocrinological aspects of low natural fertility in highland New Guinea. J Biosoc Sci 17:57–79

Worthington-Roberts BS (1989) Nutrition in pregnancy and lactation, 4th edn. Times Mirror/Mosby, St Louis

Wright GA, Dirks JD (1973) Myth as an environmental message. Ethnos 3/4:160–176

Wu J, David JL (2002) A spatially explicit hierarchical approach to modelling complex ecological systems: theory and applications. Ecol Model 153:7–26

Wynne-Edwards VC (1962) Animal dispersion in relation to social behaviour. Oliver & Boyd, Edinburgh

Yesner DR (1987) Life in the "Garden of Eden": causes and consequences of the adoption of marine diets by human societies. In: Harris M, Ross EB (eds) Food and evolution. Toward a theory of human food habits. Temple University, Philadelphia, pp 285–310

Zink AR, Grabner W, Nerlich AG (2005) Molecular identification of human tuberculosis in recent and historic bone tissue samples: the role of molecular techniques for the study of historic tuberculosis. Am J Phys Anthropol 126:32–47

Zvelebil M (1996) The agricultural frontier and the transition to farming in the circum-Baltic region. In: Harris DR (ed) The origins and spread of agriculture and pastoralism in Eurasia. University College, London, pp 323–345

Zvelebil M, Rowley-Conwy P (1984) Transition to farming in Northern Europe: a hunter-gatherer perspective. Norw Archaeol Rev 17:104–128

Zwölfer H (1987) Grundlagen der Bevölkerungsentwicklung aus ökologischer Sicht. In: Herrmann B, Sprandel R (eds) Determinanten der Bevölkerungsentwicklung im Mittelalter. VCH, Weinheim, pp 37–54

Zwölfer H (1991) Das biologische Modell: Prinzipien des Energieflusses in ökologischen Systemen. Saeculum 42:225–238

Subject Index

Ecological Studies

Volumes published since 2001

Printing: Krips bv, Meppel
Binding: Stürtz, Würzburg